The Maryland Amphibian and Reptile Atlas

The
Maryland Amphibian
and Reptile Atlas

Edited by *Heather R. Cunningham and Nathan H. Nazdrowicz*

Johns Hopkins University Press | *Baltimore*

This book was brought to publication with the generous assistance of the Maryland Department of Natural Resources and the Natural History Society of Maryland.

Johns Hopkins University Press
2715 North Charles Street
Baltimore, Maryland 21218-4363
www.press.jhu.edu

Library of Congress Cataloging-in-Publication Data
Names: Cunningham, Heather R., 1976– , editor | Nazdrowicz,
    Nathan H., editor.
Title: The Maryland amphibian and reptile atlas / edited by
    Heather R. Cunningham and Nathan H. Nazdrowicz.
Description: Baltimore : Johns Hopkins University Press, 2018. |
    Includes bibliographical references and index.
Identifiers: LCCN 2017054172 | ISBN 9781421425955 (hardcover :
    alk. paper) | ISBN 1421425955 (hardcover : alk. paper) | ISBN
    9781421425962 (electronic) | ISBN 1421425963 (electronic)
Subjects: LCSH: Amphibians—Maryland—Atlases. | Reptiles—
    Maryland—Atlases.
Classification: LCC QL653.M3 M34 2018 | DDC 597.909752—dc23
LC record available at https://lccn.loc.gov/2017054172

A catalog record for this book is available from the British Library.

Illustration credits: p. ii: Marbled Salamander, photo by Robert T. Ferguson II; p. iii: Smooth Greensnake, photo by Scott McDaniel.

Special discounts are available for bulk purchases of this book. For more information, please contact Special Sales specialsales@jh.edu

To the memory of Ronald L. Gutberlet, Jr., PhD.
Ron served as the Wicomico County Coordinator
for the field data collection portion of the Maryland
Amphibian and Reptile Atlas project and wrote
one of the species accounts prior to his premature
passing. He was a great naturalist, an enthusiastic
educator at Salisbury University, and a dedicated
herpetologist, whose commitment to biodiversity
conservation has left us a lasting legacy.

Contents

Acknowledgments

Successful completion of the Maryland Amphibian and Reptile Atlas (MARA) project was the result of a collaborative effort by state and county employees, private citizens, and conservation organizations in Maryland. The MARA project was developed through the efforts of the project's planning committee. In the earliest stages, the committee developed the MARA pilot study, determined the atlas coverage goals, and refined the methods of the project. We thank Lance Benedict, Robert Bull, David Burman, Lynn Davidson, Charles Davis, Matthew Evans, Don Forester, Matthew Grey, Rachel Gauza Gronert, Wayne Hildebrand, William Lattea, Eugene Meyer, Nathan Nazdrowicz, Mike Quinlan, Kyle Rambo, Chuck Saunders, Bill Sipple, David Smith, Scott Smith, Christopher Swarth, Mark Tegges, Glenn Therres, David Walbeck, Linda Weir, and Jim White for their work on the MARA planning committee.

The MARA project was guided by a steering committee throughout the project and the preparation of this book. The collective knowledge and experience of the committee members greatly benefited the project. Charles Davis, David Smith, Christopher Swarth, and Glenn Therres chaired the committee. Steering Committee members who served throughout the project were Lynn Davidson, Rachel Gauza Gronert, Wayne Hildebrand, Nathan Nazdrowicz, and David Walbeck. Lance Benedict and June Tveekrem graciously agreed to serve on the committee midway through the project. Committee members who initially served on the Steering Committee were Matt Evans, Don Forester, Kyle Rambo, Scott Smith, Joel Snodgrass, and Linda Weir. Thank you all for your commitment and dedication to the project.

The Maryland Department of Natural Resources (MDNR) and The Natural History Society of Maryland (NHSM) sponsored the project. The MDNR provided most of the financial support for the project through State Wildlife Grants, which were allocated to state wildlife agencies by the US Congress, administered here through the MDNR Wildlife and Heritage Service. We also thank the Maryland Coastal Bays Program for its financial support to the MARA project.

Data were collected primarily by atlas project volunteers. Additional data were provided by the MDNR Maryland Bio-logical Stream Survey, the Herpetological Education and Research Project, HerpMapper, and the Marine Mammal and Sea Turtle Stranding Network. These groups submitted to the MARA project amphibian and reptile location data that were collected by their respective projects. The *Maryland Amphibian and Reptile Atlas (MARA) Training Handbook* provided guidance on data collection techniques. We appreciate the efforts of Rachel Gauza Gronert and David Smith in developing the handbook. We thank Lynn Davidson and Linda Weir for their work in developing mapping resources for volunteers to use—specifically, digital quad and block map layers and associated online and smartphone applications. The MDNR managed and maintained data collected during the project. The NHSM hosted the official MARA website, which disseminated educational and project resources. We extend our gratitude to Dan Brellis and Joe McSharry, who developed and maintained our website.

We express our gratitude to a number of people who facilitated data management, analysis, and presentation. From the MDNR, Robert Swan, Laura Bowne, and Lynn Davidson developed and maintained the MARA database, analyzed data, and formatted data for visual presentation. Jason Poston and Nathan Nazdrowicz developed and maintained a separate database for the record verification process that held all identification evidence submitted to the project. Don Becker hosted the verification website and provided additional web management assistance. A committee of Maryland herpetofauna experts, chaired by Nathan Nazdrowicz, verified all photographs and recordings submitted to the MARA project. The MARA Verification Committee members were Lance Benedict, Don Forester, Rachel Gauza Gronert, Kyle Loucks, Nathan Nazdrowicz, Brandon Ruhe, Scott Smith, Scott Stranko, Ed Thompson, Jim White, and John White. Additional verification team members were Lynn Davidson, Wayne Hildebrand, Todd Pierson, Jason Poston, Christopher E. Smith, David Smith, Mark Southerland, Glenn Therres, and David Walbeck. Grover Brown and David Mifsud assisted with species-specific identifications. Heather Cunningham and Nathan Nazdrowicz coordinated the verification process.

County coordinators were integral to the success of the

MARA project. County coordinators facilitated data collection in their county through a variety of means, such as recruiting and training volunteers, directing surveys, and promoting the project through local conservation organizations. County coordinators also managed data collected in their county, assisted their volunteers, and submitted data to the MARA database. We are grateful for the tremendous effort put forth by the MARA county coordinators. Listed by county, the county coordinators were: Allegany—Ed Thompson; Anne Arundel—David Walbeck; Baltimore City and County—Don Forester and Joel Snodgrass; Calvert—Andy Brown; Caroline—Scott Smith; Carroll—David Smith and June Tveekrem; Cecil—Nathan Nazdrowicz and Jim White; Charles—George Jett; Dorchester—Lynn Davidson; Frederick—Wayne Hildebrand; Garrett—Seth Metheny and Amo Oliverio; Harford—Bob Chance, Brian Goodman, and Scott McDaniel; Howard—Sue Muller; Kent—Nathan Nazdrowicz; Montgomery—Lance Benedict and Rachel Gauza Gronert; Prince George's—Tasha Foreman, George Middendorf, and Mike Quinlan; Queen Anne's—Glenn Therres; Somerset—Douglas Ruby; St. Mary's—Kyle Rambo; Talbot—Kelsey Frey, Scott Smith, and Glenn Therres; Washington—Andrew Landsman and Dave Weesner; Wicomico—Lance Biechele, Ron Gutberlet, and Aaron Hogue; Worcester—Roman Jesien, Jim Rapp, and Dave Wilson.

We appreciate the support shown to the MARA project by organizations that allowed the use of their facilities for meetings. Jug Bay Wetlands Sanctuary, the NHSM, and the MDNR provided facilities for the monthly MARA Steering Committee meetings. Washington College provided facilities for meetings of the MARA Verification Committee. Facilities for the annual meeting of the MARA county coordinators were provided by Howard County Department of Recreation and Parks.

We appreciate the work of the species account authors who researched each species, compiled information about the species specific to Maryland, and wrote a narrative account. Species account authors were Lance Benedict, Heather Cunningham, Lynn Davidson, Tasha Foreman, Don Forester, Rachel Gauza Gronert, Brian Goodman, Ron Gutberlet, Wayne Hildebrand, George Jett, Andrew Landsman, Seth Metheny, Nathan Nazdrowicz, Kyle Rambo, Douglas Ruby, Brandon Ruhe, David Smith, Scott Smith, Joel Snodgrass, Christopher Swarth, Glenn Therres, Ed Thompson, David Walbeck, and Jim White. Color photographs accompany each species account. A photo contest was held to select photographs for inclusion in the book, and photographers of all skill levels were encouraged to participate. Lynn Davidson, Charles Davis, and Glenn Therres generously donated their time to judge each photograph. The photographers chosen were Andrew Adams, Lance Benedict, Jonathan Chase, Robert Ferguson, Don Forester, Wayne Hildebrand, George Jett, Matthew Kirby, Michael Kirby,

Scott McDaniel, Nathan Nazdrowicz, Bonnie Ott, Scott Smith, Kevin Stohlgren, Hugh Vandervoort, and Jim White.

More than 2,000 individuals and organizations contributed to the MARA project. Contributions included sharing observations and photographs, directing field surveys, collecting data, leading workshops, and granting access to private property. Without these dedicated volunteers and cooperators, the MARA project would not have been possible. Thank you all.

Patricia Aaron
Charlie Abeles
Stan Abremski
David Ack
Scott Ackerson
Joe Acord
Mike Adami
Andy Adams
Caitlin Adams
Julia Adams
Joe Addicks
Piers Ady
Dominic Aeschlimann
Christopher Agard
Robert Aguilar
Ed Airey
Scot Aitkenhead
Daniel Akwo
Sherry Alanis
Roberta S. Alberding
Wendy Alberg
Lori Alders
Rita Allan
Steve Allan
Michael Allen
Terry Allen
Patty Allen
Mike Allera
Annette Allor
Marci K. Altman
Mary Alves
Carolyn Anderson
Charlie Anderson
Dave R. Anderson
Gary M. Anderson
James Anderson
Jesse M. Anderson
Katie Anderson
Mariko Anderson
Matt Anderson
Sandy Anderson
Taylor Anderson
Keith Andres
Claudia Angle
Phil Angle
Max Anthony
Stephen F. Appell
Steve Appell
Nix Applin
Brad Arbogast
Greg Arbutus
Philip Ardanuy
Harry Armistead
Joan Armistead
Michael Arndt
Stan Arnold
Steve Arslanian
Jean Artes
Lori Ashbaugh
Sara Ashby

Matt Ashton
Terry Auble
Mike Auer
Paul August
Linda Augustine
Jason Avery
Kevin Aviles
Maggie Babb
Jason Babcock
Marsha Back
David Bacon
Thomas Baden
Holly Badin
Shirley Bailey
Beth Baker
Chris Baker
Joyce Baker
Linda Baker
Mike Baker
Steve Baldo
Doug Baldridge
Stan Baldwin
Brynn Bales
Bob Balestri
Marcia Balestri
Michael Balgley
Dawn Balinski
Susan Ballentyne
Ralph Ballman
Betsy Bangert
Jana Barberio
James Barnes
Brady Barr
Samantha Barrett
Marty Barron
Lisa Barry
Samuel Barry
Polly Bart
Greg Bartles
Mary Jo Bartles
Rochelle P. Bartolomei
Chris Basham
Mark Baskeyfield
Emily Bates
Scott Bates
Gary F. Battel
Ken Baxter
Jessica Baylor
Zachary Beal
James Beale
Mark Beals
Sandy Bearden
John Beatty
Bridgette Beaumont
Elizabeth Beck
Andy Becker
Paula Becker
Marie Beckey
Cadien Beckford
Karsten W. Beckmann

J. K. Beckstrom
Cindy Beemiller
Shannon Beliew
Jessy Bell
Ryan Bell
Tyler Bell
Alan Belles
Deb Bello
Hannah Bello
Joe Bello
Lance H. Benedict
Michael Benedict
Becky Benjamin
Donna Benjamin
Deborah Bennett
James Bennett
Dave Bent
Steve Berberich
K. Berenozen
Judy Berger
Ginny Bern
Seth Berry
Bonnie Bezila
Joann Biechele
Lance Biechele
Victoria Billings
David Bird
Shannon Bisese
Dawn Bisson
Garrett Black
Jim Black
Ed Blacka
Tom Blair
Chris Blakely
Jay Blakely
Carolyn Blakeney
Mike Blanchette
Peter Blank
Scott Bleakney
Harrison R. Block
Lisa Blottenberger
Lauren Blundin
Steven M. Boas
Richard Bochniewicz
Forrest Bogan
Shaun Bogan
Jorge A. Bogantes Montero
George Boker
Ray Bolger
Reed Bolger
Val Bolger
Susan Bollinger
Dan Bolton
Lydia Bonifant
Jerry Bonnie
Chris Bontrager
Sonja Boocatt
Elease Booker
Lisa Booze
Mary Ellen Borland
Shannon Borowy
Amy Bortz
Audrea Bose
Ray Bosmans
Lenore Boulet
Ed Boulware
Ray Bourgeois
James M. Bovis
Dan Boward
Scott Bowen
Mark Bowermaster
Deb Bowers
Lloyd Bowling
Laura Bowne

Samuel S. Bowser
Amanda Boyer
Jason Boyer
Melissa Boyle
Wade Bradford
Marie Brady
Michael Brady
Dan Brandewie
Larry Braun
Mary Braun
Bob Brehl
Gregory Brennan
Gwen Brewer
Sharon Brewer
Christy Bright
Jim Brighton
Christopher Briles
David Brinker
Sarah Brinker
Karin Britt
Ben Brockmeyer
Judy Broersma
Anita Bronzert
Andy Brookens
Lenore Brooks
Nymbus Broome
Tom Brosnan
Andy Brown
Jessica Brown
Kathy Brown
Lamont Brown
Mike Brown
Pat Brown
Tracilyn Brown
Tyler Brown
Michelle Browning
Dawn Brownlee-Tomasso
Rob Brownlee-Tomasso
Tom Brubaker
Cory Bruegman
Otto Bruegman
Jennifer Bruza
Karl Bryan
Dave Bryant
Julie Buchanan
David Buck
Nora Bucke
Alexander Buckler
Holly Budd
Zach Buecker
Bill Buettner
Andrew Bullen
Deborah L. Buppert
Hester Burch
Mark F. Burchick
Robyn Burger
Gwen Burkhardt
Rod Burley
David Burman
Lisa Burner
Jennifer Burnham
Angela D. Burns
Brad Burns
Elizabeth Burns
Mary Burton
Craig Busby
Paul Buse
Westley Bush
Maricel Bustos
Meghan Butasek
Garfield Butler
Bernard Butrim
Nathan Byer
Lori A. Byrne

Danny Bystrak
Paul Bystrak
Stephanie Bystrak
John Cabala
Jared Cain
P. Cain
Andrea G. Calderon
Collin Calhoun
Mike Callahan
Aaron Callis
Kathy Calvert
Brenda Campbell
Brian Campbell
Cynthia Campbell
Jeff Campbell
Shannon Campbell
Tim S. Campion
Felicia Candela
Andrew Cane
Trish Cantler
Larisa-Nichole Capers
Lorraine Cardamone
Nina Cardin
Nate Carle
Bjorn Carlson
Pam Carney
Tim Carney
John Carr
Nick Carter
Gloria Catevenis
Joseph R. Catoe
Jennifer Cedeno
Jason Cessna
Marion Chambers
Bob Chance
Rob Chapman
Aric Charowsky
Verena Chase
Judy Chatfield
William Cheeseman
Norine Chester
Jacob Chestnut
Shiehan Chou
F. Christopher
Leslie Churilla
Jack Ciesielski
Jeff E. Claffy
Donnie Clark
Katherine Clark
Ken Clark
Roxanne Clark
Tom Clark
Ty Clark
Dianne Clarke
Suzanne Clarke
Steve Claus
Denise Clearwater
Craig Clemens
Brandon Clement
Marion Clement
Rebecca Cline
Lisa A. Clossin
Virginia Cobler
Mary Catherine Cochran
Dan Cockerham
Caryn Code
Brian Coe
Krystal Coffey
Curtis Cohen
Matt A. Cohen
Wendy Cohen
Tyler J. Colaw
Brett Cole
Chap Cole

Jonathan Collins
Kathy Colston
Chris Colyer
Tom Conard
Robert Condon
Eleanor Cone
Trevor Conley
Dylan Cooper
Laura Cooper
Paula Cooper
Daniel Corcoran
Marlin Corn
Sandra Cornell
Jack Cover
Jessica M. Cox
Kelley Cox
Mike Cox
Sharon K. Cox
Patty A. Craig
Erin Crawley
Tom Creedon
Jennifer Crews-Carey
Marty Cribb
Kevin Crocetti
Susan Cromwell
Charles H. Crone
Andrew Cronin
Joe Cross
David E. Crow
Doug Crow
Wendy Crowe
Bonnie Cubbage-Lain
Jeff Culler
Bill Cumberland
Heather Cunningham
Kevin Cunningham
Mike Currie
Jeff Curry
Ken A. Curry
David Curson
Kara Curtain
James Curtis
William B. Curtis
Kristy Cusic
Brandon Custer
Matt Custer
Lee Custis
Terry Cutshaw
Roy Cutts
Rachelle Dagnault
Harvey Dail
Hilary Dailey
Donald Dale
Sharon Daley
Kaitlyn Dalrymple
April Danchik
Gerry D'Angelo
Julie Daniel
William Daniel
Curtis Daniels
Daniel Danko
Betty Darby-Glime
Karen Darcy
Jackie Darrow
Michelle Daubon
Brian David
Colin David
Michele David
Phil David
Lynn Davidson
Brittany A. Davis
Charles A. Davis
Jean Davis
Joe Davis

Kent Davis
Linda Davis
Shannon M. Davis
Tom Davis
Jon Davy
Nevin Dawson
Sierra Dawson
Larry Daye
Becky Deafenbaugh
Al Dean
Mike Dean
Samantha Dean
Scott Dearinger
Erik Deatherage
Tom Decker
Patricia Defendoch
Katherine Degerberg
Hal Delaplane
L. Dembrow
Patrick Dempkin
Brian DeMuth
Otilewa Denloye
Scott Deosaran
Dave DeRan
Francois Derasse
Deidre DeRoia
Scott Derringer
Francine DeSanctis
Vincent DeSanctis
Maria Dickhoff
Janet M. Diem
Andrew Dietrich
Debbie Dietrich
John Dilenno
Carl Diliberto
Tamene Dilnesahu
Grant Disharoon
Jen Dittmar
Beth C. Dobbins
Kevin Dodge
John Doe
April Doherty
Alyssa Domzal
Ed Donahue
Judy Donaldson
Lindsey Donaldson
Kevin Donovan
Rocky Donovan
Tara D'Orazio
Marilyn Dorfman
Charles L. Dorian
Carl Dorr
Fred Dorsey
Gina Dotterweich
Ron Douglas
Chris Drazba
Cindy Driscoll
John David Driver
Jeff Drury
Theresa D'Souza
Rhonda Duasman
Woody Dubois
Suzanne Duclos
Glenn Dudderar
Jim Duffy
Wendy Duke
Darian Dunn
Tim Dunn
Lisa D'Urso
Jesse Dustin
Nancy Dutton
Erik Eagle
Susan Earp
Wes Earp

Dave Eby
Elizabeth Echevarria
Harvey Edelson
Bruce Edwards
Djuna Edwards
Mary Edwards
Michael L. Ehrlich
Mike E. Ehrlich
Nor Eldridge
Eric Elenfeldt
Steve Elinsky
Terry Elkins
Greg Elliot
Bill Ellis
James Ellis
Walter Ellison
Bob Elmer
Jennifer Elmer
Jennifer Emanart
Marc Emond
Bill Engle
L. C. Englehardt
Eric Epstein
Lori Erb
Merrill Erichsen
Jeff Errington
Ted Ersek
Bob Escobido
Adam Eshleman
Kate Estler
Caren Euster
Janet Evans
Malachia Evans
Middleton Evans
Mark Everett
Corey Everhart
Chris Evers
Claire Ewing
Brian Eyler
Charlotte E. Eyring
Valerie Fair
Michael Faith
Holly Fallica
Terry Farman
Symantha Faulkner
Scott Faunce
John Fauntleroy
George Fear
Roy Fedders
Joe Fehrer
Dan Feller
Linda Fennell
J. R. Fenwick
Gracie Ferber
Kathryn Ferger
Bob Ferguson
Mickey Ferguson
Paula Fernandes
Mike Ficalora
Temperance Field
Alan Fink
Harry Fink
Kelly Fiore
Jake Fisher
John Fisher
Mark Fisher
Sherry L. Fisher
David Fitzpatrick
John R. Fleetwood
Kathy Fleming
Kelly Flint
Kelly Flore
Bob Flores
Barbara Fogg

Kevin Foley
Seely Foley
Mark Fondren
Shirly Ford
Tasha Foreman
Don Forester
Penny Forester
Andy Forte
Barry W. Fortune
Donna Fossaceca
Melvin Foster
Cathy L. Foutz
Kelly Fox
Pam Fox
Dave Francis
Lisa Franke
Pamela Franks
Jackie Frazee
Mary Frazier
Richard A. Freas
Gandalf French
Kelsey Frey
Robert Frezza
Rosemary C. Frezza
Elaine Friebele
James Frost
Chris Frye
Jennifer Frye
Sarah Funck
Don Furbee
Chuck Gait
Richard Gallant
James Gallion
John Gallo
Nan Gangler
Scott Gardiner
Matthew Gardner
Kriste Garmen
Michelle Garrett
Katie Garst
Barrett Garvin
Aundre Gaston
Patrick Gates
Sam J. Gayhart
Peg Gelhard
Robert Gelhard
Robert P. Gelner
Ross Geredien
Leslie Gerlich
Jonathan Germroth
Bernard Gershan
Naomi Gershan
Terence Ghee
Lisa Marie Ghezzi
Tessa Giannini
Ashley Gibbon
Rob Gibbs
Dave Gigliotti
Julie Hutto Gilbert
Teddy Gill
Cathy Gilleland
David Gillum
Teresa L. Gilmore
Maren Gimpel
Christina Ginsberg
Ben Giraldo
Alfred Girard
Christina Gladmon
Corliss Glennon
Les Glick
Mike Glynn
Stephanie Goddard
Lara Goeke
Ai Goetzke

John Goetzke
Mary Goldie
Brandon Goldner
Heather L. Goldsborough
Anne Gonnella
Samantha Good
Brian Goodman
Melissa Goodman
Scott Goodman
Carl Gordon
Carl Nate Gordon
Barbara Gottleib
Richard Gourley
Belinda A. Grabrick
Carolyn Graessle
Grant Graessle
Faye Graff
Gary Graff
Kevin Graff
Kathy Gramp
Katherine Grandine
Evan Grant
Alexis Gray
Alan S. Green
Helene Green
Jim Green
John Green
Josh Green
Kaite Green
Amy Greenwood
Benton Greenwood
Jennifer Gregory
RuthAnne Gregory
Matthew Grey
Douglas Griffith
Mike Griffith
Sharon Griffith
Damion Griffiths
Eric Grivel
Elizabeth Grogan
Keith Gromen
Rachel Gauza Gronert
Gene Groshon
William Gross
Barbara Grotheer
Steve Grudzien
Iris Grundler
Sharon Grzlick
Barry Guard
Karen Guglielmo
Cindy L. Guion
Ann Gummerson
Carol Gutberlet
Ron Gutberlet
Joey Gutkoska
Danielle Guy
Deb Gwynne
Jonathan Hakim
Craig Hall
Linus Hall
Stephanie Hall
Rick Hallsworth
Neal Halsey
Harold Haltaman
John Hamill
Jason Hamilton
Sue Hamilton
Eric Hammack
Don Hammerlund
Chris Hamrick
Peter Hanan
Patty Handy
Tom Hane
Joe Hanfman

Wendy Hanley
Charlotte Hanson
Joe Hanyok
Bill Harclerode
Chelsea Hardies
Susan Hardinger
Rebecca Hardy
Elizabeth Hargrave
Matthew Harper
Paul Harrell
Cheyenne Harris
Clarissa Harris
Hayward Harris
Herb Harris
John Harris
Laurel Harrison
Maria A. Harsanyi
Curtis Hart
Tom Harten
Margie Hartman
Stephanie Hartman
Peggy Hartnett
Kayla Hartung
Kyra Harvey
Maureen F. Harvey
Mike Harvey
William Harvey
Candice Haskin
Ammie Hassell
Audrey Hatry
Sean Hauber
Hans H. Haucke
Michael Havrilla
Monica Hawse
Harris Hayward
Jodie A. Hayward
Noah J. Hayward
Brian Heacock
Richard Hearn
Melinda Heasley
Brian Hebb
Justin Hebb
Todd Heerd
Kate Heilman
Dave Heilmeier
Theressia A. Hein
Billy Heinbuch
Stacy Helgason
K. Hellmann
Bruce Henderson
Jennifer Henderson
Vickie Henderson
Bill Henry
Lauren Hensley
Lloyd Hepburn
Bud Herb
Tiffany Hergett
Darcy Herman
Rebecca Herman
Rebecca Jean Herman
John Hernandez
John Herrick
William Hershberger
Chris Heyrend
Kendall Heyrend
Erin Hickey
Teresa Hickey
Charlie Hickman
Kenneth Hiebler
Debbie Hildebrand
Elisabeth K. Hildebrand
Jacqueline T. Hildebrand
Wayne G Hildebrand
Chris M. Hill

Russ Hill
Rebecca Hiller
Artie Hinamin
Dan Hinder
Larry Hindman
Eric Hines
Tiffany Hinson
Tim Hoen
Amy Hoffman
Brad Hoffmaster
Aaron Hogue
Todd Holden
Emy Holdridge
Chad Holdsworth
Vaughan A. Holland
Mary Hollinger
Lindsay Hollister
Shawntavia Holmes
Phil Holzbauer
Philip Holzbauer
Donna Home
Josh Homyack
Susan Hood
Mark Hooper
Samantha Hopkins
Lori Horne
Edith Hoschar
Ed Hose
Hans Houke
Scott Housten
Jane Howe
Hunter Howell
Shawn Hrotic
Sue Hu
John Hubbell
Bill Hubick
Heather Huddle
Cathy M. Hudson
Steven Huettner
J. C. Hull
Bill Hulslander
Maria N. Hult
Bud Humbert
Ben Humphrey
Denise Humphrey
Tom Humphrey
Linda J. Hunt
Becky Hunter
John Hurt
Christy Huseman
Julia Hutto Gilbert
Anna Hutzel
Steve Huy
Michael D. Hyde
Angela Hysan
Tami Imbierowicz
Marc Imlay
Adam Int Veldt
Sharon Int Veldt
Holly Ireland
Pete Ireland
Janet Ishimoto
Danielle Iuliucci
Tom Jack
Kaylee Jacobs
Frode Jacobsen
Sandra Jaffe
Donald Jameson
Anna Jansen
Teresa Jansen
Camila S. Januario
Ray Jarboe
Carl Jarvis
Peter Jayne

Bev Jeffas
Becka Jeffers
Carol Jelich
Joseph Jelich
Emily Jelick
Laura Jenkins
Matt Jennette
Nick Jennings
Ted Jensen
Roman Jesien
George Jett
Joshua Johansen
Kay John
Amanda Johnson
Danny Johnson
Derek K. Johnson
Heather Johnson
Lisa Johnson
Michelle Johnson
Mike Johnson
Paula Johnson
Wayne E. Johnson
Emily Johnston
Lynne Marie Johnston Soper
David Joiner
Peter Jolles
Alan Jones
Anita S. Jones
Kara Jones
Matt Jones
Skip Jones
Stephanie Jones
Travis Jones
Ben Jordan
Glenda Jordan
Betsy K. Kadow
Jennifer Kahl
Ted Kahn
Jack Kaiser
Grant Kalasunas
Melissa Kalb
Shawn Kalicka
Sharon Kaltwasser
Stephen Kaltwasser
Nancy G. Kaplan
Melissa Kapper
Michael Kashiwagi
Morgan Kaumeyer
David Kazyak
David Keane
Craig Kehoe
Collen Keifer
Emily Keith
W. Kellam
Bradley Keller
Cindy Keller
Fred Kelley
Greg Kellner
Tim Kelly
Patty Kendrick
Bradley Kennedy
Shannon Kennedy
Mike Kepler
S.T. Kerchner
Arlene Kerish
Stuart Kerr
Bill Kidwell
Peter Kilchenstein
Jay Kilian
William Killen
Jenifer J. Kim
Eric Kimmel
Eric Kindahl
Margo Kindig

Mollie King
Randy King
Betty A. Kipphut
Christopher Kirby
Matthew Kirby
Michael Kirby
Sarah Kirby
Ed Kirk
Ian Kirk
Tim Kirschner
Megan Kisser
Stephen Kleindienst
Leslie Kline
Gerry Klinedinst
Alan Klotz
Kyle Klotz
Gary Klue
Barbara B. Knapp
Heather Knapp
Wesley Knapp
Carrie Knauer
Anita Knight
Eric Knight
Sarah Knight
Michelle Knotts
Ashley Kobisk
Frimpong Kodua
Loraine Koepenick
Carol Koh
Greg Kolarik
Stan Kollar
Leonard Kolman
Michelle R. Kopp
Donald Koranda
John Kosko
Jane Kostenko
Meagan J. Kratz
Paul Kreiss
John H. Kresslein
Luke Kresslein
Jill Kries
Millie Krimelmeyer
Kyle Kroll
Caroline Kryger
Katrina Kugel
Sue Kullen
Crystal Kunst
Lukas Kurdziolok
Adrienne Kurtz
Laura Labella
Randy Lacey
Darrell Lack
Liz Laher
Krispen Laird
Allen Lamb
Stephanie Lamb
Jeffry Lambert
Jean Lancaster
Andrew Landsman
Jerry Landsman
Karen Lane
Leyla Lange
R. E. Langford
Mark Lanham
Linda Lanzisera
Maureen M. Larkin
Tim Larney
Angie Latham
Judy Lathrop
Mike Lathroum
Will Lattea
Cory Lavoie
Keith Lawrence
Megan Lawrence

Othalene Lawrence
Chris Lawson
Stephanie Lazarus
Helen Leahy
Heath Leaverman
John Lefebure
Lynn Leibig
Sandra Leitner
Charles B. Leonard
Kenneth Leonard
Elaine L. Leonard-Puppa
John Leskinen
Alan W. Leslie
Susan E. Leslie
Brandon Levesque
Anne Levine
Sandy Levy
B. Lewis
Krysten Lewis
Mike J. Lewis
Rob Lewis
Thomas Lewis
M. Lichtey
Eric Liebgold
Ryan Lilley
Lori A. Lilly
Dana Limpert
Rachel Lind
Tracy Lind
Bill Link
Jessica A. Lipka
Tom Lipka
Kathy Litzinger
Laina Lockett
Steve Lofgren
Bob Long
Jim Long
B. Longley
Megan Lord
Crystal Loring
Kyle Loucks
Chris Louzon
Jessi Love
Felicia C. Lovelett
Patricia Lovellette
Adrienne Lowe
Alexandria Lowe
Candace Lower
Ilene Lubeck
Katie Luczak
Frank Ludwig
Robert Lukinic
Debby Luquette
Richard Luquette
Jo Lutmerding
Mikey Lutmerding
Lloyd Lutz
Monica Lynn
Rachel M. Lynn
Keith Lyon
Rick MacDonald
Allen Mackey
Michael Mackiewicz
Patty Mackin-Dowd
Barbara MacNeil
Nancy Magnusson
Bill Mahoney
Brian Malat
Ben Malatt
Seth Malek
T. Malloy
Joan Maloof
Roy Man
Jerold Mande

Bryce Manges
Clara Mankowski
Judy Mansfield
Anne Margro
Nancy Marin
Sandra Marine
Erin Markel
Tiffani Marks
Lynne Marquess
Frank Marsden
Tommy Marsellas
Ruth Marsh
John Marshall
Joyce Marstaller
Kathy Martin
Peter Martin
William H. Martin
Woody Martin
Troy Martinez
Vince Martino
Geoffrey Mason
Kevin Mason
Lori Mason
Rick Mason
Amanda Matheny
Jason Mathias
Susan Mattews
Nichole Mattheus
Kathryn Mattingly
Sharon Mattingly
Dawn Maurer
Larry Maurer
Marianna Max
Wayne McBain
Bill McCaffery
Krista McCall
Sean McCandless
Jim McCann
Katherine McCarthy
Elizabeth McCartney
Chuck McCausland
Bruce McClelland
Grazina McClure
Mike McClure
Marlene McConnell
Scott McConnell
Tammy McCorkle
Tammy McCormack
Scott McDaniel
Tom McDaniel
Peter McDevitt
Ambria McDonald
Karen McDonald
Liz McDowell
Earyn McGee
Vincent McGettrick
Bill McGiffin, Jr.
Scott McGill
Michael McGlynn
Jeff McGoldrick
Joseph McGovern
Kate McGraw
Jennifer McIntyre
Katherine L. McKearn
Kirk McKee
Colin McKeown
Megan McKewen
Michael McKewen
Sean McKewen
Antoine McKnight
Donald McKnight
Jen McKnight
Jonathan McKnight
Hugh McLaurin

Tina McMillan
Dan McNamara
John McNamara
Maureen McNulty
Tim McNutt
Wes McReynolds
Joe McSharry
Eileen McVey
Jim McVey
Tara Meadows
Jack Meckley
Mike Medinski
Lewis Megginson
Brian Meinhart
Damodara Rao Mendu
Ellen V. Mering
Woody Merkle
Whelden Merritt
Jon Merryman
Bill Merza
Heather Messick
Seth Metheny
Chris Metzbower
Helen Metzman
David Meyer
Hannah Meyers
Erik Michelsen
Michael Mickiewicz
Katie Mickletz
Steve Mickletz
George Middendorf
Frank Middleton
Mike Miedzinski
Anne Mihalick
Myron Milbourne
Candace Milcarzyk
Adrienne W. Miller
Caitlin A. Miller
Chrissy Miller
Christina Miller
Dawn Miller
Harmony Miller
Kimmy Miller
Nate Miller
Robert Miller
Sam Miller
Sharon Miller
Sue Miller
Tom Miller
Will Miller
Sadie Mills
Gale Minnich
Andrew Minnick
Bobby Minnick
Joe Mitchell
Laura Mitchell
Robin Mitchell
Ruth Mitchell
Stuart Mitchell
Erica Mitrano
Aisha Mix
Karen Mobley
Robert Moeller
Laura Mol
Karyn Molines
Connor Molloy
Craig Monroe
Christie Montgomery
Bryan R. Moody
Gloria Moon
Dward Moore
Rory Moore
Sherita Moore
Wayne Moore

Penny Moran
Janet Morehouse
Ashley Moreland
Alana Morgan
Gary Morgan
Raeth Morgan
Shelley Morgan
Robert Moriarty
Rick Morin
Steve Moritz
Jim Morris
Tyler Morris
Susan Morrison
Merle Morrow
Ernie Morse
Mary Morton
Kevin Moses
Heidi Motz
John Moulis
Joseph D. Mowery
Flor Moya
Gordon D. Mueller
Jenny Mulhern
Tim Mullady
Jim Mullan
Sue Muller
Chelsea Mummaugh
Katherine Munsen
James Murduck
Brendan Murray
Judy Murray
Nandini Muthusubramanian
Cinthia Myers
Hannah Myers
E. Z. Naibert
Maggie Nantz
Molly Nantz
Felipe Nazario
Amy Nazdrowicz
Nathan Nazdrowicz
Bernie J. Nebel
Kelly Neff
George M. Neighbors
Amy Nelson
Katey Nelson
Linda Nelson
Rick Nelson
Alice Nemitsas
Sue Neubauer
Jeff Nibali
Renee Nicholson
Jana Nietmann
Connie Nissley
Paul Nitz
Debbie Nizer
Doug Nizer
Susan Nobile
Mark Noble
David D. Nodar
Cynthia Norman
Jane M. Norman
Janet Norman
Phil C. Norman
Billie Norris
Ginger North
Renee North
John Nudelman
Brenda Nuse
Tom Oakley
William O'Brien
Jessica L. Oby
Janice C. Odonnell
Bill Offutt
Olav Oftedal

Amo Oliverio
Barbara Olsh
Dawn Olson
Don Olson
Brooks Onley
Chris Ordiway
Phil Orndorff
Richard Orr
Aydin Örstan
Michael Osmun
Mike Ostrowski
Robert Ostrowski
Bonnie Ott
Anne Overall
Kevin Oxenrider
Sharon Page
Stephen Page
Moira Pairo
Monica Palko
Angela Pan
Bernie Panzenhagen
Matt Pape
Nick Paraskevas
Keith Pardieck
Kay Paris
Labro Parish
Ted Parker
Carolyn Parsa
Patty Parsley
Paul Parzynski
Sue Patrice
Craig Patterson
Donald Patterson
Matt Patterson
Michael C. Patterson
Nicole Patterson
Anthony D. Paul
Bethany Pautrat
Petar S. Pavlov
Ashley Payne
Eric Peach
Tami Pearl
Nick Peditto
Paul Peditto
Ed Peffly
John Pellock
Barb Penny
Kia Penso
Siobhan Percey
Andrea Perez
Hannah Perrin
Bryan Perry
Matthew C. Perry
Demeta Peterbark
Robert Peters
Stephanie Peters
David Petersen
Todd Petersen
Paul Peterson
Stephen Peterson
Stuart Peterson
Paul Petkus
Emil Petruncio
Paul Petzrick
Mary Pfafffko
Alan E. Pflugrad
Autumn Phillips
Karen Phillips
Kevin Phipps
Paul Piavis
Catherine Pichler
Bob Pickett
Elizabeth Pierce
Margaret Pierre-Nanan

Katie Pilcher
Rebecca Pilgrim
Mary Piotrowski
E. L. Pitney
Tori Pitts
Drew Pizzala
Eli Platnick
Mandy Plummer
Julie Poehlman
Danny Poet
Sean P. Poholsky
Susan Poiniaszek
Russell Poivesan
Julian Polak
Maggie Polak
Oren Polak
Mark Pollitt
Fran Pope
Jeff Popp
Anne Porkney
Frank Porkney
Ben Porter
Kathleen Porter
Amanda Poskaitis
Genie Posnett
Kari Post
Nancy Post
Robert Post
Jason Poston
Amelia Potter
Bill Potter
Erin Potter
Charlie Poukish
Bradford Powell
Christen Powell
Michael Powell
Rick L. Powell
Tim Powell
Ian Priestly
Niles Primrose
Dawn Pritchard
Tony Prochaska
Doug Proctor
Elizabeth Prongas
Terry Psaltis
Carolyn Puckett
Cliff Puffenberger
Liz Purcell-Leskinen
Kyle Putman
Christopher F. Puttock
Dominic Quattrocchi
Elsa Quillin
Kim Quillin
Lila Quillin
Mike Quinlan
Karen Quint
Juan Quiroga
Chris Radack
Gemma Radko
Elaine Raesly
Sean Rafferty
Jennifer Rafter
David E. Rager
Currin Raimer
Pezhmon Raissi
Johanna L. Rambo
Julie Rambo
Kyle Rambo
Ryan Rambo
Brittney Ramirez
Mario A. Ramirez Castelo
Paul Ramsey
Fiona Randle
Jim Rapp

Glenn Rash
Matthew Ratcliffe
William E. Ray
Christina Raymond
Theresa Raymond
Gus Redlinger
Jan Reese
Ken Register
James Rehbein
Zachary Reichold
Art Reid
Dawn Reid
Elissa Reineck
Ann Reinecke
Dave Reinecke
Chris Reinke
James Reise
Richard Reise
Karen Rennich
Mary Renoll
Armin Rest
Maria R. Reusing
Carla Rhode
Sue Ricciardi
Charles Rice
Janna Rice
Teddy Rice
Jared Richards
Teal Richards
Carson Richel
Amy Riddle
Dan Rider
Vickie Ridgeway
Marina Riese
Jason Riesner
Jim Riley
Peter Ring
Lindsay Ringgold
Robert F. Ringler
Gia Marie Ristvey
Casie Ritter
Mike Rivera
Charlie Roach
Krista Roberson
Lewis Robert
Linda Roberts
Jennifer Robertson
T. Robinette
Ken Robinson
Todd Robinson
Lynne Rockenbauch
Jim Rode
Andy Rodgers
Derek Rodgers
Elaine Rodrigues
Alonso Rodriguez
Donald Rohrback
Tom Roland
Vann Rolfson
Beverly Rollins
Willem Roosenburg
Charlotte Rose
Karen Rose
Dave Roseman
Gretchen Rosencrantz
Ryan Rosencranz
Marylee Ross
Sally Ross
Clarissa J. Rous
Jim Rousey
Bill Rouston
Christopher Rowe
Spencer Rowe
Sujata Roy

Frank Roylance
Andy Royle
Maks A. Rozenbaum
Jay Rubinoff
Doug Ruby
Sally Ruby
Melissa Ruck
Mike Rudy
Russ Ruffing
Gabriel Ruiz
Evan Rule
Anne Rullman
Dana Rupard
Ronald C. Ruse
Justin Rutledge
Bill Rymer
Greg Saba
Brian David Sadler
Greg Safko
Sarah Sagalow
Nick Sagwitz
Saki Sakakihara
Christian Sakoian
Hunter H. Sakoian
Timothy Sala
Grayson Saldana
Matt Salo
Jean Salvatore
Jim Sanborn
Tierra Sandeen
Amy Sanders
Beth Sanders
Carol Sanders
Adrianna Sante
Z. Sarrez
Matt Sarver
Chuck Saunders
Elisabeth Saunders
Jason Savage
Jan Saxton
Fred Saylor
Karen Sayre
Stephen Sayre
Susanna Scallion
Jack Scanlon
Tyler Schaeberle
Justine W. Schaeffer
Mike Scheffel
Sonja Scheffer
Greg Schenck
Irv Schindler
Mike Schindler
Kathleen Schlappal
Jeri Schlenoff
Jake Schneider
Paul Schneider
Steve Schneider
Jamie Schofield
Jeff Schramek
Thomas Schreiber
Bill Schrodel
Phil Schubert
Kay Schultz
Laura Schultz
Margaret Schultz
Gretchen N. Schwartz
Kurt Schwarz
Sara Schwarz
Sharon Schwemmer
Rosie Schwier
Julia A. Schwierking
Don Schwikert
Jake Scott
Mike Scott

Chris D. Scroggins
Savanna Scroggins
Carol Seigel
Fred Seitz
Matt Sell
Missy Burton Sellers
Phil Serafinas
M. Sgambati
Khawaja Shahrukh
Brian Shallcross
Charles A. Sharp
Teresa Shattuck
Zach Shattuck
Desiree Shelley
Jeff Shenot
Nathan Shepard
Jay M. Sheppard
Nancy Sherertz
Rob Sheridan
Dan Sherman
Frances Sherman
Georgianna Shertzer
Andrew Shifflett
Jeremy Shifflett
Kate Shoup
Gerrit Shuffstall
Bobby Siater
Jane Sidebottom
Jeff Sidebottom
Jeremy Sidebottom
Alex Siegel
Mary Sies
Christine Silberberg
Eric Silva
Susan Silver
Chris Simmers
Connie Simmons
Desiree Simmons
Mark Simmons
Michael Simmons
Susan Simpson
Evan T. Sims
Bill Sipple
Mike Sipple
Sean Sipple
Garrett Sisson
Emily Skelly
Terry Skinner
Marti Skogebo
Peter Skylstad
Edward Slade
Dylan Slagle
Bobby Slater
Rich Slavik
Edwin Sligar
Dan Small
Adam Smith
Christine N. Smith
David R. Smith
Ester Smith
Irene Smith
Jackie C. Smith
Jacob Smith
Kathy Smith
Katrina B. Smith
Korey Smith
Kristin Smith
Lisa Smith
Mandy Smith
Mike Smith
Peter Smith
Sara Smith
Scott Smith
Shelby Smith

Teresa Smith
Tim M. Smith
Tracy R. Smith
Wayne E. Smith
Mike Smolek
Chris Smolinsky
Margaret Smyles
Joel Snodgrass
Amy Snyder
Bill Snyder
Nancy Snyder
David Sohns
Mary Sokol
Bob Solem
Joanne Solem
Rylinn Sorini
James Soule
Mark Southerland
Pat Spalding
Tyler Spalding
Matt Spannare
Clarence Sparks
Darrick Sparks
James Speicher
Micheal Speicher
Anitra Speight
Marty Spellman
Mary K. Spence
Tracey Spencer
Nick Spero
Steve Spielman
Beth Spiker
Harry Spiker
John P. Spinicchia
Paul Spitzer
Ben Springer
Kristen Springer
Mark Spurrier
Achyuthan Srikanthan
Michael St. Denis
Jennifer E. St. John
Jesse St. John
Stephen Staedtler
Harold Staley
Luke Staley
Mark W. Staley
Peter Stango
Emily Stanley
Richard Stanley
Andy Stansfield
Taylor Stark
Lynette Starke
David Staton
Kristen Stauff
Chelsie Stecher
Michele Steele
Shannon Steele
Trish Steele
Marisa Steiger
Michele Steiger
Janet Steinberg
Ryan Steiner
Michele Steinitz
Ken Stepanuk
Julia Stephens
Susan Sterling
Jayne Stevens
Mary Stevens
Charles Stevland
Asli Stewart
Nancy Stewart
Jason Stick
Chuck Stirrat
David Stokes

Mike Stone
Karina Stonesifer
Nick Stonesifer
Tyler Storm
Matt Storms
Amy Strahl
Robert Strahl
Scott Stranko
Lynn Strauss
Xanthia Strohl
Susan Stuchell
Adam Stull
Rebecca Stump
Linda Subda
Kathy Sullivan
Lori Summer
Alexandra Sutton
Connie Sutton
Ingrid Swan
Ray Swank
Taylor Swanson
Chris Swarth
Jim Swift
Mac Syphax
Fumika Takahashi
Kanji Takeno
Sharon Tamburello
Winny Tan
Peggy Tanner
Ryan Tant
Gary Tarbuk
Adina Tarshish
James Tate, Jr.
Juanita Tate
Pat Tate
Adam Tatone
Heather Tatone
Christopher Tawney
J. E. T. Taylor
Matthew Taylor
Bonnard Teegarden
Benjamin Teich
Sandy Teliak
Jason Tesave
Jamie Testa
Barbara Thaler
Casey Therres
Cindy Therres
Eric Therres
Glenn Therres
Jessica Therres
Joe Therres
Leah Therres
Stephen Therres
Christy Thomas
Clayton Thomas
Ebony Thomas
Gayton J. Thomas
Mark Thomas
Meagan Thomas
Ed Thompson
Jesse Thompson
Mike W. Thompson
Christine Thurber
Ryan Thurston
Jennifer Tidd
Maggie Tieger
Matt Tillett
Rachel Tillinghast
George Timko
Paul Tingler
Josh Tiralla
Robin Todd
Chris Todd

Stan Tomaszewski
Larraine Tompkins
Lucy Tonacci
Alex Torrella
Reggie Townsend
Mary Toy
Wilson Trabal
Jason Traband
Lorraine Trabing
Mike Tracey
Kate Traut
Shannon Travers
Patrick Treece
Angela Trenkle
Scott Trexler
Joshua Trimble
Tina Tripp
Jennifer Troy
Joe Truit
Casey D. Trump
Kate Tufts
Kathy Tull
Catherine Tunis
James Turek
Jim Turek
Suzanne Turek
Mike Turner
Nicolas Turner
June Tveekrem
Will Twupack
Ronald Tydings
Kelly N. Tyree
Chris Tyson
Keith Underwood
Joyce A. Utmar
Kirsti Uunila
Carol Vaeth
Evan Vaeth
David Van Houton
Keith Van Ness
Hugh Vandervoort
Gregory Vandervoot
David Vanko
Mike VanValen
Gary Varesko
Maria Vargas
Jerry Veek
Theodore Verbich
Jillen Vest
Polly Victor
Kim Viernes
Bart E. Viguers
Lee Vines
Don Vogel
Sandra Vogel
Eddie Voorhaar
Matt Voorhaar
Martha Voorhees
Sally Voris
Chuck Wachter
Eric Waciega
Diana Wagner
Grant Wagner
Holly Wagner
JoAnne Wagner
Wes Wagner
Angus Walbeck
David Walbeck
Hutch Walbridge
Clare Walker
Julie Walker
Marie Walker
Donald W. Wallace
Mark Wallace

Garrett M. Waller
Jennifer Walsh
Mia Walsh
Amy Walters
Kent Walters
Patrick G. Wamsley
Norman Wang
Daniel Ward
Donna Ward
Ronald Ward
Doug Warner
Jordan Warner
Paris Washington
Lesley Wasilko
Meredith Watters
Alex B. Waugh
Ann Wearmouth
Peter Web
Caroline Webb
Ronald Weber
Donald Webster
Benjamin Wechsler
Dave Weesner
Linda Weir
Peter Weiser
Chuck Weinkam
Glenn Welch
Shannon Welford
Tina M. Wellington
Katelin Wells
Kyle Welsh
Melissa Wentz
Vinny Wentz
Michele Werner
Amanda Werrell
Robert J. Werrlein
Leeandra Wesley
Steve Westre
Janet Wheatley
David Wheeler
Lynne Wheeler
Michael Wheeler
Dennis Whigham
Marguerite Whilden
Matt Whitbeck
Alexa B. White
Jim White
John White
Caitlin Whitlock
Ricky Whitney
J. B. Whitten
T. B. Whitten
Sylvia Whitworth
Terry Whye
Bart Wickel
Chris Wicker
Becca Wier
Hal Wierenga
Andrea Wiggen
Jim Wilkinson
Jonathan Wilkus
Pam Willett
Guy Willey

Jonathan Willey
Levin Willey
Dwight Williams
Guy Williams
Molly O. Williams
Steve Williams
Sarah Williamson
Stephanie Williamson
Ernest J. Willoughby
Michael Wilpers
Dave Wilson
Greg Wilson
Jane Windley
Rob Windsor
Gerald Winegrad
Alex Winter
Gil Winters
Eileen Wise
Cathy Wiss
Kerry Wixted
Nancy Wolfe
Kara Wolf-Pitts
Nicholas Wood
Roy Wood
Brian Woodcock
Dave Woodcock
Peter Woodside
Bruce Woodward
Michael Woolford
Henry Woolley
Travis Wray
Christy Wright
Cynthia Wright
Leslie Wright
Terry Wright
Tommy Wright
Jim Wrigley
Albert Wurm
Mariel Yarbrough
Carol Yates
Alex Youmans
Amy Young
James Young
Megan Young
Mike Young
Vance Young
Fran Younger
Laura Younger
Garrett Younkins
Debbie Zalesak
Ed Zamaitis
Kathleen Zanoni
Robert Zappalorti
Zeke Zarnosky
Mary Zastrow
Alex Zerphy
Jeni Zerphy
Matt Zerphy
Charles E. Zimmerman
Mark Zimmerman
Aiyana Zinkand
John Zuke
John D. Zyla

Organizations

Anita C. Leight Estuary Center Herpetology Team
Audubon Naturalist Society, Woodend Sanctuary
Audubon Society of Central Maryland
EA Engineering, Science, and Technology, Inc.
Howard County Recreation and Parks
Izaak Walton League of America, Bethesda Chevy Chase
    Chapter
Jug Bay Wetlands Sanctuary
Marine Mammal and Sea Turtle Stranding Network
Maryland Department of Natural Resources
Maryland State Highway Administration
Masonville Cove Environmental Education Center
The Natural History Society of Maryland, Inc.
The Nature Conservancy
North American Field Herping Association
Sandy Spring Friends School, fourth grade class
Second Chance Wildlife Center
Smithsonian Environmental Research Center
US Department of Defense
Washington College

Acronyms and Abbreviations

ac	acre(s)	MARA	Maryland Amphibian and Reptile Atlas
C	Celsius	MASL	Maryland Academy of Science and Literature
CITES	Convention on International Trade in Endangered Species of Wild Fauna and Flora	MDNR	Maryland Department of Natural Resources
		MET	Maryland Environmental Trust
cm	centimeter(s)	mi	mile(s)
DOD	US Department of Defense	mi^2	square mile(s)
ESA	Endangered Species Act of 1973	MNHP	Maryland Natural Heritage Program
F	Fahrenheit	NHSM	Natural History Society of Maryland
ft	foot (feet)	NMFS	National Marine Fisheries Service
ha	hectare(s)	NOAA	National Oceanic and Atmospheric Administration
in	inch(es)		
IUCN	International Union for Conservation of Nature	NPS	National Park Service
		NWR	National Wildlife Refuge
km	kilometer(s)	ppt	parts per thousand
km^2	square kilometer(s)	TNC	The Nature Conservancy
m	meter(s)	USFWS	US Fish and Wildlife Service
m^2	square meter(s)	USGS	US Geological Survey

The Maryland Amphibian and Reptile Atlas

Introduction

Globally, amphibian and reptile species are among the most threatened of animal groups (J. W. Gibbons et al. 2000; Stuart et al. 2004). In the United States, fully one-third of plants and animals are of conservation concern, based on an assessment of more than 20,800 species, including 36% of amphibians and 18% of reptiles (B. A. Stein et al. 2000). Habitat alteration and loss, invasive species, emerging diseases and pathogens, environmental pollution, commercial collection, and climate change are serious threats to herpetofaunal diversity (Gibbons et al. 2000; J. P. Collins and Storfer 2003). Extinctions and population declines of herpetofauna have steadily increased since the 1970s (Alroy 2015). At least 3% of frog and toad species have disappeared globally, and scientists estimate that another 7% could be lost within the next century (Alroy 2015). The current extinction rates of amphibians and reptiles are four orders of magnitude higher than expected background rates (Alroy 2015). These pervasive threats and global population declines point out the need for increased attention to the conservation and proper management of amphibians and reptiles.

Overview of Amphibians and Reptiles in Maryland

The United States ranks within the top 10 countries worldwide for herpetofaunal diversity (Young et al. 2004). That diversity is concentrated in the eastern and southeastern regions of the country. In fact, the southeastern United States is the world center of salamander diversity. Maryland ranks second among the states of the mid-Atlantic region for amphibian and reptile diversity (B. A. Stein 2002). Currently, 89 native and naturalized species of amphibians and reptiles are found in Maryland (table 1). These species include 42 amphibian species: 22 salamander and 20 frog and toad

(i.e., anurans) species. The remaining 47 species are reptiles. Snakes are the most diverse reptile group, with 23 species, followed by turtles with 18, and lizards with 6 species. The diversity of Maryland's herpetofauna is, in part, influenced by the location of the state on the continent and the state's five distinct physiographic provinces. While many of Maryland's amphibian and reptile species are found in multiple provinces, others are restricted to particular provinces within the state.

The Maryland Amphibian and Reptile Atlas Project

In 2008, Maryland herpetologists, naturalists, and state wildlife agency personnel met to discuss herpetological issues in the state. One of the foremost concerns arising from that meeting was a need to determine current distributional patterns of Maryland's amphibians and reptiles. At the time, the collective knowledge on the distributions of Maryland's herpetofauna was based mostly on distributional maps compiled in 1975 by Herbert Harris, Jr., in "Distributional Survey (Amphibia/Reptilia): Maryland and the District of Columbia." These were point-based maps, using dots for each location a specimen had been collected. This work proved valuable for depicting county-level occurrences and for visualizing the locality data compiled at the time. However, the statewide distributional maps compiled by Harris in 1975 were not derived from a systematic effort, therefore leaving some areas of the state under-surveyed or neglected altogether. The Maryland Amphibian and Reptile Atlas (MARA) project was designed to meet the need for updated distributional data for the state's herpetofauna and comprehensive coverage of all regions of Maryland by using citizen scientists to collect data throughout the state. The Maryland

Table 1. Native and naturalized amphibian and reptile species known to occur in Maryland

Common Name	Scientific Name
AMPHIBIANS	
Salamanders	
Jefferson Salamander	*Ambystoma jeffersonianum*
Spotted Salamander	*Ambystoma maculatum*
Marbled Salamander	*Ambystoma opacum*
Eastern Tiger Salamander	*Ambystoma tigrinum*
Hellbender	*Cryptobranchus alleganiensis*
Green Salamander	*Aneides aeneus*
Northern Dusky Salamander	*Desmognathus fuscus*
Seal Salamander	*Desmognathus monticola*
Allegheny Mountain Dusky Salamander	*Desmognathus ochrophaeus*
Northern Two-lined Salamander	*Eurycea bislineata*
Southern Two-lined Salamander	*Eurycea cirrigera*
Long-tailed Salamander	*Eurycea longicauda*
Spring Salamander	*Gyrinophilus porphyriticus*
Four-toed Salamander	*Hemidactylium scutatum*
Eastern Red-backed Salamander	*Plethodon cinereus*
Northern Slimy Salamander	*Plethodon glutinosus*
Valley and Ridge Salamander	*Plethodon hoffmani*
Wehrle's Salamander	*Plethodon wehrlei*
Mud Salamander	*Pseudotriton montanus*
Red Salamander	*Pseudotriton ruber*
Mudpuppy	*Necturus maculosus*
Eastern Newt	*Notophthalmus viridescens*
Frogs & Toads	
American Toad	*Anaxyrus americanus*
Fowler's Toad	*Anaxyrus fowleri*
Eastern Cricket Frog	*Acris crepitans*
Cope's Gray Treefrog	*Hyla chrysoscelis*
Green Treefrog	*Hyla cinerea*
Barking Treefrog	*Hyla gratiosa*
Gray Treefrog	*Hyla versicolor*
Mountain Chorus Frog	*Pseudacris brachyphona*
Spring Peeper	*Pseudacris crucifer*
Upland Chorus Frog	*Pseudacris feriarum*
New Jersey Chorus Frog	*Pseudacris kalmi*
Eastern Narrow-mouthed Toad	*Gastrophryne carolinensis*
American Bullfrog	*Lithobates catesbeianus*
Green Frog	*Lithobates clamitans*
Mid-Atlantic Coast Leopard Frog	*Lithobates kauffeldi*
Pickerel Frog	*Lithobates palustris*
Southern Leopard Frog	*Lithobates sphenocephalus*
Wood Frog	*Lithobates sylvaticus*
Carpenter Frog	*Lithobates virgatipes*
Eastern Spadefoot	*Scaphiopus holbrookii*
REPTILES	
Turtles	
Spiny Softshell	*Apalone spinifera*
Loggerhead Sea Turtle	*Caretta caretta*
Green Sea Turtle	*Chelonia mydas*
Hawksbill Sea Turtle	*Eretmochelys imbricata*
Kemp's Ridley Sea Turtle	*Lepidochelys kempii*
Leatherback Sea Turtle	*Dermochelys coriacea*
Snapping Turtle	*Chelydra serpentina*
Painted Turtle	*Chrysemys picta*
Spotted Turtle	*Clemmys guttata*
Wood Turtle	*Glyptemys insculpta*
Bog Turtle	*Glyptemys muhlenbergii*
Northern Map Turtle	*Graptemys geographica*
Diamond-backed Terrapin	*Malaclemys terrapin*
Northern Red-bellied Cooter	*Pseudemys rubriventris*
Eastern Box Turtle	*Terrapene carolina*
Pond Slider[a]	*Trachemys scripta*
Eastern Mud Turtle	*Kinosternon subrubrum*
Eastern Musk Turtle	*Sternotherus odoratus*
Lizards	
Eastern Fence Lizard	*Sceloporus undulatus*
Coal Skink	*Plestiodon anthracinus*
Common Five-lined Skink	*Plestiodon fasciatus*
Broad-headed Skink	*Plestiodon laticeps*
Little Brown Skink	*Scincella lateralis*
Six-lined Racerunner	*Aspidoscelis sexlineata*
Snakes	
Scarletsnake	*Cemophora coccinea*
North American Racer	*Coluber constrictor*
Eastern Kingsnake	*Lampropeltis getula*
Northern Mole Kingsnake	*Lampropeltis rhombomaculata*
Eastern Milksnake	*Lampropeltis triangulum*
Rough Greensnake	*Opheodrys aestivus*
Smooth Greensnake	*Opheodrys vernalis*
Eastern Ratsnake	*Pantherophis alleghaniensis*
Red Cornsnake	*Pantherophis guttatus*
Common Wormsnake	*Carphophis amoenus*
Ring-necked Snake	*Diadophis punctatus*
Rainbow Snake	*Farancia erytrogramma*
Eastern Hog-nosed Snake	*Heterodon platirhinos*
Plain-bellied Watersnake	*Nerodia erythrogaster*
Common Watersnake	*Nerodia sipedon*
Queensnake	*Regina septemvittata*
Dekay's Brownsnake	*Storeria dekayi*
Red-bellied Snake	*Storeria occipitomaculata*
Eastern Ribbonsnake	*Thamnophis saurita*
Common Gartersnake	*Thamnophis sirtalis*
Smooth Earthsnake	*Virginia valeriae*
Eastern Copperhead	*Agkistrodon contortrix*
Timber Rattlesnake	*Crotalus horridus*

[a]The subspecies Red-eared Slider, *Trachemys scripta elegans*, is naturalized in Maryland.

Department of Natural Resources (MDNR) and the Natural History Society of Maryland (NHSM) agreed to cosponsor the MARA project.

Citizen Science and Herpetological Research

Citizen science is often a practical way to collect ecological data on broad spatial scales. Citizen science-based projects can greatly extend the geographic area covered in a study. For example, privately owned land is often more accessible through volunteers' relationships with landowners. Citizen science projects have been used successfully across an array of disciplines. However, there are some challenges in using citizen science projects in herpetofaunal research, often stemming from two main factors. The first factor is the cryptic nature of most amphibians and reptiles. Without targeted surveys with the express purpose of documenting

Wehrle's Salamander is found in the Appalachian Plateaus Province in Maryland. Photograph by Ed Thompson

Timber Rattlesnakes are one of the most persecuted snakes because they are venomous and perceived to be dangerous to humans. Photograph by Lance H. Benedict

amphibians and reptiles, many of these animals go unnoticed (Price and Dorcas 2011). The second factor could be considered a public relations dilemma. Developing effective citizen science projects focused on animals that are misunderstood can be challenging, and amphibians and reptiles are an often maligned and misunderstood group of wildlife. Myths and misconceptions about these organisms are pervasive in society. Unfortunately, these misunderstandings may result in the ruthless targeting of some species. For example, snakes are frequently reviled and feared. The saying "the only good snake is a dead snake" persists in present-day society. Low appeal to volunteers for projects focused on these organisms can greatly slow the development of new citizen science projects that study herpetofauna (Price and Dorcas 2011).

Despite these challenges, the number of citizen science projects that focus on amphibians and reptiles has grown since the 1990s. Monitoring projects such as the North American Amphibian Monitoring Program, the Association of Zoos and Aquariums' FrogWatch USA, and the Global Amphibian BioBlitz are a few such examples. The types of citizen science projects that have seen the largest increase in number are statewide atlas projects, also known as ecological or natural history atlas projects.

An ecological atlas is a compilation of spatially explicit data on species occurrences (A. M. Dunn and Weston 2008). The distributional data of these projects are collected during a set period of time; 5 or 10 years are common durations. The first grid-based ecological atlas was published by the Botanical Society of the British Isles in 1962. The *Atlas of the British Flora* (Perring and Walters 1962) was an impressive achievement requiring a 10-year effort by 1,500 botanists to map the distribution of approximately 2,000 species of plants. Hundreds of ecological atlas projects have since been developed and conducted all over the world. By the 1970s, North American atlas projects had started in the Northeast and quickly spread across the continent. To date, a number of successful breeding-bird atlas projects have been conducted in North America under the guidance of the North American Ornithological Atlas Committee (NAOAC 1990). In fact, two breeding-bird atlases have been successfully compiled in Maryland. The first project was conducted from 1983 to 1987 (Robbins and Blom 1996) and the second from 2002 to 2006 (Ellison 2010).

Successful ornithological atlases served as templates for the later development of amphibian and reptile atlases. In 1992, one of the first herpetofaunal atlases was published: "The Amphibians and Reptiles of Maine" (Hunter et al. 1992). It was a township-based effort that spanned a five-year period. Since publication of the Maine atlas, the number of herpetological atlas projects has steadily grown. Other amphibian and reptile atlas projects have been conducted or are underway in multiple states across the United States, including Georgia (Jensen et al. 2008), Massachusetts (Jackson et al. 2010), New Jersey (NJENSP 2002), New York (Gibbs et al. 2007), North and South Carolina (Price and Dorcas 2011), Pennsylvania (PARS 2017), Vermont (Andrews 2013), and Wisconsin (Casper 1996). While the goals of these atlas projects are similar, the methods for collecting distributional data differ. In some cases, species locality data are collected as point localities across the state. Often, the resulting distributional maps show species as present or absent by county, yielding a county-based distributional range map. An alternative method uses a statewide grid, such as the US Geologic Survey (USGS) quadrangle map framework, as the basis for collecting and depicting distributional data.

The New Jersey and New York atlases were among the earliest amphibian and reptile atlases to use grid-based systems. An advantage of a grid-based approach is that the distribution of survey effort can be monitored to ensure coverage throughout the state, rather than emphasizing the most species-rich or popular sites. By establishing fixed survey areas, changes in herpetofaunal distribution can be detected through comparison with future surveys of the same areas.

Goals of the Atlas

The primary purpose of the MARA project was to collect distributional data on the state's amphibian and reptile species over a five-year duration, using a systematic and repeatable approach. The data collected during the project will serve as a reliable baseline with which future distributional patterns can be compared. Additionally, the data provide a robust measure of the present herpetofaunal diversity of Maryland, as well as an assessment of the state's endangered and threatened amphibians and reptiles, which the MDNR and others can use while managing and protecting Maryland's herpetofauna.

Finally, perhaps the most significant goal was the hope that raising public awareness about this group of often misunderstood organisms would contribute to the long-term protection and conservation of the state's herpetofaunal diversity. During the MARA project, thousands of citizens across Maryland learned about the state's native herpetofauna through the project's social media campaign and educational outreach initiatives. Educating citizens about native amphibian and reptile diversity and its ecological benefits is an important step in creating an informed society that actively participates in the long-term conservation and protection of Maryland's natural heritage.

Organization of the Atlas

In organizing this atlas, we first provide information that is helpful to understanding the context of the survey and the methods we used to conduct the survey. We then present

The Eastern Ribbonsnake is a semiaquatic snake that is native to Maryland. Photograph by Hugh Vandervoort

the results of our efforts, including details for each of the amphibian and reptile species we documented.

First, in "History of Herpetofaunal Distributions in Maryland," we provide background information about the history of herpetological surveys in Maryland, particularly noting the people involved, the historical stages in assembling distributional data from throughout the state, and the historical and potential pitfalls of assembling accurate distributional information. Next, in "Maryland's Environment," we summarize the environmental conditions within the state, including the geologic basis of landforms, how these conditions influence and are influenced by climate, and the resulting types of habitat that support Maryland's herpetofauna. In the chapter "Conservation of Maryland's Herpetofauna," we characterize the land uses that occur throughout the state and summarize the history of public efforts—through laws, regulations, and public programs—and private efforts to conserve herpetofaunal biodiversity in Maryland.

Preparing this initial atlas was a complex and complicated process that involved many people. In "Designing and Implementing the Atlas Project," we document our methodology for organizing the core volunteers and for assessing and processing data, managing data quality, ensuring statewide coverage, and acquiring public participation.

Lastly, we summarize the survey effort in "Results of the Atlas Project." We note our degree of success in reaching our effort and coverage goals, show the distributional patterns and counts for each Maryland county, summarize statistics for herpetofaunal taxon groups and rare species, present statistics for the accuracy of participants' identifications of species, and note the relative confidence of the supporting

The Coal Skink is Maryland's only state-endangered lizard. Photograph by Ed Thompson

The Eastern Fence Lizard is one of six species of lizards that are native to Maryland. Photograph by Robert T. Ferguson II

data by quantifying reported occurrences and confirmed observations.

Following these chapters, we present the species accounts, with maps for each species, prepared by our many authors. We hope that through the comprehensive information in this atlas, we have provided a baseline for future atlas surveys to be used as meaningful tools for conserving Maryland's herpetofauna.

History of Herpetofaunal Distributions in Maryland

CHARLES A. DAVIS

Where are the amphibians and reptiles? The answer to this question is basic to our understanding of these animals in Maryland. Our knowledge of the historical and recent distributions of herpetofauna in Maryland is the culmination of centuries of observations. The path to understanding the distribution of amphibians and reptiles has included a series of practices: discovering, collecting, and describing; documenting species ranges; assembling locality-based checklists; creating statewide lists and maps; surveying rare species; and organizing crowd-sourced surveys, which ultimately led to the MARA project. Each activity benefited from an increasing accumulation of and access to distributional information, more skilled observers, more directed surveys, and advances in transportation and information technology. These factors allowed interested people to organize and extend their potential for discovery.

Discovery

The precolonial inhabitants of Maryland were certainly aware of the surrounding herpetofauna, but little of this knowledge transferred to the early explorers. Food was the exception. In 1588—before Maryland was chartered as a colony—the first account of the New World noted the abundant "Tortoyses bothe of lande and sea kinde" that were consumable as meat and eggs (Heriot 1588). Perhaps the earliest recorded observations in the Chesapeake Bay region were those by John Smith, who first visited from 1607 to 1609. Smith described eating turtle that was offered by the native people and incidentally noted reptile parts incorporated in the body ornamentation of the native people (Warner 2006). On one head ornament, he noted a "small rattle growing at the tayles of their snaks" (Warner 2006), and he

In each eare commonly they haue 3 great holes, whereat they hange chaines, bracelets, or copper. Some of their men weare in those holes, a small greene and yellow coloured snake, neare halfe a yard in length, which crawling and lapping her selfe about his necke often times familiarly would kiss his lips.　　—John Smith (1607–1609); quoted by Arber (1884)

observed a small live snake that, during an encounter, some native men used as an ear ornament (see above) (Arber 1884). The identities of species in early observations like these are full of uncertainty, since the early explorers had no names to distinguish among the diverse and abundant new species that they encountered (Branch 2004). And without a scientific name, this early knowledge is unreliable.

The early naming process was fraught with perils. There was no common compilation of names with reference specimens or drawings, and species were repeatedly "discovered" and assigned a different name. Holbrook (1836) expressed this concern in the preface to his monumental illustrated work *North American Herpetology* (see next page). Much confusion existed at least until 1905, when International Nomenclatural Rules were adopted (Melville 1995). The process of clarifying the meaning of names continues to the present; a current iteration relative to the MARA project is *Scientific and Standard English Names of Amphibians and Reptiles of North America North of Mexico, with Comments Regarding Confidence in Our Understanding* (Crother 2017). But what do these names mean? What did early Maryland observers and researchers see? Even with names, their application can be inconsistent and confusing without corresponding physical evidence for comparison. Future scientists require more than a name to repeat the observations of the original ob-

With an immense mass of materials, without Libraries to refer to, and only defective Museums for comparisons, I have constantly been in fear of describing animals as new that have been known to European Naturalists. In no department of American Zoology is there so much confusion as in Herpetology.   —John Holbrook, Preface, in *North American Herpetology*, vol. 1 (1836)

servers. To accomplish this we require descriptions, drawings, preserved specimens, or, in recent years, photographs and recordings.

Collections and Descriptions

Repeatable observations are a fundamental characteristic of scientific study. Collecting specimens is the ultimate practice of documentation because it allows a future researcher to repeat the original observation of the discoverer. Historical collections are some of the earliest scientific evidence of the past presence of particular amphibian and reptile species in Maryland. Documentation of the amphibian and reptile distributions began through arrangements between European collectors and agents sent to the New World. Collecting began in Maryland when the Reverend Hugh Jones was sent to the Maryland Colony in 1696 with an additional obligation to collect natural specimens for James Petiver, an apothecary in England (Stearns 1953). Petiver provided directions to those collecting for him (see below).

On 1 March 1697, Reverend Jones sent to Petiver a small turtle, among other biological specimens, which Petiver (1698) later described in a presentation to the Royal Society

Brief Directions for the Easie Making, and Preserving Collections of all Natural Curiosities. For Iames Petiver Fellow of the Royall Society London.
All small Animals, as Beasts, Birds, Fishes, Serpents, Lizards and other Fleshy Bodies capable of Corruption, are certainly preserved in Rack, Rum, Brandy, or any other Spirits; but where these are not easily to be had, a strong Pickle, or Brine of Sea Water may serve; to every Gallon of which, put 3 or 4 Handfulls of Common or Bay Salt, with a Spoonful or two of Allom powderd, if you have any, and so send them in any Pot, Bottle, Jarr, &c. close stopt, Cork'd and Rosin'd N.B. You may often find in the Stomachs of Sharks, and other great Fish, which you catch at Sea, divers strange Animals not easily to be met with elsewhere; which pray look for, and preserve . . .
—Iames Petiver (1767)

Testudo terrestris Americana, dorso elato.
I do not find this certainly described by any Author. Its Shell an Inch and a quarter long, and one broad, the Scales about the Edges are Quadrangular, those above Pentangular; he is guarded along the Back with a round Ridge; his Head about the Bigness of our Horse bean; the Orbits of the Eyes very large; his Snowt not very unlike a parrot's Bill, his upper Jaw including the under; each Foot has four sharp Claws like a Mouse.

His Belly is made up of several thin Scales; whose middle Pair are long and quadrangular, that next the Head and Tail triangular, the rest irregular; his Tail taper, and about half an Inch long.

His whole Body exceeds not the half as a large Walnut.
—James Petiver (1698)

of London (see above right). Petiver (1767) also described from his later acquisitions a "Maryland blue tailed lizard" (possibly a *Plestiodon* skink) and a "Maryland blue chin lizard" (possibly a male Eastern Fence Lizard). His descriptions and drawings are the known beginning of the documentation of Maryland herpetofauna.

During the 1700s, natural history interests were concen-

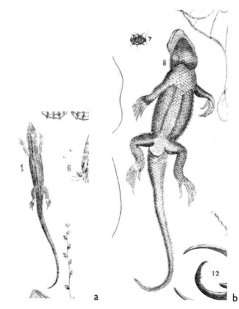

Drawings of two Maryland specimens in the collections of James Petiver (1663-1718). From Petiver (1767)
a. "*Lacertos MARIANUS minor cauda caerulea*. Particular in having a blew thinning Tail, found about Trees in MARYLAND." Jacobus Petiver Gazophylacii Naturae & Artis, Decas Prima, p. 1, table 1, fig. 1.
b. "*Lacertos mariana undata, subtus caerulea*. This is a rough fealed lizard, with blackish waved lifts on the back; blue on the chin, and each side of the belly. From Maryland." Jacobus Petiver Gazophylacii Naturae & Artis, Decas Prima, p. 3, table 14, fig. 8.

trated in places outside Maryland, specifically Charleston, Philadelphia, New York, and Boston (Smallwood and Smallwood 1941). But interest also existed outside these centers of natural history study, and corresponding naturalists started societies for promoting studies of nature, the museum concept, and scientific journals (Smallwood and Smallwood 1941). Early activities in Maryland originated in the vicinity of Philadelphia and later in Washington, DC, where there was a concentration of natural history interest and expertise.

By 1740, the Maryland settlement frontier had reached the vicinity of the future Cumberland (Jordan and Kaups 1989). From then until the late 1800s, public interest in amphibians and reptiles continued to focus on their values as food and on safety concerns. Diamond-backed Terrapins collected from the salt marshes of coastal counties and Snapping Turtles from throughout the state were consumed by the thousands. In 1880, about 30,000 pounds of Diamond-backed Terrapin were harvested in Maryland (True 1887), an amount then considered too small to be of commercial importance (Earll 1887). At least one attempt was made to introduce a non-native species to establish a harvestable population. In 1883, John Peck Dukehart, then manager of the Oakland Hotel in Garrett County, transferred 18 Spiny Softshells from the Ohio River to below the dam at Cumberland, and he planned to place more at Woodmont, near Club House, in 1884 as an attempt to establish this edible species in the Potomac River (Dukehart 1884).

Organizing for Discovery and the Early Amphibian and Reptile Collections

In the late 1790s, there were attempts in Baltimore to establish an organization to promote the study of natural science, but these attempts were unsuccessful (J. G. Morris 1907). Charles Willson Peale, a painter and keen observer of nature from Queen Anne's County, Maryland, established the nation's first natural history museum in Philadelphia in 1784 (Stone 1899; Smallwood and Smallwood 1941). And later, in 1813, his son, Rembrandt Peale, started a similar venture in Baltimore (Colton 1909); that museum was a crossroads for academics and other prominent people of Baltimore to meet about their common interest in natural sciences. From these interactions, a group formed to establish the Maryland Academy of Science and Literature (MASL), which remained in existence from 1822 to 1844 (Steuart 1931).

Members of the MASL observed and collected amphibians and reptiles as part of their activities in the early to mid-1800s. During this period, the publication of Holbrook's (1836, 1838a, 1838b, 1840, 1842a, 1842b, 1842c) *North American Herpetology* helped facilitate interest in amphibians and reptiles because these were the first books to clarify nomen-

clatural synonyms, thereby making a clearer exchange of information possible. Also, by listing known localities of occurrences, the books allowed naturalists to begin to envision the ranges of species. Dr. Eli Gettings, one of the prominent members of the MASL and a friend of Holbrook, contributed three observations from Maryland to *North American Herpetology*: Common Five-lined Skink (Holbrook 1838b), Rough Greensnake (Holbrook 1842c), and Eastern Milksnake (Holbrook 1842a). Julius Ducatel (1837), who was active in the MASL, noted the following animals in a summary of Maryland geography: Diamond-backed Terrapin of Somerset County, Snapping Turtle of the Patuxent River drainage, and Timber Rattlesnake of the far western mountains. Instructions for collecting animal skins and associated data were available to the general public through the *American Farmer* (Skinner 1819), which was published in Baltimore. How many of these observations were backed by specimens is uncertain. Little is known of the collections that were accumulated during the period when the MASL was active, for two reasons. First, collections and corresponding records made before 1835 were destroyed in a disastrous fire at the academy's building (J. G. Morris 1907). To replace the specimens lost in the fire, the MASL prepared a pamphlet describing collecting procedures (Coale et al. 1838) and made an appeal to the public (Macaulay and Fisher 1836) to collect and donate specimens, including reptiles, with locality information (MASL 1837a). Among the donations was a "corn snake" (*Coluber eximius*—now: Eastern Milksnake, *Lampropeltis triangulum*) from Dr. A. H. Bailey of Easton and several specimens of *Rana* and *Coluber* (current species identity indeterminate) from Dr. W. E. Coale (MASL 1837b). The second reason we know little about the collections of the academy is that when it ultimately dissolved in 1844, the new collections that had been assembled were divided and sold (J. G. Morris 1907). Even if we were to find specimen lists of the amphibian and reptile collection, the actual identity of the taxa would be uncertain because of the persistent nomenclatural confusion and the lack of specimens to verify the identifications.

The Smithsonian Institution was founded in 1846, and by 1853, some information about snakes collected in Maryland was documented by the enumeration of Maryland specimens, as part of periodic reporting by the staff (Baird and Girard 1853) (table 2). Included in this list is the only amphibian or reptile species first described from Maryland. Kent County is the type locality (Stejneger and Barbour 1923) for the Smooth Earthsnake (*Virginia valeriae* [Baird and Girard 1853]), which was collected by and named after Spencer Baird's cousin, Miss Valeria Blaney.

After 1863, the MASL was revived as the Maryland Academy of Sciences and, again, zoological collections were assembled; ultimately, these collections were given to Johns Hopkins University (P. Uhler 1888). But whether the trans-

Table 2. Collections of snakes from Maryland in the holdings of the Smithsonian Institution, Washington, DC, in 1853

Current Species Name	Species Name in Baird & Girard (1853)	Maryland Location	Collector
Common Gartersnake			
Thamnophis sirtalis (Linnaeus, 1758)	*Eutainia sirtalis* B.&G.	Centreville	S. F. Baird
Common Watersnake			
Nerodia sipedon (Linnaeus, 1758)	*Nerodia sipedon* B.&G.	Centreville	S. F. Baird
North American Racer			
Coluber constrictor (Linnaeus, 1758)	*Bascanion constrictor* B.&G.	Anne Arundel Co.	J. H. Clark
Rough Greensnake			
Opheodrys aestivus (Linnaeus, 1766)	*Leptophis aestivus* Halbe	Anne Arundel Co.	J. H. Clark
Eastern Smooth Earthsnake			
Virginia valeriae valeriae Baird & Girard, 1853	*Virginia valeriae* B.&G.	Kent Co.	Miss V. Blaney (#1962, Type)
Eastern Smooth Earthsnake			
Virginia valeriae valeriae Baird & Girard, 1853	*Virginia valeriae* B.&G.	Maryland	Prof. C. B. Adams

Source: Baird and Girard 1853.

fer contained amphibians and reptiles is unclear; if so, their condition is uncertain. Uhler (1888) stated that amphibians and reptiles were among the natural science collections sent to the World's Industrial and Cotton Centennial Exposition in New Orleans, Louisiana, in 1884–1885 as part of the Maryland State exhibit; many specimens were not returned, however, and those that were returned were in such poor shape that they were disposed of and not transferred to Johns Hopkins University. Strangely, the guidebook for the exposition (Perkins 1885) does not list any amphibians or reptiles as part of the Maryland exhibit.

Not all specimens collected in Maryland have remained within the state. Collections documenting historical distributions occur in various museums around the country and the world (VertNet 2016), and they can be just as vulnerable to loss elsewhere. For instance, John van Denburgh, while completing a medical degree at Johns Hopkins Uni-

versity in 1899, collected amphibians and reptiles in the Baltimore community of Windsor Hills and in Catonsville. He returned with 141 specimens to California, where he was curator of the Herpetological Department of the California Academy of Sciences. Those collections were later lost in the earthquake and fire of 1906 (VertNet 2016).

Edward Drinker Cope (1900) provided another measure of the extent of collections of snakes and lizards from Maryland at the Smithsonian Institution National Museum. In 1900, Smithsonian holdings contained 20 species of reptiles (16 snakes and 4 lizards) collected by at least 35 people or institutions.

Collections play an essential scientific role for confirming the identities of amphibians and reptiles claimed to have been observed during the beginning of herpetofaunal distributional research in Maryland. Merely having collections, though, is not enough for them to be scientifically valuable.

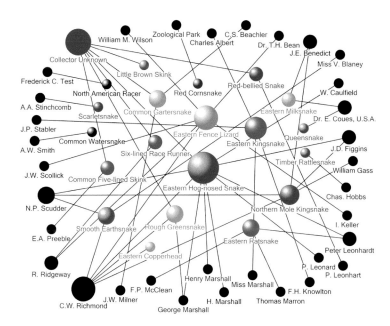

Collectors of Maryland lizard and snake specimens in the Smithsonian Institution, Washington, DC, collections in 1900. Circle diameter is in proportion to the number of collections. From Cope (1900)

They must be reexamined and compared to ensure that organisms are correctly identified. A purported Bog Turtle was collected from along the Potomac River north of Washington, DC (Brady 1924), but only after multiple authors had cited this record numerous times did someone actually look at the specimen and discover that it was a young Wood Turtle (Mitchell 1989). This erroneous record still appears periodically from authors who miss the correction during their research (Mitchell 1989). For lost specimens, it is difficult to know whether they still exist or were just misidentified and are now filed under a new name in a collection. Fortunately, current efforts to establish an international collections database allow researchers to locate some of these specimens and determine whether they still exist.

Limitations and Challenges of Assessing Historical Ranges: Where Were the Rattlesnakes?

Venomous snakes had a special place in the psyche of colonists and later settlers. Strong myths and attitudes generated then still persist today. Well before Hopkins (1890) documented the life history and distribution of Eastern Copperheads in Maryland, the general public was well aware of venomous snakes and their hazardous potential. By the 1800s, rattlesnakes had been through at least a century of attack by humans. Slaughter (1996), in his biography of naturalist John Bartram (1699–1777), recounts the common attitude toward these snakes: "Whacking a rattler was something of a male rite of passage in colonial America—a moral obligation, a public service, and dangerous fun all in one." At first, the death of rattlesnakes was so common it was not newsworthy (Slaughter 1996). Bartram (1734), living six miles outside Philadelphia, noted that the rattlesnake had "now become a rarity so near our settlements." Cope (1873) indicated that the Timber Rattlesnake "originally inhabited the whole country . . . but they are . . . exterminated from the best-settled regions."

In Maryland, the western counties of Allegany and Garrett were the battlefront in the slaughter of rattlesnakes in the 1800s. From newspaper reports, we know that in one season at least 230 were killed at three locations in Allegany County (Sun 1838): John Gruber killed 17 in western Allegany (Sun 1856); Joshua Summers killed 111 near Oakland (Sun 1877a); and James Davis and Owen Brady killed 16 near Dan's Rock (Sun 1882). Further to the east, in Washington and Frederick counties, there were fewer encounters. Rufus Wilcoxen killed a large rattlesnake in Frederick County, "the largest rattlesnake that has been killed in this county for some time" (Sun 1867). A woman killed a rattlesnake near Mt. Zion Church, Frederick County (Sun 1877b), and another woman later killed one elsewhere in that county (Sun 1890).

William Smith was bitten while catching 26 rattlesnakes in Garrett County (Sun 1897b). In March 1886, State Senator Richard T. Browning introduced a bill authorizing a bounty of 50 cents per rattlesnake in Garrett County (Sun 1886).

Based on earlier writings, rattlesnakes were more widely distributed throughout Maryland during the colonial period. The nature of these sighting records exemplifies the uncertainty of interpreting historical distributions of amphibians and reptiles where no verifiable evidence is preserved. The historical abundance of rattlesnakes in the eastern part of Maryland is unknowable with certainty because it is based mostly on writings, possible piggyback reporting, with few complementary specimens or other means to currently confirm the identities of these past observations. The evidence is mainly circumstantial. Rev. Hugh Jones (1699), living in what is now Calvert County, highlighted rattlesnakes as the reptile of interest in the Maryland Colony. The Baltimore County Land Records indicate that one was killed in 1700 at the confluence of Deer Creek and Thomas Run, in what is now Harford County (Maryland Historical Magazine 1908). Based on Bartram's assessment, by this time rattlesnakes had become so rare that an occurrence was now newsworthy. In 1738, John Greer was bitten by one at Long Green Run in Baltimore County (M. Howard 1923). Marye (1921) noted that by 1738, "rattlesnakes . . . had been so long extinct in that part of the country to which John Greer referred in his deposition that most of the inhabitants could hardly be convinced that they ever existed there." In 1739, Dr. Andrew Scott sent a pair of live rattlesnakes to Dr. Hans Sloane in England, with the hope of indirectly obtaining favor in appointment as sheriff of Prince George's County (Frick et al. 1987). We do not know the source of Dr. Scott's rattlesnakes. Andrew Burnaby listed rattlesnakes occurring in the Washington, DC, vicinity in 1759 (McAtee 1918), and later, Warden (1816) noted their occurrence just several miles upriver from Washington at the Great Falls of the Potomac. Some reports also included references to the rattlesnake's former range (M. Howard 1923). In 1840, the Baltimore Sun reported the first instance of rattlesnakes occurring in Pikesville in Baltimore County, wherein one specimen of the four snakes killed that season was displayed at the "Museum of the Maryland Seminary" (Sun 1840). In 1845, the Patriot reported that one was observed several miles east of Baltimore in the woods of the Ridgely estate (as reported in Sun 1845). About 1898, Julian Smith observed a rattlesnake at Loch Raven in Baltimore County (M. Howard 1923). In the early 1900s, multiple rattlesnakes were killed in Rayville and Freeland in Baltimore County (Sun 1905, 1906), and in 1909, a young boy was bitten in Monkton (Sun 1909). Howard (1923) noted that the occurrences of rattlesnakes along the Gunpowder River drainage (including reports in Sun 1845, 1909) were possibly individuals washed down by floods from a larger population known to exist in northern Baltimore County.

The northern Baltimore County population persisted until relatively recent years (J. E. Cooper and Groves 1959; H. S. Harris 2007a).

Similar stories have been recorded on the Eastern Shore. In 1749, the *Maryland Gazette* reported that a rattlesnake killed James Taylor, who lived on the North Fork of the Nanticoke River (now Marshyhope Creek) in Dorchester County (Johnston 1922), but the report does not clarify whether he was bitten at his home. In 1878, a rattlesnake was killed in Harmony, Caroline County (*Denton Journal* 1878). In 1892, James Jones killed one with seven rattles and a button in Wicomico County, noting that rattlesnakes hadn't been seen in that vicinity for some time but were "plentiful" in another section of the county (*Sun* 1892). They were said to be numerous in the swamps of Wicomico County (*Sun* 1897a). In 1897, another was killed in Dorchester County, near Vienna, by Rev. L. T. McLain, which was the "first rattler to be killed here in 30 years" (*Sun* 1897a), suggesting that another had been killed sometime in the 1860s. This report also influenced political commentary. Because of a then commonly held myth that alcohol was an effective remedy for a rattlesnake bite (*New York Daily Times* 1854; H. A. Kelly 1926), someone in Dorchester County editorialized that the governor should call a special legislative session to eliminate the local-option laws of prohibition, so that alcohol would be readily available within the county should anyone be bitten. In good debating form, the commenter preemptively acknowledged that some may attribute the original sighting of the rattlesnake to consumption of the medicine prophylactically (*Sun* 1897c).

Some of the stories associated with these encounters contain details that make them more credible than others—for instance, sightings by multiple people, counts of rattlesnake rattle segments, multiple stories from the same vicinity, public display of the captured snakes, or death of a person from the bite. Despite these seemingly true accounts, there is no way to repeat these observations and independently verify the claims. In addition, the report of a rattlesnake does not necessarily mean that a breeding population was present.

A complicating factor in interpreting historical distributions is the practice of intentionally and unintentionally moving animals to other locations and the possibility of their release or escape, which has occurred for some time in Maryland and continues today. For example, in the case of rattlesnakes, Mr. Stabler, a ticket agent in Baltimore for the B&O Railroad, received a surprise gift from Garrett County of a box with a live rattlesnake (*Sun* 1859). In 1875, Prof. C. Johnson received a large living "Florida rattlesnake" from Capt. William Kelly and displayed it at the Maryland Academy of Sciences (*Sun* 1875). In 1886, in Baltimore City, a man entered City Hall with a box containing two defanged rattlesnakes, including a "California rattle-

snake," looking for a permit to display them to the public (*Sun* 1886). In 1895, George W. Bishop died from the bite of a "Florida rattlesnake" that he was exhibiting in Quantico, Wicomico County (*Sun* 1895). Three Timber Rattlesnakes were sent from northern Pennsylvania to Woodstock College, Baltimore County (McClellan 1916). And rattlesnakes also rode on trains unassisted (*Sun* 1916). In 1957, two Massasaugas killed in east Baltimore City were suspected to have arrived by lumber truck from Indiana (*Sun* 1957; F. Groves 1991). A note in the Virginia Herpetological Society's *Bulletin* (VHS 1960) mentions a Timber Rattlesnake from Kent County, Maryland, that was "believed to be an escapee." In 1991, a "Canebrake" Timber Rattlesnake was killed by a resident mowing a yard in Howard County (*Sun* 1991). None of these rattlesnakes, if released, would represent a native rattlesnake population in the places where they arrived. Additionally, many of these observations are encounters that provide no information about the origin of the rattlesnakes. And for those that were not collected, we are left to judge the accuracy of these observations based on the fallible judgment and technical reputations of the reporters, with no supporting evidence. The lack of proper care and tracking of voucher collections is an ongoing theme in the uncertainty of published lists of Maryland amphibians and reptiles (R. W. Miller 2009b). This is in part because Maryland has had no designated natural history museum to officially conserve collections for the long term.

In more recent times, this type of uncertainty in our knowledge about the historical distribution of amphibians and reptiles has led to disputes about appropriate resource management practices—for instance, management actions regarding the Eastern Pinesnake (*Pituophis melanoleucus*), for which there is no current scientific evidence of a viable population having existed in Maryland, but for which stories of encounters exist or are implied (Holbrook 1842a; Fisher 1932; H. A. Kelly et al. 1936; McCauley 1945; Kirkley 1949; Lee 1972a; CREARM 1973; Grogan 1973; H. S. Harris 1975, 2007b; Ashton et al. 2007). Similarly, these same factors have led to questions in interpreting the historical reports of other species in Maryland, such as Wood Turtle (Henshaw 1907; Shufeldt 1919; A. H. Clark 1930; W. Norman 1939b; Simmons and Hanzley 1952; Reed 1956h; Norden and Zyla 1989; R. W. Miller 1993; Proffitt 2001); Eastern Spadefoot (Mansueti 1947; Reed 1956j; R. W. Miller 1977, 2013; H. S. Harris and Crocetti 2008; Hildebrand 2009); Eastern Tiger Salamander (R. W. Miller 2010); Scarletsnake (H. S. Harris 2013; R. W. Miller 2016); Mudpuppy (R. W. Miller 2015c); and various other species records from the Baltimore vicinity by Mansueti (R. W. Miller 2015a, 2015b). Some of these researchers reexamined prior publications to reinterpret the reported species distributions—basing the new conclusions on observations that are truly repeatable based on voucher collections.

Early Geographic Lists

In the 1800s, as public interest in local amphibian and reptile species increased, amateur and professional herpetologists began assembling lists for specific geographic locations in Maryland. This may have been stimulated in part by Baird (1850) at the Smithsonian Institution, who called for people to submit observations of periodical natural phenomena, including hibernation of reptiles, and the "utterance of characteristic cries of reptiles," as well as detailed lists "of all the animals and plants of any locality throughout the continent. These, when practicable, should consist of the scientific names as well as those in common use; but when the former are unknown the latter alone may be employed. It is contemplated to use the information thus gathered in construction of a series of species, showing the geographical distribution of the animal and vegetable kingdoms of North America."

Some researchers attempted lists for local habitats or local areas; others focused on larger regions and, ultimately, statewide lists. Early lists were included in descriptions of Maryland's geography. Cope (1873) provided several paragraphs listing the diversity of amphibian and reptile species in his description of Maryland zoology. Scharf (1882) gave a glimpse of the popular understanding of the herpetofauna of western Maryland in his *History of Western Maryland*, the preparation of which was aided by Philip Uhler, president of the Maryland Academy of Sciences. The listing was very likely based on the knowledge that had been assembled by the academy. Scharf listed at least 6 turtles, 2 lizards, 27 snakes, and 9 frogs (see above right). Frederick C. Test began a checklist for the Washington, DC, vicinity in 1890, which was continued by Hay (1902) and his students and friends, and later by McAtee (1918). In 1907, Keim (1914) prepared a short list for Jennings, Garrett County, which was later expanded countywide by McCauley and East (1940). Henry Fowler (1915) assembled a checklist for Cecil County and also made the first attempt at a partial statewide list (Fowler 1925). In Baltimore County, Rev. William H. McClellan, SJ, established a live and preserved snake collection at Woodstock College, the first Jesuit Seminary in the United States, located along the Patapsco River. McClellan established the collection to educate the seminarians, who would sometimes return from walks along the river and ask him to identify the baby Eastern Copperheads that they were holding. The reference collection that he assembled resulted in what he believed to be a complete list of snakes occurring in the Patapsco River Valley (McClellan 1916).

When researchers prepare checklists that include observations by others, they weigh their confidence in a reporter's ability to accurately identify before adding the observations to their list. This doesn't guarantee that errors are prevented. Errors are more likely to occur when expert

The reptiles and fish likewise comprise numerous species of curious appearance of value for food. Among the former, the great snapping-turtle, the slider, two kinds of mud-terrapins, the musk turtle, the land tortoise, the gray swift, and six-lined skink may be mentioned as conspicuous and well known creatures. Of the worm-shaped reptiles, the dreaded rattlesnake and copperhead still occur among the low rocks in the wilder places of the backcountry, besides which three kinds of water-snakes, four varieties of garter snakes, the blowing viper, the chain and milk snakes, the great horse-runner and common black snake, the delicate green snake, and a dozen other species affect most parts of the region where vegetation grows thickly. Of frogs, most of the kinds common to the Atlantic region occur in moderate numbers. Thus two forms of toad, two tree-toads, the bull-frog, leopard frog, woods frog, savannah cricket, and spring frog are numerous in most of the low grounds and wet meadows.

—Thomas Scharf (1882)

opinion is substituted for scientific evidence. Historical compilations of records are particularly challenging because often the observer(s) become increasingly difficult to contact, until we are ultimately left to interpret cryptic notes tied to a specimen—if specimens or notes exist at all.

More Efforts to Organize

In 1929, the Maryland Academy of Sciences renewed attempts to establish a public museum (Maryland Academy of Sciences 1928). Frustrated with the direction of the academy's new museum efforts—which were neither based primarily on Maryland collections nor research-oriented—life member and treasurer of the academy Edmund B. Fladung resigned from his position. He, along with Wallace W. Coleman, F. Stansbury Haydon, Herbert C. Moore, and Eugene R. Polacek, established the Natural History Society of Maryland, Incorporated, in 1929. Because the mission of the society included the vision of a prominent natural history museum, a library for reference, and research and public education activities, its creation was a key action in establishing an incubator for many individuals who would contribute significantly to herpetology in Maryland throughout the twentieth century.

In 1930, the NHSM established a Department of Herpetology, headed by Gilbert C. Klingel (NHSM Trustees 1930). Aware of conservation concerns, the society emphasized capturing distributional records through photography to assemble a reference collection of glass photographic lantern slides of all species known in Maryland (Klingel 1930, 1932a). The activities of the department included collect-

Gilbert Klingel photographing frogs, 1932. Archives of the Natural History Society of Maryland, photographer unknown

ing and photographing specimens, reporting discoveries through short notes in the in-house journals, and sharing findings through public presentations. Because Klingel was also preoccupied with his interests in marine sciences and herpetological expeditions to the Caribbean (Klingel 1929, 1932b; Noble 1931), Harry C. Robinson became curator of the department in 1933. During his tenure, the society started a Junior Division with a stated purpose of training future scientists. This single action cultivated a cohort of young herpetologists who significantly contributed to her-petological discovery in Maryland between the late 1930s and early 1960s. Among them were Elias Cohen, Francis Groves, Romeo Mansueti, John Norman, Robert Simmons, James Fowler, Jerry Hardy, and John Cooper.

Meanwhile, William H. Fisher, an officer of the Maryland Academy of Sciences, was active in listing the animals of Maryland. In 1885, he prepared lists of mammals (Lee 1988); then he helped Kirkwood (1895) prepare the first Maryland bird list. Later he prepared a list of the snakes of Maryland (Fisher 1932). This unpublished list was shared with Dr. Howard Kelly.

Howard Kelly arrived in Baltimore in 1889 as a gynecol-ogist and one of four founding surgeons of Johns Hopkins Hospital (Robison 2010). He traveled annually to Florida and Ontario to vacation and to study snakes, mushrooms, and other natural history topics of medical interest (Or-tenburger and Ortenburger 1933). Dr. Kelly kept a private live snake collection in the basement of his summer home, "Liriodendron," in Bel Air. He was a member of the Mary-land Academy of Sciences and also a financial supporter of the NHSM, where he shared common interests, particu-larly herpetology, with young and old colleagues. Kelly had been mentored as a child in Philadelphia by Edward

Drinker Cope, a prominent herpetologist (E. B. Kelly 1949). In 1936, Kelly became the honorary curator of the Depart-ment of Herpetology at the NHSM (NHSM Trustees 1937). Also in 1936, Howard Kelly et al. (1936) prepared a short book for the general public entitled *Snakes of Maryland*, based on Fisher's (1932) list and field notes and on collec-tion records compiled by the department under the prior curator, Harry C. Robinson. The book was available in li-braries—at least in the Baltimore region—in part because its production was advised by Kelly's friend and chief librarian of Enoch Pratt Library in Baltimore, Dr. Joseph J. Wheeler. Kelly paid the library's contracted printer for its production. It was the first statewide listing of an amphibian or reptile taxon that also included information for identification, in this case about snakes.

Kelly's publication has been criticized for containing er-rors (McCauley 1945; J. E. Cooper 1966; R. W. Miller 2015a, 2015b, 2016). These errors included six taxa now considered never to have been part of the resident fauna of Maryland. McCauley (1945) assessed each taxon to explain why that was so. Such errors highlight the potential perils of assem-bling a regional checklist. Some of the errors appear to be misapplied names inherited from other published or cred-ible sources, particularly for species having color pattern variability. Kelly included those species based on what he considered to be credible evidence or claims. For example, Kelly et al. (1936) reported some of these extralimital taxa based on Fisher's (1932) compiled notes. Dr. Raymond L. Ditmars (1907), curator of reptiles at the New York Zoo-logical Park, reported that Brown Watersnake (*Nerodia tax-ispilota*) occurred from "Maryland to the Gulf States." The Eastern Pinesnake was included based on Fisher's (1932) an-notation that R. V. Truitt, professor of zoology at the Univer-sity of Maryland, saw one taken in a heavily wooded thicket in the back beach at the North Beach Life Saving Station near Ocean City, Maryland.

Usage of common names may have added to the mis-reported species. Fisher's (1932) list did not include Glossy Swampsnake (*Liodytes rigida*), but the confusion here may have resulted from an additional common name, "Striped Water Snake," which Fisher listed for Queensnake. Ditmars (1907) used the common name "Striped Water Snake" for Glossy Swampsnake and claimed, erroneously, a distribu-tion of "Pennsylvania to the Gulf." This confusion over dis-tribution began when Holbrook (1842b) reported the snake from as far north as Philadelphia, apparently based on a mis-reading of Say's (1825) original description and comparison with other Pennsylvania species. This claimed northern range was repeated by Cope (1875) and A. E. Brown (1901) and in one statement was expanded to New York by S. Gar-man (1883). Stejneger and Barbour (1923) later corrected the distribution of Glossy Swampsnake to the "Southern States," but by then the confusion had resulted in some Queensnakes in the northern part of their range being reported erroneously as Glossy Swampsnakes (McCauley 1945).

Inclusion of Louisiana Milksnake (*Lampropeltis triangulum amaura*) was a misidentification while attempting to name the variation in color patterns that occurs within Milksnake in Maryland. Kelly used "Red Milk Snake" as the common name for this morphotaxon, following Ditmars's (1907) book, which also contains a photo from Maryland of "Red Milk Snake." Ditmars (1907) called these snakes *Lampropeltis triangulum clericus*, but only two subspecies of Milksnake were recognized in the eastern United States in the official checklist by Stejneger and Barbour (1923): *Lampropeltis triangulum triangulum* and *Lampropeltis triangulum amaura*. The range of *L. t. amaura* was stated as "Eastern United States, west to Texas, south to the Gulf," despite research by Blanchard (1921) showing that it was restricted to the lower Mississippi Valley. But for Kelly et al. (1936), this was probably "identification by geographic default." Kelly et al. had access to the detailed drawings and description of Milksnake patterns that appeared in Cope (1900), but given the two apparent milksnake taxa in Maryland and only two taxa in the Stejneger and Barbour (1923) checklist, the identities might have seemed obvious. Current research by Ruane et al. (2014) shows that the delimitations between these taxa are not readily apparent by morphology alone.

"Yellow Ratsnake" (*Elaphe quadrivittata*, now considered a color variant of Eastern Ratsnake [Burbrink et al. 2000]) was caught twice—two years apart in the same vicinity—in St. Mary's County. This is unlikely to be a misidentification because the second snake was apparently kept in captivity for at least four years (H. A. Kelly et al. 1936). Kelly et al. doubted it was indigenous to Maryland given its southern range, but included it in their book as they did other native species for which there were few records. Undetected introductions are easiest to determine in hindsight.

The selection process for including species in *Snakes of Maryland* was not without evaluation. Fisher's list also included "Gray Ratsnake," based on a 1916 photograph in *American Forestry* by Dr. R. W. Shufeldt (1921), claiming it to be that species. Kelly rejected that report, and the inclusion of "Gray Ratsnake" in the book was instead based solely on one purportedly collected in Harford County by Richard von Hagel (H. A. Kelly et al. 1936), assistant curator of the NHSM's collections in 1933. The draft of *Snakes of Maryland* contained an additional claim from Fisher's notes of a Rough Earthsnake (*Haldea striatula*), based on Holbrook's claim of its presence in Maryland (as *Calamaria striatula* L. [Holbrook 1842a, 124]). But this was rejected before publication because a review of Fisher's supporting field journals could find no specific note about it ("Note to Miss Davis" 1935).

Although *Snakes of Maryland* refers to specific collectors, there are no references to specific voucher specimens to verify the observations (McCauley 1945). Kelly et al.'s book

is a cautionary example of why researchers should not rely solely on unsubstantiated reports—even from recognized experts and published reports—as the basis for accurate distributional claims. Despite these scientific shortcomings, *Snakes of Maryland* played a critical role in supporting the interests of young naturalists in their quest to learn about and discover snakes.

After 1937, George William Maugans, Jr., became curator of the Department of Herpetology at the NHSM, and under his leadership the Junior Division members flourished. These younger members could participate in the department's activities but were expected to publish in the in-house newsletters and journals. From this, a flurry of amphibian and reptile listings occurred, starting in the Baltimore vicinity. Mansueti (1939b, 1941c) prepared a list for Patapsco State Park, and later for the Baltimore City environs (Mansueti 1941a, 1941b). Buxbaum (1942) prepared a list for the Gwynns Falls. William Norman (1939a) inventoried McManon Quarry, and John Norman (1939) listed Curtis Wright Airport, Bonnie View Golf Club, and Western Run Parkway. For the Cherry Hill community in Baltimore, M. F. Groves (1940) prepared a list of herpetofauna that included Mansueti's (1940) list of amphibians. More distantly, H. C. Robertson (1939) began listing herpetofauna for Catoctin Mountain Park, and Mansueti (1939a) for Frederick County. James Fowler (1944) began listing salamanders of Allegany County. Mansueti (1942) compiled notes for Calvert County, which later led to a county checklist (J. D. Hardy and Mansueti 1962).

In the Washington, DC, vicinity, Brady (1937) assembled a checklist for Plummers Island in the Potomac River, which was later updated by Manville (1968). James Fowler (1943b) revisited and expanded the lists of amphibians and reptiles

Members of the Natural History Society of Maryland search for amphibians and reptiles. Left to right: John (Jack) E. Norman, [unknown], Joseph A. Bures, Conrad P. Kenny, [unknown], Romeo J. Mansueti. ca. 1942. Archives of the Natural History Society of Maryland, photographer unknown

Members of the Herpetology Department of the Natural History Society of Maryland, 22 March 1946, at the annual meeting of the Department. Back row: H. V. Stabler, F. Groves, H. F. Howden, G. Maugans (Curator), B. Bennett, E. B. Fladung, and R. Mansueti (Assistant Curator). Front row: J. A. Fowler, D. Oler, J. Cooper, J. Leake, and J. D. Hardy, Jr. Archives of the Natural History Society of Maryland, photographer unknown

for Washington, DC, and vicinity in a master's thesis that highlighted the role of physiography in distribution. Later he prepared a list for the National Capitol Parks and the District of Columbia (J. A. Fowler 1945). McCauley transformed his PhD dissertation (McCauley 1940) into *The Reptiles of Maryland and the District of Columbia* (McCauley 1945). This publication was a significant milestone because it was the first statewide listing of all reptiles that included, in addition to life history information, dot maps of distributions. These maps challenged the herpetologists of the NHSM to refine the boundaries of ranges or to fill gaps within ranges. McCauley applied a scientifically conservative approach to inclusion of records—preferring voucher specimens where possible and otherwise accepting locality records from observers recognized as knowledgeable, but only if he felt confident in their skills. Even for recognized observers, McCauley required a voucher specimen for claims of a range extension of a species. Despite his conscientious effort, the current whereabouts of some vouchers are unknown (R. W. Miller 2009a).

Roger Conant (1945), meanwhile, assembled a list of amphibians and reptiles for the Delmarva Peninsula, later expanding the list to the northeastern states (Conant 1947b). Conant explored Delmarva at every possible chance. In April 1947, while traveling to his wedding in Pocomoke City, Conant and his fiancée, Isabelle dePeyster Hunt, visited Gum Swamp in Dorchester County and discovered the first Carpenter Frog in Maryland (Conant 1993). During the same period, Mansueti (1953) assembled observations of turtles and terrapins on the Eastern Shore.

Efforts to locate amphibians and reptiles and to assemble lists sometimes had a public education motive. In August 1949, the NHSM led a three-day expedition to Hooper Island, Barren Island, and Blackwater National Wildlife Refuge (NWR) in Dorchester County. Researchers were aware of Conant's recent nearby discovery of Carpenter Frog, but were only able to detect its call during this expedition (Kirkley 1949). Surveyors collected numerous other species of snakes, lizards, and amphibians, which were displayed first at the Fresh Air Camp in Bel Air, Harford County, then later at the Baltimore Zoo (Kirkley 1949). The expedition was featured in the Sunday Magazine section of the *Baltimore Sun*—giving it wide public exposure. John Norman (1949) prepared a summary of turtles of Maryland that was distributed as a "nature leaflet" to the general public. Later, F. J. Schwartz (1961, 1967), at the Chesapeake Biological Laboratory, assembled a booklet about the turtles of Maryland that included basic life history facts for each species, shaded maps of the likely distributions within Maryland, drawings of the turtles by Mrs. Alice Jane Mansueti (Alice Jane Lippson), a key to the species, notes on commercial consumption, and a list of 10 species of fossil turtles from Maryland. The booklet proved popular and went through several printings.

Prince et al. (1955) listed amphibians and reptiles found in the Broad Creek / Deep Run watershed of Harford County. Reed prepared lists for Harford County (Reed 1956a) and Somerset County (Reed 1957c). One year after John Cooper (1956) listed amphibians and reptiles of Anne Arundel County, Reed (1957b) published a list for that county that incorporated voucher specimens from Ralph Daffin and Donald Linzey. Cooper (1953) shared additional notes on herpetofauna of southern Maryland.

Capturing an Eastern Ratsnake in Dorchester County, Maryland, 1949. From Kirkley (1949). Photograph by staff of the *Baltimore Sun*, 28 August 1949. Reprinted with permission from the Baltimore Sun Media Group

Between 1956 and 1958, Reed assembled a series of publications about Maryland herpetofaunal distribution. First, he prepared a bibliography of published reports of amphibian and reptile distribution in Maryland (Reed 1956b). From this effort, he prepared a series of published papers and privately published lists of the amphibian and reptile subgroups of Delmarva and Maryland: lizards (Reed 1956c), turtles (Reed 1956d), snakes (Reed 1956e), frogs and toads (Reed 1956f), and salamanders (Reed 1957a). He synthesized a Delmarva list of all amphibian and reptile taxa (Reed 1956g). Reed's lists of amphibians were the first for Maryland, but the lists were not widely available. From these lists, he published comparisons of the relative distributions of amphibians and reptiles of Delmarva, compared on larger spatial scales—for instance, noting typically Piedmont species occurring on the Atlantic Coastal Plain (Reed 1958a) and southeastern species reaching their northern limit on the mid-Atlantic Coastal Plain (Reed 1958b). Arnold Norden of the NHSM was able to recover from Reed's estate some of the vouchers that Reed possessed at the time of his death in 1999. Many of the specimens were dried in jars and in poor condition; some specimens were incorporated into the NHSM collections.

The publication of Roger Conant's (1958a) *A Field Guide to the Reptiles and Amphibians of the United States and Canada East of the 100th Meridian* was a major stimulus for herpetofaunal discovery (Lee 1993; Mitchell 2013). The field guide included range maps for each species, and young naturalists felt challenged and empowered to discover extensions to species' ranges (Lee 1993).

In 1959, the Virginia Herpetological Society established a DC-Maryland section in its *Bulletin* (VHS 1959) for notes on Maryland species occurrences. Several Maryland members were granted a "staff" membership status within the society because of their breadth of knowledge about herpetofaunal distributions: Clyde Reed, Romeo Mansueti, Robert McCauley, Jr., Charles Wierer, William F. Stickel, and Lee D. Schmeltz. Many Maryland participants were interested in the adjacent geographic distributions of lower Delmarva and the Potomac River environs near Washington, DC. For instance, John Cooper shared information on the occurrence of venomous snakes in the region (VHS 1960). The Virginia Herpetological Society *Bulletin* (VHS 1959) was a vehicle for requesting data and exchanging observations (e.g., Tobey 1962).

John Cooper (1960b) compiled known observations and records of Maryland amphibians and reptiles into two statewide matrixes, one based on county and the other on physiographic region. He did not indicate which of the records were backed by preserved specimens. The Cooper paper is the first published state list to identify county distributions for all of Maryland's herpetofauna.

The Maryland Herpetological Society (1965), an unincor-porated club within the NHSM, was established in 1965 with the purpose of guiding "the individual and aesthetic education of the young and uninitiated in the field of herpetology while attempting to aid the more advanced herpetologist in his work of locating, cataloging information and specimens of the state of Maryland." The *Bulletin of the Maryland Herpetological Society* was also established in 1965. The first volume contained a republished version of John Cooper's (1960b) paper, with the tables amended with new distributional records by Herbert Harris, Jr. (J. E. Cooper 1965b). During this period, there was a flurry of field activity and library research to document all historical and current occurrences of each species throughout Maryland. To help focus some of these searches, Herbert Harris et al. (1967) assembled a list of questions about species' distributions for field surveyors to ponder as they explored Maryland. Vouchers were collected for new occurrences and stored in the basement of the NHSM's offices. Several additions to the Cooper (1965b) list were published (H. S. Harris 1966a, 1968, 1969a) as new information accumulated, leading to a compilation, "Distributional survey: Maryland and the District of Columbia," which was the first attempt to present dot distributional maps for all of Maryland's amphibians and reptiles (H. S. Harris 1969b). Survey efforts continued for several years and resulted in an updated compilation, "Distributional survey (Amphibia/Reptilia): Maryland and the District of Columbia" (H. S. Harris 1975). These two published surveys included historical reports in addition to the then current observations and collections. As in past surveys, some records were included based on the author's knowledge of the credibility of the reporter. Harris added occasional notes within the publications to explain the basis of his decision to accept or reject some past observations. This was the clearest way to handle the uncertainty of historical information with no vouchers.

Efforts to Update the Distributional Survey

After 20 years, Herbert Harris (1995) initiated an update to the 1975 survey. Many additional new and old records had been discovered since the earlier survey: additional historical collections (e.g., R. W. Miller 1982); new state and county records (e.g., Thompson and Chapman 1978); habitat-based records (e.g., Bascietto and Adams 1983); life history studies with incidental distributional data (e.g., D. M. Hillis 1977); stranding records of sea turtles (e.g., Norden et al. 1998); site checklists (e.g., R. H. Johnson and van Deusen 1979); field guides (e.g., White and White 2007); rare species studies (e.g., Thompson and Taylor 1985); bioblitzes (e.g., Swarth et al. 2010); and anuran call surveys (e.g., Hildebrand 2005). New dot maps were published for the following: Mud Salamander (Chalmers 2006; R. W. Miller 2011,

2014), Long-tailed Salamander (R. W. Miller 1984b), Eastern Spadefoot (R. W. Miller 2013), Eastern Narrow-mouthed Toad (J. D. Hardy 2009), and Scarletsnake (Grogan and Close 2008; H. S. Harris 2013; R. W. Miller 2016). Herbert Harris (2009) again announced a renewed effort to publish an updated survey and began organizing the update by enlisting volunteers. That effort expanded and morphed from a survey of distribution to the current MARA project.

Challenges and Practices for Gathering Distributional Data

Sources of information for checklists of Maryland amphibians and reptiles were derived from three types of activities: personal discoveries and observations of single species, checklists of small locales or habitat types, and life history studies indicating source locations of the subject animals.

Challenges for Discovering New Occurrences

The listing and mapping efforts for amphibians and reptiles depended on the persistence of individuals in searching for and reporting new occurrences, particularly if these were a new state or county record or an extension of a species' range. Numerous amateur and professional herpetologists have been exploring the state. Herbert Harris (2009) lists the many people who contributed observations for his 1975 distributional survey.

Documenting past discoveries is not always simple. Reconstructing or rediscovering past claims can be puzzling even when a voucher is available. James Fowler (1969) and John Cooper and Thomas Hunt (1970) wondered about the potential occurrence of the Pine Woods Treefrog (*Hyla femoralis*) in southern Maryland, which was documented there in 1937 when Carl and Laura Hubbs purportedly collected it from the Battle Creek Cypress Swamp (J. A. Fowler and Orton 1947). It took almost 30 years to conclude that we have not confirmed the Three-lined Salamander (*Eurycea guttolineata*) in Maryland (Ireland 1979; R. W. Miller 1980, 2009a), despite the species being reported on several lists.

Nomenclatural confusion of cryptic species can contribute to erroneous claims of occurrence. Noble and Evans (1932) in their study of Northern Dusky Salamander mistakenly also reported the Carolina Mountain Dusky Salamander (as *Desmognathus fuscus caroliniana*) from Garrett County. This was based partly on their mistaken claim of the northerly extent of its range. Pope (1924) probably contributed to this confusion by renaming the taxon as a subspecies of the Northern Dusky Salamander, rather than as originally described by E. R. Dunn (1916) as a subspecies of Allegheny Mountain Dusky Salamander. Eventually, Tilley

and Mahoney (1996) resolved the distribution of the Carolina Mountain Dusky Salamander through genetic studies and showed that the taxon was limited to the mountains of Tennessee and North Carolina. The Northern Dusky Salamander and Allegheny Mountain Dusky Salamander were the two species within the Maryland range of Noble and Evans's experiments and observations.

Similarly, names can change as our understanding of species relationships improves. Highton (1962b) reported "Northern Ravine Salamander" (as *Plethodon richmondi richmondi*) from western Maryland based on a morphological analysis of members of the genus. Angle (1969) studied this salamander's ecology on the Ridge and Valley Province in Maryland. But Highton (1972), using protein studies, later proposed that the forms at the northern end of the range are a different species and named it the Valley and Ridge Salamander (*Plethodon hoffmani*). Consequently, older occurrence records of "Northern Ravine Salamander" on Maryland's Ridge and Valley Province now represent the Valley and Ridge Salamander.

Non-native species have also been encountered in the search for new amphibian and reptile occurrences. Although the apparent attempt to introduce softshell turtles from the Mississippi Basin into the upper Potomac Basin (Dukehart 1884) failed, Spiny Softshells have been encountered at the head of the Rhode River (Mansueti and Wallace 1960) and at Patuxent River Naval Air Station (Rambo 1992). Chinese Softshell Turtles (*Pelodiscus sinensis*) have been encountered in the tidal Potomac (H. S. Harris 2004; Mitchell et al. 2007). Other reported non-native turtles include the False Map Turtle (*Graptemys pseudogeographica*) (J. A. Fowler 1943a); Red-eared Slider (Nemuras and Sparhawk 1966; J. D. Groves 1972); Yellow-bellied Slider (*Trachemys scripta scripta*) (J. E. Cooper 1959; Reed 1956d; H. S. Harris 2006; Norden and Browning 2013); Blanding's Turtle (*Emydoidea blandingii*) (Carver 1970); Coastal Plain Cooter (*Pseudemys concinna floridana*) (F. J. Schwartz 1961); Mississippi Map Turtle (*Graptemys pseudogeographica kohnii*) (F. J. Schwartz and Dutcher 1961; Hildebrand 2011); and Three-toed Box Turtle (*Terrapene carolina triunguis*) (Polley 1989); as well as claims of Reeve's Turtle (*Mauremys reevesii*) (VHS 1962) and Cumberland Slider (*Trachemys scripta troostii*) (VHS 1962; H. S. Harris 1975). Many of these are believed to have been escaped or abandoned pets. Mediterranean Geckos (*Hemidactylus turcicus*) in Baltimore most likely arrived in shipping containers from Spain (Norden and Norden 1989). While it is generally easy to recognize non-native species when they are encountered, cryptic introduction of native species, based on the extensive trade in native amphibians and reptiles (see Bridges et al. 2001), has probably occurred through pet release and passive and incidental transport by people. However, these types of introductions are very difficult to detect.

Local and Habitat Checklists

A second practice that was helpful in synthesizing state-wide lists was the assembling of local geographic and habitat checklists. Many additional regional and local checklists have been put together. Some emphasize many species in one geographic area (table 3), while others attempt to show the extent of one species across its Maryland range. Other surveys and studies have emphasized habitats, such as: Maryland caves (Schreiber 1952; J. E. Cooper 1960c, 1965a; Franz 1964, 1967a) and stormwater management basins (Bascietto and Adams 1983). The scientific quality of these lists varies. Some are supported by voucher specimens or photographs, and others are unverifiable. Even lists apparently vouchered at the time of publication may become less certain over time as owners or museums lose, discard, destroy, abandon, or transfer specimens, or, as in some cases, specimens were never deposited in an appropriately curated museum.

Life History Studies

Other distributional data have been amassed incidentally as part of life history field studies, but these vary in quality depending on the specificity of the locations stated in the publications and on whether voucher specimens were collected as part of the study protocol. Life history studies in Maryland that incidentally contributed distributional data began with a study of the Eastern Copperhead (Hopkins 1890). Hopkins's research focused on various life history traits, but incidentally documented occurrence sites throughout Maryland. He mentions occurrences in Baltimore, Montgomery, Prince George's, Washington, and Worcester counties, and near Harpers Ferry. Researchers have studied many aspects of amphibian and reptile life history within Maryland. For instance, reproductive cycles have been studied in the Little Brown Skink in Queen Anne's County (J. D. Groves and Norden 1995), Eastern Red-backed Salamander throughout Maryland (Sayler 1966), and the Valley and Ridge Salamander (Angle 1969). Cordero (2009) noted the response of Spotted Turtles to the displacement effects of flooding of the Patuxent River. Norden (1999) pondered on the mortality effects of flood events on Wood Turtles he had observed in Allegany and Washington counties. McClellan et al. (1943) assembled life history information for the lizards of central and southern Maryland. Noble and Evans (1932) studied the brooding habit, larval life, and habitat selection of Northern Dusky Salamander in Garrett County. Periodic movements during annual life cycles have been documented for Eastern Box Turtle (Sipple 2008; Hagood and Bartles 2010), Eastern Mud Turtle (Cordero et al. 2012b), and Long-tailed Salamander (Franz and Harris 1965). Hagood (2009) studied the effect of habitat fragmentation on the genetic diversity of Eastern Box Turtles in Montgomery

Table 3. Regional and local herpetofaunal checklists for Maryland

Location	Authors
Anne Arundel County	J. E. Cooper 1956; Reed 1957b
Assateague Island, Worcester County	Lee 1972a, 1973a
Broad Creek/Deep Run, Harford County	Prince et al. 1955
Calvert County	J. D. Hardy and Mansueti 1962
Camp Moshava, Harford County	P. Martin 1995
Catoctin Mountain Park (salamanders)	Randolph 1984
Chesapeake and Ohio Canal (amphibians)	Thompson 2000; Forester 2000
Chesapeake Bay Center for Environmental Studies, Anne Arundel County	Lynch 1980
Chesapeake Bay Region (reptiles)	J. D. Hardy 1972; Musick 1972
Coastal Plain of Maryland	Musick 1972
Cove Point, Calvert County	Bushmann 2000; McCann et al. 2002; Norden 2005
Delmarva Peninsula	Conant 1945; Reed 1956g
District of Columbia region	J. A. Fowler 1943b, 1945
"The Elms," St. Mary's County	van Deusen and Johnson 1980
Harford County	Reed 1956a
Irish Grove Sanctuary, Somerset County	Lee 1973b
Jug Bay Wetlands Center, Anne Arundel County	Swarth 1998a; Smithberger and Swarth 1993; Swarth et al. 2010
Kent Island, Queen Anne's County	Grogan and Bystrak 1973a
Moyaone Reserve, Prince George's County	Geddes 1981
National Capital Region	Pauley et al. 2005
Naval Surface Warfare Center at Indian Head, Charles County	Forester and Miller 1992
Plummers Island, Montgomery County	Manville 1968
Potomac Gorge	Gibson and Sattler 2008
Severn Run Natural Environmental Area	H. S. Harris 2015
Somerset County	Reed 1957c
Southern Maryland	J. E. Cooper 1953
Thomas Stone Site	Mitchell 2008
Vienna power plant site, Dorchester County	R. H. Johnson and van Deusen 1979, 1980

County. Forester and Lykens (1991) documented the population age structure of Eastern Newts near a pond on the Blue Ridge Province. Many other studies include incidental information about the distribution of Maryland herpetofauna. But here, as in other cases, not all have preserved evidence of the identity of the researched species.

Additional Institutional Efforts That Document Amphibian and Reptile Distributions

In the late twentieth century, academic and research institutions played a significant role in the distributional and life history research of amphibians and reptiles in Maryland.

Starting in 1977 or thereabout, much herpetofaunal fieldwork in western Maryland was facilitated by Dr. J. Edward Gates at the University of Maryland Center for Environmental Science (UMCES) at Frostburg University. In addition to basic life history and distributional studies, the research there has also emphasized resource management (Thompson and Gates 1979; Thompson 1980; Thompson et al. 1980; R. D. Williams et al. 1981a; R. D. Williams et al. 1981b; Gates and Thompson 1981; Davis-Chase 1983; Gates et al. 1984; Gates et al. 1985a; Gates et al. 1985b; Durner 1991; Durner and Gates 1991, 1993; McLeod 1995; McLeod and Gates 1998). At about the same time, Towson State University (now Towson University) began building its collections, and research has been undertaken by Dr. Donald Forester, Dr. Joel Snodgrass, and Dr. Richard Seigel and their students (Forester and La Pasha 1982; Forester and Lykens 1991; Grogan and Forester 1998; Forester 2000; Snodgrass et al. 2000; Forester et al. 2003; Forester et al. 2006; Massal et al. 2007; Snodgrass et al. 2008; Brand and Snodgrass 2010; Brand et al. 2010; Farnsworth and Seigel 2013; K. P. Anderson 2014; Gallagher et al. 2014). Dr. William Grogan and his students and associates at the University of Maryland, College Park, and Salisbury University contributed to field research mostly on the Eastern Shore of Maryland (Grogan and Prince 1971; Grogan 1973, 1974a, 1974b, 1975, 1985, 1994; Grogan and Bystrak 1973a, 1973b; Grogan and White 1973; Grogan and Williams 1973; Grogan and Forester 1998; Grogan and Close 2008).

The Jug Bay Wetlands Sanctuary in Anne Arundel County, a National Estuarine Research Reserve, has been the site of concentrated herpetological research, particularly between 1995 and 2011. During this period, numerous reports were produced on the behavior and life history of amphibians and reptiles on the site. Much of the work focused on native turtle species at the sanctuary. Observations generated at least 15 technical reports for the sanctuary (Linebaugh 1995; Nuebert 1996; Crumrine and Bartimo 1997a; Parks 1998; Swarth 1998a; Bulte 2000; K. Capps 2001; Curless 2001; M. Bennett 2002; J. Capps 2002; DeGregorio 2002; Lentz 2004a; Thurston 2005; Matthews 2007; Swarth and Friebele 2008); at least five popular articles (Crumrine and Bartimo 1997b; Molines 1997; Swarth 1998b, 2005a; Teece and Swarth 1998); three academic theses (Dreher 1999; Lentz 2004b; Woodward 2005); four presentations at conferences (Molines and Swarth 1997, 1999; Fogel et al. 2002; Szlavecz et al. 2007); and 15 peer-reviewed articles, books, and book chapters (Smithberger and Swarth 1993; Smithberger 1995; Marchand et al. 2003; Quinlan et al. 2003; Swarth 2004, 2005b, 2005c; Teece et al. 2004; Swarth and Hagood 2005; Somers and Mathews 2006; Giery and Ostfeld 2007; Costanzo et al. 2008; Dickey et al. 2009; Savva et al. 2010; Cordero et al. 2012b). Each of these publications or unpublished reports indirectly confirms the presence of the researched species at Jug Bay Wetlands Sanctuary.

Studies at Patuxent Wildlife Research Center have included herpetological investigations related to heavy metal and pollution accumulation in amphibians and reptiles. Hall and Mulhern (1984) investigated heavy metal accumulation in anurans. Birdsall et al. (1986) specifically studied lead accumulation in Green Frogs and American Bullfrogs. Others studied the effects of water quality on Spotted Salamander egg survival (Albers and Prouty 1987) and of pollutants on Snapping Turtles (Albers et al. 1986). The extent and permanence of the Patuxent Wildlife Research Center provide a relatively natural site for long-term field studies within the highly developed East Coast corridor. Studies have featured Eastern Ratsnake (Stickel et al. 1980), Eastern Box Turtle (Stickel 1978, 1989; Hallgren-Scaffidi 1986; Stickel and Bunck 1989), and other turtles (Gotte 1988, 1992).

In 1992, the National Park Service (NPS) established the Inventory and Monitoring Program (Monahan and Gallo 2014), which fostered several herpetofaunal surveys on National Park lands in Maryland (Forester 2000; Thompson 2000; Pauley et al. 2005; Mitchell 2008). The results of surveys of NPS properties are available through the NPSpecies online database, with data fields indicating the type of evidence supporting each claimed presence. The very first "bioblitz" was organized by the NPS and took place at Kenilworth Park and Aquatic Gardens in Washington, DC, and in Prince George's County. It included searches for amphibians and reptiles, as did the later bioblitz at Potomac Gorge (Gibson and Sattler 2008).

Rare Species Surveys

By the 1950s, enough information about amphibians and reptiles in Maryland had been assembled to determine which species were less common. Beyond range-wide studies of these species, searches and studies were begun on suspected rare species: Hellbender (J. A. Fowler 1947a), Eastern Tiger Salamander (Stine 1953; Stine et al. 1954), Green Salamander (H. S. Harris and Lyons 1968a, 1968b; Lee and Norden 1973), Jefferson Salamander (Netting 1946; Stine and Simmons 1952; Stine 1953), Eastern Narrow-mouthed Toad (Noble and Hassler 1936; J. A. Fowler and Stine 1953; Conant 1958b), Coal Skink (Lemay and Marsiglia 1952), Rainbow Snake (McCauley 1939b; J. E. Cooper 1960a), Mountain Earthsnake (J. E. Cooper 1958), and Bog Turtle (McCauley and Mansueti 1943; J. E. Cooper 1949; Campbell 1960). Information generated about sea turtles generally consisted of spot distributional data (H. S. Harris 1969b), except for comments on the Kemp's Ridley Sea Turtle (J. D. Hardy 1962) and Leatherback Sea Turtle (J. D. Hardy 1969).

In 1971, Maryland passed the state's endangered species act. The next year, the Maryland Herpetological Society established the Committee on Rare and Endangered Amphibians and Reptiles of Maryland to provide recommendations for species to be included in regulations. The committee consisted of John E. Cooper, L. Richard Franz, Frank Groves, Jerry D. Hardy, Jr., Herbert S. Harris, Jr., David S. Lee, Robert G. Tuck, Jr., and Peter Wemple. By 12 October 1972, the aforementioned rare species were included under the protection of the Maryland act.

Concurrently, at the national level, a protocol was needed to standardize how distributional data were collected so that sense could be made of rare species' ranges, given that many ranges cross state boundaries. In 1976, the Nature Conservancy (TNC) began an effort to standardize the collection of species distributional information by establishing joint programs with states (Hoose 1981). The Maryland Natural Heritage Program (MNHP) began as a cooperative program with TNC in 1979 and was ultimately staffed by the MDNR in 1981 (Chipley et al. 1984). Establishment of the MNHP and the Maryland endangered species program (G. J. Taylor 1984) facilitated surveys of many species considered rare in the state: Bog Turtle (Dawson 1984; Lee and Norden 1996); Hellbender (Gates et al. 1984); sea turtles (J. D. Groves 1984); Jefferson Salamander, Green Salamander, Mountain Earthsnake, and Coal Skink (Thompson 1984a); Wehrle's Salamander (Thompson and Chapman 1978; Thompson 1984b); and Eastern Tiger Salamander (Stine 1984). Regulations enacted to protect rare species resulted in additional surveys related to development proposals, state government actions, or expenditure of public funds. All of these efforts resulted in additional understanding about amphibian and reptile distributions. The results of these early efforts were presented at a public conference at Towson State University (Towson University) in 1981 (Norden et al. 1984).

Since then, many additional local studies have been undertaken as part of the development of permitting requirements and project reviews that require surveys for rare and protected species. Other specific investigations, such as the sea turtle tagging study (Kimmel 2007), have contributed to our understanding of harder to detect, more mobile rare species.

Increasing Access to Herpetofaunal Information

Before 1900, most amphibian and reptile reports were brief notes in private journals, letters, land records, newspapers, privately published books, or occasional government reports. Between 1930 and 1960, many of the sighting records were published in the in-house publications of the NHSM: *Bulletin of the Natural History Society of Maryland*; *Maryland, a Journal of Natural History*; *Maryland Naturalist*; *Junior Division Bulletin*; *Junior Society News*; and *Junior Naturalist*.

Since November 1965, when the *Bulletin of the Maryland Herpetological Society* was established by Herbert S. Harris,

Jr. (editor), it has been the single most dominant instrument for the publication of notes and observations by amateur and professional herpetologists on the distribution of Maryland's amphibians and reptiles. By providing a place for publishing both incidental observations and formal research, it became, as intended, a critical repository for distributional information. In addition, distributional knowledge has been published in *Proceedings of the Biological Society of Washington*; *Transactions of the Maryland Academy of Sciences*; *Copeia*; *Annals of the Carnegie Museum*; *Journal of Herpetology*; *Herpetological Review*; *Herpetologica*; *Proceedings of the Pennsylvania Academy of Sciences*; *Bulletin of the Chicago Herpetological Society*; *American Midland Naturalist*; *SSAR Herpetological Review*; *Maryland Naturalist*; and *Biological Conservation*. Unvouchered reports within all these publications, though, create uncertainty for future researchers.

Over the course of assembling amphibian and reptile distributional data, the technology for capturing evidence has steadily progressed. For images, we transitioned from Petiver's (1698, 1767) descriptions and sketches, to flash photography using glass plates (e.g., Klingel 1932a), to film cameras, then digital cameras, and now the convenience of cellphone cameras (Selleck 2014). Similarly, how we gather, store, and access other herpetological data has transitioned from paper-and-pencil lists, mimeograph processes, and copiers, to computers for electronic assembly and dissemination of data. The current state of technology allows the real-time gathering of and access to herpetological data from any point on Earth accessible to the internet. Numerous projects sponsored by organizations and individuals now collect distributional records: Sea Turtle Stranding and Salvage Network (National Oceanic and Atmospheric Administration), FrogWatch USA (Association of Zoos and Aquariums), iNaturalist (California Academy of Sciences), Project Noah (National Geographic support), Nature's Notebook (National Phenology Network), Global Amphibian Bioblitz (iNaturalist project), iSpot (Open University), Global Reptile Bioblitz (iNaturalist project), Amphibian Research and Monitoring Initiative (US Geological Survey), Maryland Biodiversity Project (database managed by Bill Hubick and Jim Brighton), Herpetological Education and Research Project (unincorporated nonprofit managed by its board members), and HerpMapper (incorporated nonprofit with a database administered by Don Becker, Mike Pingleton, and Christopher E. Smith). Improving technologies have had two effects. First, the assembly and dispersal of herpetological distributional information can occur in near real time, allowing immediate public access. Second, increasing access to information and expertise allows consumers of the information to more readily learn about their local amphibians and reptiles. This fosters public expertise and the probability that people will contribute to the ongoing accumulation of herpetofaunal knowledge and conservation in Maryland.

Maryland's Environment

Maryland is the ninth smallest state in the United States. It encompasses 25,317 km^2 (9,775 mi^2) of mountains, valleys, rolling hills, flatlands, and beaches (Frese 1996). Within this varied topography lie diverse forests, a variety of wetlands, and abundant streams and rivers. The state boasts more than 14,800 km (~9,200 mi) of freshwater streams and approximately 12,400 km (~7,700 mi) of shoreline along the Chesapeake Bay and Atlantic Ocean (Hennessee et al. 2003). Traversing the state from west to east, mountains transition to rolling hills that level out to mixed flatwoods with swamps and Delmarva bays, and barrier islands buffering the shores of the Atlantic Ocean.

Physiographic Provinces

Despite its size, Maryland has a broad geographic reach that crosses five major physiographic provinces (Reger and Cleaves 2008). Within a province, the landscape is similar in relief, topography, geologic structure, and geomorphic history (A. E. Godfrey and Cleaves 1991; Reger and Cleaves 2008). Maryland is part of the Appalachian Plateaus, Ridge and Valley, Blue Ridge, Piedmont, and Atlantic Coastal Plain physiographic provinces (Reger and Cleaves 2008; Schmidt 2010). These provinces encompass a diversity of aquatic and terrestrial habitats. Maryland's geography also places the state in two major climate zones (Peel et al. 2007): the Humid Continental Climate Zone in the far west and the Humid Subtropical Climate Zone in the remainder of the state. Maryland's geography, habitat, and climate influence the distribution of its diverse amphibian and reptile fauna.

In Maryland, the Appalachian Plateaus Province includes western Allegany County and all of Garrett County. The eastern edge of the Appalachian Plateaus in Maryland is demarcated by the Allegheny Front, a steep escarpment extending from southern Pennsylvania to eastern West Virginia. In Allegany County, the Allegheny Front is continuous with Dans Mountain and Piney Mountain. The Appalachian Plateaus Province accounts for 8% of Maryland's land area. This is the highland region of Maryland, with the highest point reaching 1,024 m (3,360 ft) above sea level at Hoyes Crest, located on Backbone Mountain in southwestern Garrett County (MGS 2004). The province is underlain by sedimentary rocks that lie relatively flat or have been folded slightly and uplifted (Schmidt 2010). The steep ravines and valleys have been carved out by stream systems, forming a dendritic, or irregular branching, topographic pattern in the landscape. Stream systems here are part of the Ohio River watershed and flow generally north, except those along the eastern edge of the Appalachian Plateaus, which are part of the Chesapeake Bay watershed. The landscape is heavily forested, being too steep for agriculture in many areas. Peat bogs, swamps, and glades form in low-lying areas between hillsides and over bedrock, which traps water at the surface as it flows from springs through cracks and fissures in the ground.

To the east, the Ridge and Valley Province covers 7% of Maryland, stretching from the lower limit of the Allegheny Front in western Allegany County to the western base of South Mountain in Washington County. The area is underlain by strongly folded and faulted sedimentary rocks (Edwards 1981). The long mountain ridges of this province, with southeast- to east-facing slopes, run nearly parallel and are separated by equally long valleys. One such valley, known as the Great Valley, marks the eastern limits of the Ridge and Valley Province. The recent geologic history of this province is evident, with areas of exposed rock outcrops and shale barrens. This region is also heavily forested, although the soils are less developed, resulting in drier habitats.

Abutting the eastern border of the Ridge and Valley Province is the narrow Blue Ridge Province, which accounts for only 3% of Maryland's land area. On this province, South

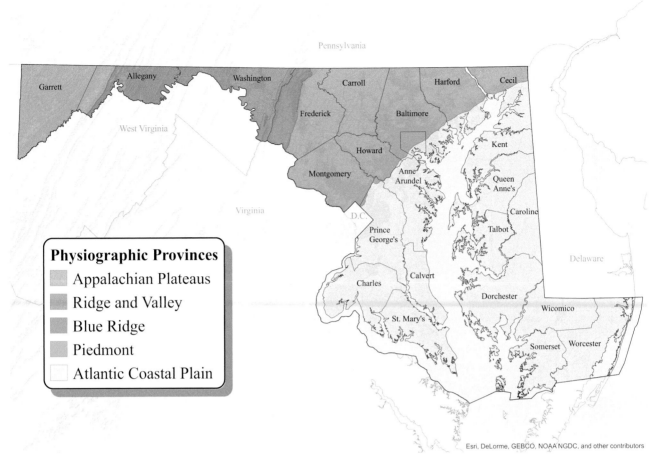

Physiographic provinces in Maryland. From MDNR; base map from ESRI, Inc., DeLorme Publishing Company, Inc., General Bathymetric Chart, NOAA National Geographic Data Center, and other contributors

Mountain and Catoctin Mountain are shaped by a large anticlinal fold of sedimentary rock. Buried between the mountain ridges is the region's oldest volcanic and gneiss rock (MGS 2015). On the South Mountain ridge, the highest point on the Blue Ridge Province is Quirauk Mountain, rising approximately 652 m (2,139 ft) above sea level. This is the eighth highest peak in Maryland (MGS 2015). Along the Potomac River in Washington County, near Harpers Ferry, West Virginia, the elevation drops to its lowest point in the province, 76 m (250 ft). Soils in the region are typically thin, with loose rock or exposed bedrock.

The Piedmont Province encompasses 28% of Maryland and is characterized by rolling hills within an agricultural landscape. The region is underlain by a variety of igneous and metamorphic rock. The province extends from the foothills of Catoctin Mountain east to the Atlantic Coastal Plain Province. Sugarloaf Mountain, in Frederick County, is the highest point in the province at roughly 391 m (1,283 ft) (MGS 2004). Like other provinces, the Piedmont contains a variety of wetland and upland habitats.

The Atlantic Coastal Plain Province is the largest physiographic province in Maryland, comprising approximately 54% of the land area. It is delineated from the Piedmont Province by the Fall Line (also known as the Fall Zone), a region of contact between the two provinces where water flow transitions from fast flowing, over hard rocky streambeds, to slow moving over unconsolidated soils (Schmidt 2010). In Maryland, the lands of the Atlantic Coastal Plain Province to the east and west of the Chesapeake Bay are referred to as the Eastern Shore and Western Shore, respectively (Scott 1801). The terrain is relatively flat, although the Western Shore shows more topographic relief than the Eastern Shore. Unconsolidated sandy or loamy soils are the result of erosion from the Piedmont and Appalachian Mountains and deposits from ancient seas (Schmidt 2010). Water flows slowly through swamps and creeks to tidal river systems of the Chesapeake Bay and, to a lesser extent, the Atlantic Ocean. Due to the flat landscape and rich organic soil, agriculture dominates in adequately drained areas, particularly on the Eastern Shore.

The Green Salamander is a species that is restricted to the Appalachian Plateaus Province in Maryland. Photograph by Ed Thompson

Habitat of the Ridge and Valley Province is well suited to species such as the Eastern Copperhead. Photograph by Robert T. Ferguson II

Eastern Box Turtle, a denizen of Maryland's forests. Photograph by Andrew Adams

Habitat Types

Forests

Forests cover about 38%, or approximately 1 million ha (2.5 million ac), of Maryland's landscape (MDP 2010; Lister et al. 2011). Seventy-four percent of the state's forests are privately owned (USDA 2013). The assemblage of tree species in Maryland's forests varies across the state. White pine (*Pinus strobus*), eastern hemlock (*Tsuga canadensis*), and northern hardwoods comprise forests in the mountainous regions of western Maryland. The rolling landscape of the Piedmont in central Maryland has been heavily cleared to accommodate agriculture and urban development; thus, forests are mostly confined to riparian areas or small woodlots. The southern region of the Western Shore, with its deeper, less rocky soils, generally supports white oak (*Quercus alba*), Virginia pine (*Pinus virginiana*), and tuliptree (*Liriodendron tulipifera*). Forests of the northern Eastern Shore contain a mixture of hardwoods, while those in the southern region are dominated by loblolly pine (*P. taeda*), sweet-

gum (*Liquidambar styraciflua*), and red maple (*Acer rubrum*) (Rider 2015).

Rivers and Streams

Maryland has more than 15,000 km (9,400 mi) of streams and rivers (Peterson and Urquhart 2006), approximately 95% of which flow into the Chesapeake Bay. The state has a variety of rivers and streams, including tidal, nontidal, fresh, and brackish. More than half of Maryland's stream-miles are first order, with no tributaries. A small number of Maryland's streams are fourth order or larger, such as the Patapsco, Susquehanna, and Potomac rivers, which have extensive drainage basins and multiple tributaries.

The physical characteristics of Maryland's rivers and streams depend on their location. On the Atlantic Coastal Plain Province, rivers and streams tend to have sand and gravel substrates, whereas in mountain and Piedmont streams the substrate is cobble, boulder, and bedrock (Roth et al. 1999). Flow velocity varies depending on gradient. Generally, the steeper stream gradients in the mountain provinces and the Piedmont result in swifter-moving water relative to those on the Atlantic Coastal Plain Province. For example, the well-known whitewater areas of the Youghiogheny River, in Garrett County, result from steep mountain gradients, exposed boulders, and bedrock in the river. In general, the temperatures of rivers and streams in western provinces are cooler than those of the Atlantic Coastal Plain Province. Streams along the Fall Line tend to have falls, where they pass from the rocky, higher elevations of the Piedmont Province to the lower, flatter Atlantic Coastal Plain Province. Many of the streams on the Atlantic Coastal Plain Province have been channelized to drain surrounding uplands. In Maryland, about 50% of stream-miles have forested riparian buffers (Versar 2011).

Wetlands

Maryland has an abundance of wetlands. Overall, nearly 10% of the state is classified as wetland or tidal (LaBranche et al. 2003). Most wetlands are concentrated on the Atlantic Coastal Plain Province, surrounding the Chesapeake Bay. The two most common wetland types in Maryland are palustrine (freshwater) and estuarine (brackish water), comprising more than 99% of the state's wetlands. Palustrine wetlands may be either tidal or nontidal, with the majority (89%) in Maryland being nontidal (Tiner and Burke 1995). Forested wetlands are the most widely distributed and abundant type of palustrine wetland in the state. These wetlands are found in riparian floodplains and upland depressional swamps. Forty-two percent of the state's wetlands are estuarine, concentrated on Maryland's tidal rivers and extending upstream to freshwater areas.

The lower Eastern Shore has the state's greatest extent of wetlands, due to its low topography, clay-rich soils, and high groundwater tables (LaBranche et al. 2003). Dorchester and Somerset counties have the highest proportion of wetlands in the state, 45% and 38%, respectively, while Allegany and Washington counties have the lowest (LaBranche et al. 2003). Tidal marshes are the most extensive type, but Delmarva bays, bald cypress swamps, and Atlantic white cedar swamps are just a few of the nontidal wetland habitats found on the lower Eastern Shore.

In Maryland, fewer wetlands occur on the Piedmont than on the Atlantic Coastal Plain. (LaBranche et al. 2003). Most Piedmont wetlands consist of isolated palustrine and riverine types, such as floodplain and upland depressional swamps. Further west, on the Blue Ridge and Ridge and Valley provinces, wetlands are uncommon and often restricted to depressions at the base of topographic slopes and narrower floodplain depressions. Wetland diversity increases on the Appalachian Plateaus Province. Wetlands of this province include wet thickets, shrub bogs, and seasonally flooded meadows and marshes (LaBranche et al. 2003).

Chesapeake Bay

The Chesapeake Bay is the largest and most productive estuary in the United States (CBP 2004). Approximately 4,470 km^2 (1,726 mi^2) of the main basin is in Maryland (Frese 1996). The shoreline of the bay is irregular and includes brackish marshes, sandy beaches, low banks, cliffs, and bluffs. On the lower Eastern Shore, the shoreline of the Chesapeake Bay is characterized by extensive marsh and estuarine beaches. The water level is affected by semidiurnal tides, wind, and precipitation. A number of islands are found within the Bay, some consisting almost entirely of salt marsh. The Chesapeake Bay includes diverse habitats that contribute to its productivity. These include open water,

Forested ephemeral wetlands are important breeding habitats for the Marbled Salamander, shown here guarding eggs. Photograph by Don C. Forester

shallow nearshore waters, oyster reefs, tidal marshes and wetlands, tidal flats, and sandy beaches.

Coastal Bays

Coastal bays are lagoon-like estuaries contained by barrier islands on Maryland's Atlantic Coast and small, distinct watersheds on the mainland. Maryland contains five coastal bays: Assawoman, Chincoteague, Isle of Wight, Newport, and Sinepuxent. These shallow bays, collectively, form a biodiverse estuary. Salt marshes, mussel beds, and submerged seagrass meadows are found in the shallowest regions.

Atlantic Ocean

Maryland has approximately 50 km (31 mi) of shoreline fronting the Atlantic Ocean (E. R. Smith et al. 2016). The highly populated Ocean City is found in the northern portion of the coastline; the southern portion contains the undeveloped Assateague Island. From the shoreline, Maryland authority extends 4.8 km (3 mi) into the Atlantic Ocean. Nearshore currents, tides, and storm events shape the coastline. Shallow depths near the shore make these waters highly productive.

Climate

Maryland's climate exerts a strong influence on the distribution and abundance of the state's amphibians and reptiles. The state's geographic position yields a temperate climate to the area. Coastal influences of the Chesapeake Bay and Atlantic Ocean in the east and the Appalachian Mountains in the west collectively moderate the state's climate. In general, Maryland experiences a mild, humid, and relatively stable climate.

Maryland has four distinct seasons throughout the year. Summers range from mild to hot and winters from very cold to moderate. The state lies within two climate zones (Peel et al. 2007). In western Maryland, the Appalachian Plateaus Province is within the Humid Continental Climate Zone, which is characterized by large seasonal temperature differences. The remainder of Maryland is within the Humid Subtropical Climate Zone, characterized by hot and often humid summers and cool to mild winters.

The average annual temperature statewide is 12.1 °C (53.8 °F) (NOAA 2017a). Although summer temperatures vary across the state, July is often the warmest month, with average temperatures exceeding 26.7 °C (80 °F). Late summer and fall are often the driest periods of the year. The winters of Maryland are typically cold. January is often the coldest month, with average temperatures reaching –7 °C (20 °F) (NCDC 2015). Annually, the state receives approximately 108 cm (43 in) of precipitation, with the greatest amounts falling in early to midsummer (NOAA 2017b). However, precipitation varies across the state. The wettest area in the state is western Garrett County. The Ridge and Valley Province is often the driest region in Maryland.

The higher elevation of the Appalachian Plateaus Province results in a much cooler climate than in the rest of Maryland. In general, this area of the state averages 150 days of subfreezing temperatures annually (MDP 1973). Here, the growing season rarely exceeds 130 days. Summers are typically warm to hot, at times humid, and winters are very cold. Precipitation is usually well distributed throughout the year. In western Maryland, the Appalachian Mountains create a rain-shadow effect that causes heavy precipitation on the western slopes with correspondingly dry conditions on the lee side. Annual precipitation exceeding 127 cm (50 in) is not uncommon in the extreme southwestern corner of the Appalachian Plateaus Province in Maryland. Conversely, the eastern side of the province may receive less than 91 cm (36 in) annually (USGS 1999).

The Ridge and Valley Province also experiences rain-shadow effects due to the Allegheny Mountain range in the west and the Blue Ridge Mountains in the east (MDP 1973). Annually, the province receives an average 91-102 cm (36-40 in) of precipitation. Growing days generally range between 160 and 170 days per year. The Piedmont Province receives, an average 102-112 cm (40-44 in) of precipitation annually, with an average of 170-190 growing days per year (MDP 1973). The Atlantic Coastal Plain Province experiences the longest growing season in Maryland, which can reach 230 days along the southernmost shores of the Chesapeake Bay. The northern region of the Atlantic Coastal Plain Province has an average annual growing season of 190 days. Annual precipitation in the province averages 112-122 cm (44-48 in).

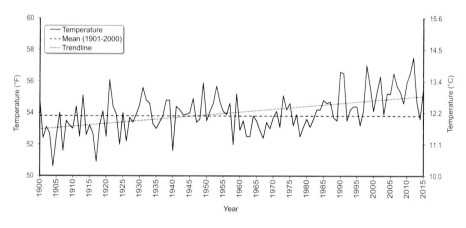

Mean annual temperature in Maryland, 1900-2015. Annual 1901-2000 mean temperature = 12.1 °C (53.8 °F); annual 1900-2015 trend = +0.11 °C (+0.2 °F) per decade. From NOAA (2017a)

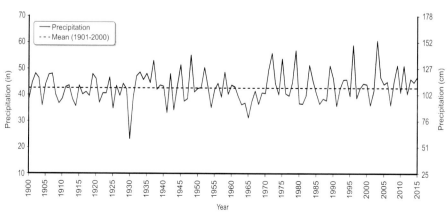

Mean annual precipitation in Maryland, 1900-2015. Annual 1901-2000 mean precipitation = 108.1 cm (42.6 in.). From NOAA (2017b)

Temperature and precipitation influence the surface activity of the state's herpetofauna. Timing of the active seasons of Maryland's amphibians and reptiles varies geographically due to variation in temperature and precipitation across the state. For example, species on the Atlantic Coastal Plain are active for longer periods of the year than those on the Appalachian Plateaus. However, peak surface activity for most herpetofauna in the state is generally observed during the spring and fall. Periods of hot, dry conditions in the late summer often result in temporary bouts of low surface activity for many amphibian and reptile species.

Maryland's climate is changing: over the past century, the state has warmed by an average of 0.5-1 °C (1-2 °F) (EPA 2016). During the MARA project, 2010-2014, the average annual temperature was approximately 13 °C (56 °F), slightly higher than the average baseline (1901-2000) annual temperature of 12.1 °C (53.8 °F) for Maryland (NOAA 2017a). For each year of the project, with the exception of 2014, annual temperatures were above average. During the project, the greatest departure from the baseline average was observed in 2012, when the annual temperature was approximately 2 °C (3.7 °F) above average (NOAA 2017a).

Over the past century, the average annual precipitation has increased by approximately 5% (EPA 2016). Annual precipitation during the MARA project was 118 cm (46 in) (NOAA 2017b). Overall, during the project, annual precipitation exceeded the state's normal average of 108.1 cm (43 in). Two years during the project, 2010 and 2012, experienced below-normal precipitation, but the greatest departure from the normal average was during 2011, when average annual precipitation was 22 cm (9 in) above normal (NOAA 2017b).

Land Use

The majority of the state's land consists of forests (38%), agricultural lands (30%), and nonforested wetlands (4%). The remaining landscape is urban development (MDP 2010),

which includes very low to high residential density, as well as commercial and industrial areas. Maryland's forested land is concentrated in the western and southeastern regions of the state, whereas agricultural lands are concentrated on the Piedmont and Atlantic Coastal Plain provinces. The Atlantic Coastal Plain Province is one of the most heavily utilized areas in the state, with urbanization dominating the northern regions, particularly along the Baltimore-Washington, DC, corridor, and agriculture and silviculture dominating the southern regions. Silviculture has a long history on the Eastern Shore and southern portions of the Western Shore. It has been practiced to varying degrees of intensity throughout the region, with the highest intensity on the Eastern Shore. On the Piedmont Province, undeveloped areas have become fragmented, as forest and agricultural land have been converted to residential use. This conversion is due, in large part, to expansion of the urban centers of Baltimore and Washington, DC (Kearney 2003). The majority of the Appalachian Plateaus, Ridge and Valley, and Blue Ridge provinces consists of forest, with agriculture and urban areas confined to the lower valleys (MDP 2010). Throughout the montane provinces, urban development is sparse and is generally located in the larger valleys.

Maryland's population has increased slightly more than 4% since the 2010 census (MDP 2016), but the rate of population growth varied across the state. Due to the state's everburgeoning population, land use has changed considerably. Since 1973, more than one million acres of land have been developed: from 1973 to 2010, low-density residential and industrial land use increased threefold and fourfold, respectively (MDP 2010). During this time, nearly 250,000 ha (62,000 ac) of forest were lost across Maryland, and agricultural land decreased by 19%; from 2002 to 2010, however, only about 6% of forested and agricultural land was lost (MDP 2010). Maryland's wetlands have also decreased by more than 405 ha (1,000 acres) since 1973, and an additional 1,620 ha (4,000 ac) are now covered by water, probably due to a combination of land subsidence and sea-level rise.

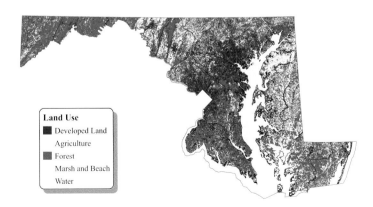

Land use in Maryland. From MDP (2010)

Conservation of Maryland's Herpetofauna

LYNN M. DAVIDSON AND GLENN D. THERRES

The list of threats and challenges to global herpetofaunal diversity is long, with habitat loss, invasive species, emerging diseases, environmental pollution, illegal collection, and climate change posing the most serious dangers. Within Maryland, native amphibians and reptiles face these same pressures. For example, a study by McLeod and Gates (1998) found that some timber practices within Maryland—specifically, disturbance of small patches of forest—resulted in decreased local diversity of herpetofauna. Maryland has lost an estimated 45% to 65% of its original wetlands (Tiner and Burke 1995), which are critical habitats for many amphibians and reptiles. The chytrid fungus (*Batrachochytrium dendrobatidis*), responsible for the often fatal disease chytridiomycosis, is a very real threat to the state's amphibians. The fungus was detected about a decade ago in the Chesapeake and Ohio (C&O) Canal National Historical Park (Grant et al. 2008). This was the first documentation of the pathogen in the United States, found in two species of stream-associated salamanders. The illicit collection and overharvesting of turtles and other species for the pet trade and for overseas markets as exotic delicacies is a concern for species such as the Bog Turtle (USFWS 2001). Climate change poses a serious risk to regional herpetofauna. For example, sea-level rise in the Chesapeake Bay threatens nesting grounds of the Diamond-backed Terrapin, Maryland's official state reptile. In Maryland's urban areas, trace metal contamination of amphibians in stormwater retention ponds is of serious concern (Casey et al. 2005). The myriad threats faced by amphibians and reptiles must be addressed through equally numerous conservation actions if Maryland's native herpetofaunal diversity is to remain intact.

Some herpetofauna, such as Dekay's Brownsnake and the Eastern Ratsnake, have adapted to alterations to their natural habitat and can persist in urban areas. Other species, however, require protection. The conservation of amphibians and reptiles in Maryland is accomplished through various means. Certain species, such as the Eastern Tiger Salamander, are protected under state legislation, and a subset of these, such as the Loggerhead Sea Turtle, are also protected under federal legislation (table 4). Some species are also benefiting from more intensive monitoring, research, and management projects. Finally, suitable habitats for herpetofauna are safeguarded through additional legislation, land acquisition and easement programs, and compatible management practices.

Legislation, Regulation, and Monitoring Programs

The Endangered Species Act of 1973 (ESA) is federal legislation that provides a means to protect imperiled species through conservation of the ecosystems on which they depend. The ESA provides programs for conservation to prevent the extinction of both plants and animals. The law is administered by the US Fish and Wildlife Service (USFWS) and the National Marine Fisheries Service (NMFS), depending on the species. Species are ranked according to the degree to which they are imperiled. A species may be listed as endangered, threatened, or candidate for listing. In Maryland, six species of turtles are protected under the ESA: three species are classified as endangered and three are listed as threatened (table 4).

Internationally, the trade in endangered species and other species at risk of exploitation is regulated by the Convention on International Trade in Endangered Species of Wild Fauna and Flora (CITES). CITES is an agreement among countries to ensure that international trade in individuals of wild animal and plant species does not threaten their survival. CITES was drafted as a result of a resolution adopted in 1963 at a meeting of members of the International Union for Conservation of Nature (IUCN).

Table 4. Threatened and endangered amphibian and reptile species in Maryland during the Maryland Amphibian and Reptile Atlas project period (2009–2015)

Common Name	Scientific Name	State Status	Federal Status
AMPHIBIANS			
Salamanders			
Eastern Tiger Salamander	*Ambystoma tigrinum*	E	—
Hellbender	*Cryptobranchus alleganiensis*	E	—
Green Salamander	*Aneides aeneus*	E	—
Wehrle's Salamander	*Plethodon wehrlei*	I	—
Mudpuppy	*Necturus maculosus*	X	—
Frogs & Toads			
Barking Treefrog	*Hyla gratiosa*	E	—
Mountain Chorus Frog	*Pseudacris brachyphona*	E	—
Eastern Narrow-mouthed Toad	*Gastrophryne carolinensis*	E	—
REPTILES			
Turtles			
Spiny Softshell	*Apalone spinifera*	I	—
Loggerhead Sea Turtle	*Caretta caretta*	T	T
Green Sea Turtle	*Chelonia mydas*	T	T
Hawksbill Sea Turtle	*Eretmochelys imbricata*	E	E
Kemp's Ridley Sea Turtle	*Lepidochelys kempii*	E	E
Leatherback Sea Turtle	*Dermochelys coriacea*	E	E
Bog Turtle	*Glyptemys muhlenbergii*	T	T
Northern Map Turtle	*Graptemys geographica*	E	—
Lizards			
Coal Skink	*Plestiodon anthracinus*	E	—
Snakes			
Rainbow Snake	*Farancia erytrogramma*	E	—
Mountain Earthsnake	*Virginia valeriae pulchra*	E	—

Source: MDNR 2016b.

E = endangered; T = threatened; I = in need of conservation; X = endangered extirpated.

The species covered by CITES are listed in the convention's three appendixes, according to the degree of protection they need (CITES 2017). Appendix I includes species threatened with extinction. Trade in specimens of these species is permitted only in exceptional circumstances. Appendix II includes species not necessarily threatened with extinction but for which trade must be controlled to avoid utilization incompatible with their survival. Appendix III lists species that are protected in at least one country that has asked other CITES parties for assistance in controlling the trade.

Eleven species of Maryland turtles are on the CITES list

Table 5. Maryland species protected by the Convention on International Trade in Endangered Species of Wild Flora and Fauna (CITES)

Common Name	Scientific Name	CITES Appendix
Loggerhead Sea Turtle	*Caretta caretta*	I
Green Sea Turtle	*Chelonia mydas*	I
Hawksbill Sea Turtle	*Eretmochelys imbricata*	I
Kemp's Ridley Sea Turtle	*Lepidochelys kempii*	I
Leatherback Sea Turtle	*Dermochelys coriacea*	I
Spotted Turtle	*Clemmys guttata*	II
Wood Turtle	*Glyptemys insculpta*	II
Bog Turtle	*Glyptemys muhlenbergii*	I
Northern Map Turtle	*Graptemys geographica*	III
Diamond-backed Terrapin	*Malaclemys terrapin*	II
Eastern Box Turtle	*Terrapene carolina*	II

(table 5). Six are listed as Appendix I species, including all the sea turtles. Four species are listed in Appendix II, and one Maryland turtle is in Appendix III.

The state of Maryland has its own endangered species legislation, called the Nongame and Endangered Species Conservation Act, established in 1971 and revised in 1973. The act authorizes the MDNR to establish and maintain a list of legally protected species. Through enabling regulations, a species may be assessed and listed as endangered, endangered extirpated, threatened, or in need of conservation. State law prohibits the direct collection of listed species. The habitat of threatened and endangered species is protected through other environmental laws that dictate considerations for state-listed species. During the MARA project period, six species of amphibians were listed as endangered, one species was ranked as in need of conservation, and one species was classified as endangered extirpated (table 4). Seven species of reptiles were listed by the state as endangered, three as threatened, and one as in need of conservation.

When the state's endangered species act was passed, a committee of the Maryland Herpetological Society evaluated the status of Maryland's amphibians and reptiles and, on 28 August 1972, recommended several for listing to the MDNR (CREARM 1973). Fourteen species were recommended for listing, including five amphibians and nine reptiles, of which five were sea turtles; all were subsequently added as the first species of amphibians and reptiles on the state's legal list. The committee acknowledged that knowledge about the distribution and ecology of Maryland's herpetofauna was still far from complete and that a great deal of time-consuming fieldwork remained to be done.

The designation of "in need of conservation" was created for species thought to be less rare but could warrant endangered or threatened status in the foreseeable future. The first list of wildlife species in need of conservation did not include any amphibians or reptiles. By 1990, four species were declared protected under this category: Wehrle's Salamander, the Carpenter Frog, the Mountain Chorus Frog, and the Spiny Softshell. After targeted field surveys, the Carpenter Frog was removed from this list in 2007, and the Mountain Chorus Frog was elevated to endangered status that same year.

Some of the MDNR's early conservation efforts for these species were to conduct intensive surveys to identify populations of these amphibians and reptiles and to further assess their population statuses. The results of many of these early studies were presented at a symposium on threatened and endangered plants and animals at Towson State University (now Towson University) in September 1981 (Dawson 1984; Gates et al. 1984; Thompson 1984a, 1984b). As a result of these studies, it was determined that two species, Jefferson Salamander and Bog Turtle, were more secure than originally thought and no longer warranted listing. They were removed from the state's endangered species list in 1982 (G. J. Taylor 1984).

In 1990, the MDNR and National Aquarium entered into a partnership and established the Maryland Marine Mammal and Sea Turtle Stranding Network. The network was designed to rescue injured sea turtles and marine mammals, as well as collect scientific data on dead animals washed ashore in Maryland's portion of the Chesapeake Bay and Atlantic Ocean. Injured animals are transported to the National Aquarium for care and eventual release back to the wild.

In the 1990s, the MDNR revisited the known locations of Bog Turtles and found that many of the sites had experienced significant habitat degradation or loss. The MDNR subsequently relisted the Bog Turtle in 1994. Shortly thereafter, the USFWS listed the Bog Turtle as a federally threatened species (J. R. Clark 1997), based on Maryland's findings and those of other states. The MDNR initiated an intensive research and monitoring program, as well as habitat restoration efforts at several Bog Turtle sites, primarily through woody vegetation control.

In the 2000s, the MDNR implemented habitat restoration efforts at several Delmarva bay and Atlantic Coastal Plain

Barking Treefrog. Photograph by Scott A. Smith

pond complexes on the Eastern Shore. Some of these habitats served as breeding sites for the state-endangered Eastern Tiger Salamander. Those restoration efforts resulted in an increase in Tiger Salamander breeding activity. Some of these sites also supported the endangered Barking Treefrog.

In addition to protection afforded to state-listed species, the MDNR adopted regulations in 1993 to protect all native species of amphibians and reptiles in the state. These regulations established limits on the number of individuals of amphibian and reptile species that may be taken from the wild. In Maryland, collecting from the wild was a concern as early as the 1960s (Roeder 1966; Saul 1966). These regulations also prohibit the taking of certain species, require a permit for the possession of certain species, and require a permit to breed certain species and control the trade of offspring produced through captive breeding. Additionally, these regulations prohibit the sale of any native species taken from the wild and require a permit to sell any captive-bred individuals of species native to Maryland, regardless of where the animal originated. For example, a person may take from the wild and possess four individuals of several common species, such as Eastern Newt and Eastern Ratsnake. However, only one individual of most turtle species may be taken and possessed, including Eastern Box Turtle and Painted Turtle. No threatened or endangered species may be taken from the wild, except for scientific purposes with appropriate permits. Certain other species of conservation concern, such as Diamond-backed Terrapin and Timber Rattlesnake, also may not be taken or possessed.

Additional protection was afforded to the Diamond-backed Terrapin in 2007, when the Maryland General Assembly passed a law prohibiting the commercial harvest of this species. Prior to that legislation, there was a commercial season on this turtle in the tidal waters of the state. The only other commercially harvested species of reptile in Maryland is the Snapping Turtle. In 2010, the MDNR set a minimum

Bog Turtle. Photograph by Kevin M. Stohlgren

size limit of 28 cm (11 in) on Snapping Turtles harvested for commercial purposes.

Several other state laws provide protection to habitat that supports Maryland's amphibians and reptiles. One such law is the 1991 Maryland Forest Conservation Act, which requires offsetting forest clearing during land development with forest protection or afforestation. Another is the Maryland Critical Area Law, passed in 1984, which regulates development and land alteration within 305 m (1,000 ft) of tidal waters and oversees local jurisdiction regulations for these same areas.

The one state law that may benefit amphibians in Maryland the most is the state's Nontidal Wetlands Protection Act of 1989. This act regulates and restricts activities that could impact nontidal wetlands or "waters of the state" and also helps to ensure "no net loss" of wetlands by requiring mitigation or compensation for any wetland losses. This law differs from federal regulations on issues of "isolated" wetlands, alteration of vegetation and hydrology, and protection of wetland buffers. For example, federal law does not protect wetland buffers, whereas Maryland law provides a protected buffer of 8 m (25 ft) that can be expanded to 30.5 m (100 ft) for "nontidal wetlands of special state concern" that have been designated by regulation as having exceptional ecological or educational value, including known habitats of rare, threatened, and endangered amphibians and reptiles.

In 2005, the MDNR developed its first Wildlife Diversity Conservation Plan (MDNR 2005) and worked with knowledgeable professional and amateur state herpetologists to develop a list of Species of Greatest Conservation Need (SGCN). This list included both legally listed species and species thought to be declining or otherwise needing conservation efforts to stabilize or improve their populations. This plan was updated in 2015 and renamed the Maryland State Wildlife Action Plan (MDNR 2015). The current SGCN list includes about half of Maryland's native herpetofauna, or 45 species: 19 amphibians and 26 reptiles. Preliminary results of the MARA project were used as a resource to help inform the development of the SGCN list. The current plan, published on the MDNR's website (http://dnr.maryland.gov /wildlife), also includes some conservation actions that could be undertaken by government agencies, researchers, or concerned citizens to help Maryland's vanishing amphibians and reptiles.

Protected Lands

Suitable habitat for Maryland's native amphibians and reptiles is found throughout the state. Among conservation lands, some properties are managed specifically for wildlife, while other areas are managed more generally for natural resources. In addition to various types of public lands, private conservation organizations also own and manage land that provides suitable habitat for native amphibians and reptiles. TNC, for example, owns ecologically valuable properties throughout the state.

Federal Land

The USFWS and NPS manage the majority of federally owned conservation land in Maryland. The US Department of Defense (DOD) holds military land that provides habitat for many of the state's amphibians and reptiles.

Within Maryland, the USFWS manages several large national wildlife refuges. Blackwater NWR, located on the Eastern Shore near Cambridge, is the largest refuge in the complex. This 10,900-ha (27,000-ac) refuge consists mainly of tidal marsh, freshwater ponds, and mixed coniferous and deciduous forests. The refuge is home to a mixed assemblage of herpetofauna. Eastern Neck NWR is a 925-ha (2,285-ac) island at the mouth of the Chester River on the Eastern Shore. The refuge is a mix of wetlands (mainly tidal marsh), small forested areas of loblolly pine and hardwoods, and croplands. Glenn L. Martin NWR, on Smith Island, is a 1,841-ha (4,550-ac) refuge that consists primarily of tidal marshes and supports only a few species of amphibians and reptiles. The Patuxent Research Refuge is located along the Patuxent River near Laurel. This large refuge comprises 5,196 ha (12,841 ac). It includes a diverse array of habitats that

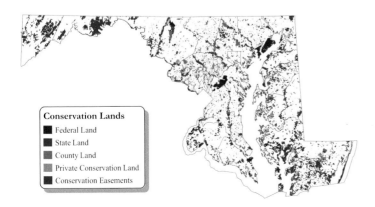

Conservation lands in Maryland. From MDNR, MDP, numerous federal and county agencies, MET and numerous local land trusts, TNC and numerous conservation organizations

Protected lands provide important habitat for a variety of species such as the Rough Greensnake. Photograph by Don C. Forester

National Park Service properties in western Maryland provide suitable habitat for species such as the Wood Turtle. Photograph by Andrew Adams

support an impressive community of amphibians and reptiles. Upland forests, with a mix of mature hardwood tree species, comprise the majority of the refuge's land and support the high level of species diversity in the refuge. Atlantic Coastal Plain river and stream habitats and floodplain forests provide vital habitat to a diverse herpetofaunal community.

The NPS owns properties in Maryland that contain important habitat for the state's amphibians and reptiles. Assateague Island National Seashore, a barrier island bordering the Atlantic Ocean, provides nesting areas for sea turtles. The 2,351-ha (5,810-ac) Catoctin Mountain Park, near Frederick, consists of hardwood and evergreen habitat and provides protection for many amphibian and reptile species of the Blue Ridge Province. The C&O Canal National Historical Park supports herpetofaunal habitat adjacent to the Potomac River along a 297-km (184.5-mi) stretch from Washington, DC, to Cumberland, Maryland. The upland and wetland portions of the canal property harbor a wide variety of amphibians and reptiles. The canal bed is an especially important breeding habitat for amphibians in the western, mountainous portion of the park.

The DOD properties that provide habitat for the state's amphibians and reptiles include Aberdeen Proving Ground, Indian Head Naval Surface Warfare Center, and Patuxent River Naval Air Station. Encompassing 29,526 ha (72,960 ac), Aberdeen Proving Ground is the largest of these properties and maintains vast forested areas and tidal wetlands near the top of the Chesapeake Bay. Additional, smaller DOD properties in Maryland support appropriate but limited habitats as well.

State Land

Most of the state-owned land is held by the MDNR. The department owns approximately 195,000 ha (482,000 ac) spread across the state's 23 counties (MDNR 2016a). The majority of these properties are managed as state forests, state parks, and wildlife management areas.

Most of the 88,200 ha (218,000 ac) of state forests are located in western Maryland and on the Eastern Shore. In the mountains, the largest state forests are Green Ridge, Potomac-Garrett, and Savage River. These forests are home to amphibian and reptile species not found anywhere else in the state. The 26,300-ha (65,000-ac) Chesapeake Forest Lands are spread across the five southern counties of the Eastern Shore.

Eighty-seven state parks comprising more than 55,800 ha (138,000 ac) are managed by the Maryland Park Service. Important habitat for the state's herpetofauna is supported in many of these state parks, including Assateague, Calvert Cliffs, Gunpowder Falls, Patapsco Valley, and Patuxent River. More than 15,700 ha (39,000 ac) of habitat are found in 31 natural resources management areas and natural environment areas, both designations managed by the Maryland Park Service.

Sixty-one state-owned wildlife management areas total approximately 50,000 ha (123,500 ac) in Maryland. These lands are managed mainly for wildlife and wildlife-oriented recreation. Across the state, these areas support important wetland and forested habitats for many of the state's amphibians and reptiles.

The Maryland State Highway Administration owns thousands of hectares of land throughout the state that are permanently protected for mitigation of wetland and stream impacts. Many of the constructed wetlands and restored streams on these lands provide habitat for amphibians and reptiles.

The Marbled Salamander may be encountered in county parks that contain ephemeral wetlands. Photograph by Robert T. Ferguson II

Wood Frogs may be encountered on public parklands that contain ephemeral wetlands. Photograph by Robert T. Ferguson II

County Land

All 23 Maryland counties and Baltimore City own parklands and other properties that support some species of amphibians and reptiles. Many of these properties are primarily sports fields with limited habitat, but many counties also own natural areas open to the public for outdoor recreation. Some of the best of these county parks are Flag Ponds Nature Park in Calvert County, Oregon Ridge Park in Baltimore County, and Jug Bay Wetlands Sanctuary in Anne Arundel and Prince George's counties. Jug Bay is home to 39 species of amphibians and reptiles (Smithberger and Swarth 1993). Although the county sports complex parks are not managed for herpetofauna and have limited habitat for these species, they often provide some habitat, such as streams, ponds, or grassy or forested peripheral areas, for common species.

Private Conservation Land

In Maryland, several nongovernmental conservation organizations own and manage land for wildlife and other natural resources. Twenty-seven properties in Maryland are owned and managed by TNC. These lands are managed to protect biodiversity. The National Audubon Society owns and manages conservation land in the state, with the largest being Pickering Creek Audubon Center in Talbot County. The Maryland Ornithological Society also owns and manages land for wildlife and natural resources in several lo-

cations throughout the state. Some corporate-owned lands are managed for wildlife, or at least compatibly for wildlife, including amphibians and reptiles.

Several state programs that protect private land from development on a voluntary basis have resulted in conservation of thousands of acres of private land through long-term and perpetual conservation easements. Most of this land is protected through the following programs:

- The Maryland Environmental Trust (MET), initiated in the 1970s, holds voluntary easements (with tax benefits to landowners). The MET typically coholds these easements with local land trusts, such as the Eastern Shore Land Trust.
- The Maryland Agricultural Land Protection Fund, initiated in the 1980s, is designed to protect the best farmland from development. Landowners are compensated for relinquished development rights, and easements are placed on entire farms, including both agricultural and forest lands.
- The Rural Legacy Program, which began in the 1990s, is specifically designed to protect properties in rural landscapes to support sustainable, natural resource-based industries through the acquisition of conservation easements. Each county defines a target area for protection, and then, over time, specific parcels in the target area are protected. Since it was enacted in 1997, this program has preserved more than 34,800 ha (86,000 ac) of farmlands, forests, and other habitats.

Designing and Implementing the Atlas Project

To distribute effort across all areas of the state and to facilitate a large network of people working together on this endeavor, the MARA project adopted a clear and objective methodology for data collection. The MARA procedures were, in large part, modeled after Maryland's two successful breeding-bird atlas projects, which were grid based (Robbins and Blom 1996; Ellison 2010). Conducting the MARA project entailed overcoming several general challenges: coordinating the project; defining the survey methods; recruiting, training, and managing a volunteer network; collecting data; and handling and managing the data.

Coordination

The MARA project was coordinated by a steering committee consisting of ecologists, professional biologists, and consultants, all of whom were associated with academia, federal and state wildlife agencies, consulting firms, or nonprofit conservation organizations (table 6). Most of the committee members were volunteers from Maryland's herpetology community, although a few were biologists employed by the MDNR. For the duration of the atlas project, the steering committee was headed by three co-chairs. Over the course of the project, the role of the steering committee evolved from planning, training, and implementing to monitoring progress, encouraging atlasers, and advancing and resolving technical issues. To complete these activities, the steering committee met monthly throughout the project.

Initially, the MARA steering committee conducted a one-year pilot project in 2009 to help develop the atlas methodology, especially to determine coverage targets. Based on the results of the pilot project (Therres et al. 2015), methodology was developed by the steering committee for the statewide effort. Next, the committee recruited at least one coordinator to facilitate and organize the collection of field data for each of Maryland's 23 counties (table 6). Some steering committee members also had a dual role as a county coordinator, database coordinator, or verification committee chairperson. County coordinators were chosen based on their interest in amphibians and reptiles, their knowledge of the county, and their connections with citizen naturalists who might help with data collection. Employees of county parks and recreation agencies coordinated a few counties. These coordinators had access to networks of volunteers interested in amphibians and reptiles and relied heavily on those volunteers to collect data. Other coordinators acted as the primary data collectors or recruited a few dedicated volunteers to collect most of the field data within their county. County coordinators of adjacent counties often worked together to survey areas along shared county lines. Each February, the steering committee held a meeting for all county coordinators to discuss results and technical issues from the preceding year and to prepare for the upcoming field season.

To aid the county coordinators and other atlasers, the steering committee developed a detailed, 50-page "how to" manual, *Maryland Amphibian and Reptile Atlas (MARA) Training Handbook* (Gauza and Smith 2010). This handbook was needed to explain data collection methods and to ensure that sightings were recorded according to the standard protocols. The MARA handbook included information on the purpose and goals of the project, description of the atlas grid system, study design and data collection procedures, atlasing ethics and vouchering methods, tips for successful atlasing, and expected distributions of species by county. The handbook also contained safety precautions for protecting habitat, animals, and surveyors. The MARA steering committee distributed a hardcopy of the training handbook to county coordinators at the launch of the MARA project

Table 6. Structure of coordinator positions for the Maryland Amphibian and Reptile Atlas project

Position	Name
Project Coordinator	Heather Cunningham
Co-chairs	Charles Davis
	David Smith
	Christopher Swarth
	Glenn Therres
Steering Committee	Lynn Davidson
	Matthew Evans
	Don Forester
	Rachel Gauza Gronert
	Wayne Hildebrand
	Nathan Nazdrowicz
	Kyle Rambo
	Scott Smith
	June Tveekrem
	David Walbeck
	Linda Weir
Verification Committee Chair	Nathan Nazdrowicz
MARA Database Coordinator	Lynn Davidson
County Coordinators	
Allegany	Ed Thompson
Anne Arundel	David Walbeck
Baltimore Co./City	Don Forester
	Joel Snodgrass
Calvert	Andy Brown
Caroline	Scott Smith
Carroll	David Smith
	June Tveekrem
Cecil	Nathan Nazdrowicz
	Jim White
Charles	George Jett
Dorchester	Lynn Davidson
Frederick	Wayne Hildebrand
Garrett	Seth Metheny
	Amo Oliverio
Harford	Bob Chance
	Brian Goodman
	Scott McDaniel
Howard	Sue Muller
Kent	Nathan Nazdrowicz
Montgomery	Lance Benedict
	Rachel Gauza Gronert
Prince George's	Tasha Foreman
	George Middendorf
	Mike Quinlan
Queen Anne's	Glenn Therres
Somerset	Douglas Ruby
St. Mary's	Kyle Rambo
Talbot	Kelsey Frey
	Scott Smith
	Glenn Therres
Washington	Andrew Landsman
	Dave Weesner
Wicomico	Lance Biechele
	Ron Gutberlet
	Aaron Hogue
Worcester	Roman Jesien
	Jim Rapp
	Dave Wilson

Note: Coordinators of each position/county, who served over the course of the project, are listed alphabetically.

during statewide training sessions, and an electronic version was made available to all volunteers on the project's website.

The committee chairs hired a statewide project coordinator in August 2010 (table 6), who was paid through a contract with the NHSM. The statewide coordinator encouraged and assisted county coordinators with recruiting volunteers and developing strategies for data collection. The statewide coordinator also provided general assistance to the MARA volunteers, served as the project spokesperson, created and distributed educational materials about the project, developed content for the MARA website, and promoted the project via monthly newsletters and social media.

To validate sightings, the MARA steering committee established a verification committee. Over the course of the project, 20 individuals experienced with identification of Maryland's herpetofauna served on the committee. Members of the verification committee had extensive identification skills and were considered experts on several species groups in Maryland. The task of the MARA verification committee was to provide an unbiased and independent review of species identifications and reach a consensus in confirming, or rejecting and subsequently determining the identity of, submitted voucher documentation. The verification committee examined all photographs, audio recordings, video recordings, and species' remains (e.g., turtle shells, snake sheds) submitted to the MARA project.

Survey Methodology

The MARA methodological foundation was based on herpetofaunal atlases conducted in other states (NJENSP 2002; Price and Dorcas 2011), two Maryland breeding-bird atlases conducted in 1983–1987 and 2002–2006 (Robbins and Blom 1996; Ellison 2010), and the pilot project conducted entirely within Anne Arundel County, Maryland, in 2009 (Therres et al. 2015). The pilot project was a test run of data collection in a grid-based system and was used to develop an estimate of species detection rates for Maryland. Based on results from the pilot project, the steering committee finalized the MARA protocols, developed minimum project targets for coverage effort and species per grid unit, and expanded the project to a statewide effort beginning in January 2010. The committee chose a five-year duration for data collection based on an estimate of expected recruitment and effort of volunteers needed to achieve the project targets. Ideally, this was the narrowest time interval possible in which the project targets could be adequately met and the results least affected by any changes in species distribution that may occur during the survey period. Data collection ended December 2014 for all counties with the exception of Garrett County, for which the end date was extended until

December 2015 due to a one-year delay in the start of data collection for that county.

The Atlas Survey Grid

The MARA project used a grid-based system rather that a GPS or point-based system to establish a benchmark of herpetofaunal distributions for Maryland. Using a grid-based system allowed for a systematic and repeatable methodology. The atlas grid provided a defined unit of area in which future survey efforts could be replicated. The units of area comprising the grid incorporated the general dispersal distances of amphibians and reptiles, yielding a robust approximation of range-wide distributions. The grid system allowed MARA coordinators to direct survey effort to defined areas, ensuring coverage of the entire state. A benefit of this approach was the ability to document the amount of survey effort that was dedicated to a particular unit of the grid, thus providing data for robust comparisons with future surveys.

The grid system of the MARA project was based on USGS 7.5-minute topographic quadrangle maps (quads). Each quad was subdivided into six equal blocks comprising approximately 25.8 km^2 (10 mi^2). The blocks were designated as Northwest (NW), Northeast (NE), Central West (CW), Central East (CE), Southwest (SW), and Southeast (SE). The state of Maryland includes all or parts of 260 quads and 1,427 blocks, of which 1,253 are "land blocks" (i.e., containing more than 40.47 ha [100 ac] of upland). For blocks that crossed state lines, volunteers collected data only from within Maryland. County coordinators were asked to cover entire quadrangles; those that straddled county boundaries were assigned based on the preponderance of total area (i.e., the quad was assigned to the county that comprised the most area in that quad).

County coordinators and their volunteers were provided with multiple tools to facilitate and plan data collection. These tools included USGS topographic quadrangle maps;

a KML (Keyhole Markup Language) file of the atlas grid system to overlay in Google Earth; an atlas grid layer on the MD iMAP interactive online mapping website, which could be used with other map layers (e.g., public lands); and an atlas grid map for use on smartphones, available through an online geographic information system (ArcGIS Online, ESRI Inc., Redlands, CA). Road maps, available in PDF (Portable Document Format), for each atlas block were made available to county coordinators and their volunteers.

Species and Effort Targets

The steering committee used the results of the pilot study to establish minimum coverage targets for number of species and search-hours of active searching to ensure a minimum effort for all blocks and all quads across the state. The minimum target at the block level was to document at least 10 species per atlas block. At the quad level, the target was documentation of at least 25 species or 25 hours of active survey time per quad. These targets were chosen to balance the probable number of species detected with the expected volunteer effort to search the state within the five-year project interval. Block and quad coverage was considered adequate once the target thresholds were attained, after which survey effort priorities were shifted to blocks and quads that had not yet met those targets. Some blocks were dominated by habitats with low species diversity (e.g., portions of Baltimore City or open water areas), which potentially did not support the minimum 10 species per block target. Under these circumstances, once the quad threshold of 25 hours of active searching was achieved, survey effort was considered adequate. Cumulative time spent searching in each block was captured as a record of effort for future comparisons. Once the minimum targets were achieved for all blocks and quads within a county, volunteers continued searching in an effort to document additional species in blocks and quads.

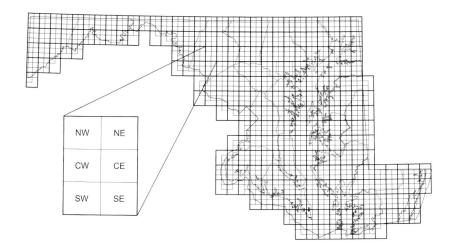

Maryland 7.5-minute USGS quadrangles comprising the Maryland Amphibian and Reptile Atlas grid. Atlas blocks are designated by the name of the quadrangle and position within the quadrangle: NW, NE, CW, CE, SW, and SE. From USGS

The Volunteer Network

The county coordinators, along with the statewide coordinator, recruited volunteers to the MARA project through various methods: personal connections, public media outlets, local conservation groups, social networking websites (e.g., Facebook), the MARA website (hosted by the NHSM), and project exhibits at wildlife- and nature-related conferences and festivals held within the state. The volunteer base for the MARA project consisted of amateur and professional herpetologists, as well as people with a general interest in nature. Volunteers were recruited from within Maryland and surrounding states. The two previous Maryland breeding-bird atlases also provided an existing network of citizen scientists experienced in atlas methodology from which to recruit volunteers. At the request of county coordinators, local newspapers and nature club newsletters published articles describing the MARA project and soliciting volunteers. Additionally, county coordinators frequently held training sessions at the beginning of the field season to offer hands-on training to volunteers.

As a means of retaining volunteers, the statewide coordinator developed two outreach products: a monthly newsletter and a MARA Facebook page. The newsletter featured photographs of interesting amphibian and reptile species encountered by volunteers and accounts of their experiences while surveying for the MARA project. The newsletter regularly contained features on species identification, surveying techniques, and guidance for finding species. The newsletter was distributed by email to county coordinators and other active atlasers and was posted to the MARA website and the MDNR website. The MARA Facebook page provided a forum for online exchange of project information and to encourage volunteers to communicate with each other. Here, volunteers also posted photographs of interesting and recent sightings of amphibians and reptiles in Maryland.

Data Collection

Volunteers documented the presence of species by two approaches within a block: "active searching" and "incidental observations." Active searching was the main source of atlas data collection and involved purposely looking for amphibians and reptiles specifically for the MARA project. Effort was documented by recording the amount of time actively searching in a given atlas block on each date of survey work. The number of participants in the survey party on each date of survey work in a block was also documented. Volunteers were asked to follow standard data-recording procedures but were not required to follow any standardized survey methodology during active searching.

Techniques used during active searching included walking through habitats; turning over logs, rocks, and debris; road cruising; visually scanning ponds and shorelines; and listening for calling frogs and toads. Incidental observations were sightings that observers made while not engaged in active searching for the MARA project. For example, if a volunteer encountered a turtle crossing the road while driving to the store, the observation was recorded as an incidental observation. Photos submitted to the MARA project or MDNR staff by someone not actively participating in the project were added to the database as incidental records. Volunteers did not record survey effort for incidental records. Most data came from encounters with live individuals of amphibian and reptile species; however, any identifiable evidence of species presence was accepted, including road-killed amphibians and reptiles, shed skins of snakes, and turtle shells.

Volunteers recorded data on a standardized MARA data sheet. This data sheet served as a yearly record of all species encountered and time spent surveying in a single atlas block. Data recorded included observer name, quad name, block, date, search time, number of surveyors, specific locality, and species encountered. A checklist of the common names of all native species (based on Crother 2008) was included on the data sheet. Volunteers recorded life stage (egg, larva, juvenile, adult), evidence observed (call heard, animal observed, remains, shed skin), and verification submitted (verification form, photo, recording, voucher) for each species encountered. Sightings of rare, threatened, and endangered species required an additional verification form. The verification form provided a place for volunteers to record additional, more detailed information, including exact locality, behavior, abundance, and description of habitat.

Sighting records for most common species were accepted without verification evidence, although the submission of record vouchers (e.g., photographs or call recordings) for all species was strongly encouraged. The MARA steering committee required verification evidence for rare, threatened, and endangered species and species difficult to identify, although if verification evidence was not obtained, the record could still be submitted. The collection and submission of voucher evidence to confirm species identifications was important to reduce potential misidentifications, which is a serious error that can occur in citizen scientist projects (M. P. Robertson et al. 2010).

In addition to data collection by volunteers, the steering committee requested and obtained data for the MARA project from other organizations that maintained databases of species occurrence. These included HerpMapper, the Herpetological Education and Research Project, the Maryland Biodiversity Project, the Maryland Biological Stream Survey, and the Maryland Marine Mammal and Sea Turtle Stranding Network.

Maryland Amphibian and Reptile Atlas Block Data Sheet

YEAR:

OBSERVER(s):

QUAD NAME:

BLOCK:
(indicate by
BOLD)

NW	NE
CW	CE
SW	SE

Document ID:
(assigned by computer during data entry)

☐　☐　☐

	Coverage			Active Search Time			
Date	Location within Block	# People	Start	End	Hours	Incidental?

County Coordinator (s): _____

Address: _____

Return to: County Coordinator; or Natural History Society of Maryland, Inc. P.O. Box 18750, Baltimore, MD 21206-0750 atlas@marylandnature.org 410-882-5376

SHEET # _____

* Species that require detailed verification, additional verification form required or the record will not be accepted.

	Age	Evidence	Verified		Age	Evidence	Verified
Turtles				**Salamanders**			
Eastern Musk Turtle				Common Mudpuppy *			
Eastern Mud Turtle				Eastern Hellbender *			
Eastern Box Turtle				Marbled Salamander			
Spotted Turtle				Jefferson Salamander *			
Bog Turtle *				Spotted Salamander			
Wood Turtle				Eastern Tiger Salamander *			
No. Diamond-backed Terrapin				Red-spotted Newt			
Northern Map Turtle *				Eastern Redbacked Salamander			
Painted Turtle				Wehrle's Salamander *			
Red-eared Slider				Northern Slimy Salamander			
Northern Red-bellied Cooter				Valley and Ridge Salamander *			
Eastern Snapping Turtle				Seal Salamander *			
Loggerhead Sea Turtle *				Northern Dusky Salamander			
Kemp's Ridley Sea Turtle *				Alleghany Mtn. Dusky Salamander *			
Green Sea Turtle *				Northern Red Salamander *			
Leatherback Sea Turtle *				Eastern Mud Salamander *			
Atlantic Hawksbill Sea Turtle *				Northern Spring Salamander			
Eastern Spiny Softshell *				Northern Two-lined Salamander			
				Southern Two-lined Salamander *			
Lizards				Long-tailed Salamander			
Eastern Fence Lizard				Four-toed Salamander			
Eastern Six-lined Racerunner				Green Salamander *			
Little Brown Skink							
Northern Coal Skink *				**Frogs & Toads**			
Common Five-lined Skink *				Eastern Spadefoot			
Broad-headed Skink *				Eastern American Toad			
				Fowler's Toad			
Snakes				American x Fowler's Toad hybrid *			
Northern Watersnake				Eastern Narrow-mouthed Toad *			
Red-bellied Watersnake *				Upland Chorus Frog *			
Queen Snake				New Jersey Chorus Frog			
Eastern Smooth Earthsnake *				Spring Peeper			
Mountain Earthsnake *				Mountain Chorus Frog *			
Northern Brownsnake				Eastern Cricket Frog			
Northern Red-bellied Snake				Green Treefrog			
Eastern Gartersnake				Unknown gray treefrog sp.			
Common Ribbonsnake *				Cope's Gray Treefrog *			
Ring-necked Snake				Gray Treefrog *			
Eastern Wormsnake				Barking Treefrog*			
Smooth Greensnake				Carpenter Frog *			
Northern Rough Greensnake				Wood Frog			
Eastern Hog-nosed Snake				Southern Leopard Frog			
Rainbow Snake *				Pickerel Frog			
Northern Black Racer				Northern Green Frog			
Red Cornsnake *				American Bullfrog			
Eastern Ratsnake							
Mole Kingsnake *				**Other Species ***			
Eastern Kingsnake							
Eastern Milksnake							
Coastal Plain Milksnake *							
Northern Scarletsnake *							
Copperhead							
Timber Rattlesnake							

Examples of "Other Species"; including status uncertain species or escaped exotic pets: Northern Pine Snake Northern Leopard Frog Mediterranean Gecko Green Iguana Boa Constrictor any python		**Age Codes**			**Evidence Codes**			**Verified Codes**
	A	Adult		**C**	Call heard (frogs & toads)		**FS**	Form Submitted
	J	Juvenile		**O**	Observed (including captured)		**PS**	Photo Submitted
	L	Larva*		**R**	Remains (carcass, bones, shell)		**RS**	Recording Submitted
	E	Egg*		**S**	Shed skin (snakes)		**VS**	Voucher Submitted

Revised: 2010-November-15

SHEET # _____

Maryland Amphibian and Reptile Atlas data sheet, page 2.

Data Management

The MDNR developed and maintained a central database to manage all the MARA data. Atlas volunteers and county coordinators entered data and submitted verification evidence (e.g., photographs, audio recordings, video recordings) through an online, password-protected database. Observer accounts were established by county coordinators or database administrators, and observer roles were assigned (e.g., incidental observer, atlaser, county coordinator, administrator) that granted increasing availability to database functions and tools. At the MDNR, data underwent quality control to check for data entry errors by comparing entries in the database to paper or electronic data sheets submitted by county coordinators and other MARA volunteers at the end of each survey year. The primary MARA database was linked to a secondary data management system, maintained on a third-party server that housed the verification evidence and was designed to support the work of the verification committee.

The MARA verification committee reviewed all submitted verification evidence (i.e., vouchers of species sightings) to confirm species identities. Digital vouchers were accessible to the verification committee via a password-protected verification website. Committee members reviewed records independently through a blind review process designed to minimize bias. On the verification website, the verification committee viewed submitted vouchers based on county of occurrence, but no additional information was provided. County of occurrence was important information for identifying some species (e.g., the different species of chorus frogs); however, identification based solely on range was discouraged. Rather, the verification committee used evidence present in photographs and/or audio and video recordings to determine species identity. Any verification committee members in a dual role, serving also as county coordinators, were blocked from identifying records from their respective county, as well as their own submissions, to prevent potential identification biases.

For each record, three opinions on species identity were required. A record was labeled as "confirmed" if the first three committee members to review the record agreed on the species' identity. In instances when the confirmed identity determined by the committee differed from the identity submitted by the atlas volunteer, the submitted species' identity was changed to reflect the committee's identification. When the first three committee members who reviewed the verification evidence did not agree on the species identity, the record was returned to the committee for additional review. The record remained under review until the committee reached a consensus. If the committee

The New Jersey Chorus Frog is one of two species of chorus frogs found during the Maryland Amphibian and Reptile Atlas project. Photograph by Andrew Adams

could not reach a consensus, the record was labeled "unconfirmed." The central MARA database was updated regularly with the decisions of the verification committee.

During active data collection for the project, four main levels of species' occurrence status were recognized: "confirmed," "pending," "accepted," and "unconfirmed." For confirmed records, the associated verification evidence successfully passed review by the MARA verification committee. For pending records, the associated verification evidence was still under review by the verification committee, or the records were of species that required verification evidence not yet submitted. Accepted records were sightings that lacked any verification evidence and were of species for which no evidence was required. Unconfirmed records either were determined by the verification committee to lack sufficient quality to confirm identification or were records for species that required verification but for which no voucher evidence was collected. Unconfirmed status allowed atlasers to know that additional verification evidence needed to be submitted for that species.

Summary statistics of the data were available online to the public (no password was needed to view the summary data). Statistics included number of species per quad and block and distribution of species (including occurrence status) per quad and block. Data were available in both map and table format. For sensitive species (e.g., Bog Turtle), only quad-level data were available publicly; however, county coordinators and steering committee members had access to block-level data for these species when logged in to the database. The real-time data system helped volunteers and coordinators plan fieldwork and continually assess their progress toward coverage targets.

Results of the Atlas Project

Overall Results

The MARA project was designed to document the distribution of Maryland's herpetofauna, using a systematic and repeatable approach. The project successfully met this objective. MARA volunteers submitted an incredible 34,910 total sightings (table 7). Volunteers documented 85 of Maryland's amphibian and reptile species, including the introduced Red-eared Slider, a Pond Slider subspecies, and both Smooth Earthsnake subspecies. A photograph or audio voucher accompanied 56% of these sightings. Protocols of the project did not preclude a species from being reported multiple times in a single quad or block; thus, redundant records are included in the total number of sightings. Most volunteers, however, did not submit multiple sightings of the same species per block. Additionally, redundancy improved confidence in the dataset for the presence of a given species in a single quad or block. Excluding redundant records, a total of 7,123 unique quad records for Maryland's herpetofauna were collected, of which 89% were confirmed by voucher evidence. At the block level, excluding redundant documentation of species, a total of 21,574 unique block records were submitted, and species identifications were confirmed for 77% of those records. Active data collection remained consistent during the atlas project, with no significant declines in the number of records submitted each year, 2010–2014 (table 7).

Effort

Sightings reported from active surveys for amphibians and reptiles were submitted along with a record of time spent surveying and number of individuals in the survey party. For the purposes of reporting hours of effort, the project used two different measures: search-hours and total-hours. Search-hours are defined as the number of hours of active searching conducted by an individual or group of individuals working together. For example, three individuals searching together for two hours equals two search-hours. Total-hours are defined by the number of individuals multiplied by the number of hours spent actively searching. For example, three individuals searching together for two hours equals six total-hours. Over the course of the MARA project, volunteers logged 14,752 search-hours while surveying for amphibians and reptiles; taking the size of survey parties into consideration, 29,180 total-hours were devoted to active searching (table 7). The reported hours do not include time contributed to data collection, such as hours accumulated while traveling to survey sites and entering data. Most sightings reported were a result of active surveys; only 23% of the total number of records were incidental, with no associated effort data (table 7).

Coverage

Coverage targets (25 species and/or 25 search-hours) were met in 78% of the 260 quads comprising the atlas grid (see appendix A). In general, quads that did not meet coverage targets were largely those dominated by open water and tidal marsh, such as those on the lower Eastern Shore in Dorchester and Somerset counties. For example, within the Terrapin Sand Point Quad in Somerset County, only one species was documented. This quad was predominantly open water, with a tiny sliver of land extending into the Chesapeake Bay. Additionally, coverage targets were not met in some quads positioned on the Maryland state line, particularly some quads in eastern Caroline County and quads in counties bordering the Potomac River. In these cases, there was often very little land within Maryland to be surveyed, and access to that land may have been

Table 7. Total number of sightings, minimum number of incidental observations submitted, and total effort-hours by year during the Maryland Amphibian and Reptile Atlas project

Table 7. Total number of sightings, minimum number of incidental observations submitted, and total effort-hours by year during the Maryland Amphibian and Reptile Atlas project

Project Year	Sightings	Incidental Observations	Search-Hours	Total-Hours
2009[a]	471	66	233	488
2010	7,044	1,800	2,876	5,418
2011	7,584	1,830	3,332	7,436
2012	6,894	1,649	3,010	6,457
2013	6,230	1,597	2,716	5,041
2014	6,086	1,001	2,398	4,072
2015[b]	601	10	187	268
Total	34,910	7,953	14,752	29,180

[a]Anne Arundel County only, pilot year.

[b]Garrett County only.

a limiting factor. For example, in the Vienna Quad, on the Maryland and Virginia state line in Montgomery County, only a small bit of land in one block was within Maryland. Here, a lack of available land to survey resulted in only two search-hours and two species. Even so, in some quads positioned on the state line, coverage targets were close to being met. One such quad was the Davis Quad in Garrett County, on the state line between Maryland and West Virginia. In this quad, 24 species were documented. Seventeen quads contain less than 40.47 ha (100 ac) of upland in the state, including quads on the state line and quads with only open water or a mixture of open water and salt marsh. Because few species were expected in these quads, valuable

search time was directed elsewhere. When these 17 quads are removed from the data analysis, 84% of quads met the coverage targets.

Coverage targets were not met in a few quads in highly urbanized areas (e.g., Baltimore City), possibly due to low species diversity, limited availability of natural areas to survey, or a combination of both. However, in many urbanized quads where the targets were not met, the results did approach the coverage targets. For example, in the Baltimore West Quad, 23 species and 24 search-hours were reported.

Seventy-two percent of quads met the target of 25 or more species (appendix A). An average of 32.6 species and 70.1 search-hours were documented in quads that met the species target. The geographic distribution of quads that met the target of 25 or more species was, for the most part, consistent across the state. Not meeting the species target in a quad did not necessarily indicate a lack of search effort. For example, 16 quads met the effort target of 25 search-hours but did not meet the species target (see appendix B). In general, quads that met the species target also met the effort target, but 19 quads did not meet the effort target despite meeting the species target.

The block target of 10 or more species was met in 1,110 (78%) of the total 1,427 blocks comprising the atlas grid (appendix A). However, 174 blocks (12%) were designated as "water blocks," consisting entirely or mostly of open water with less than 40.47 ha (100 ac) of land (upland or wetland, including salt marsh); or "salt marsh blocks," consisting

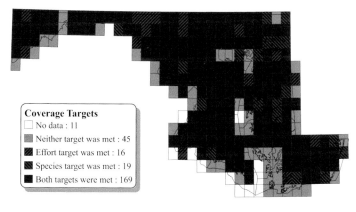

Coverage Targets
- ☐ No data : 11
- ▨ Neither target was met : 45
- ▨ Effort target was met : 16
- ▨ Species target was met : 19
- ■ Both targets were met : 169

Coverage target results. Quads with 25 or more species met the species target; quads with 25 or more search-hours met the effort target.

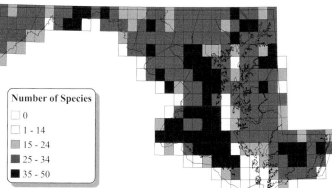

Number of Species
- ☐ 0
- ☐ 1 - 14
- ▨ 15 - 24
- ▨ 25 - 34
- ■ 35 - 50

Herpetofaunal richness in Maryland. Richness is defined by the total number of species documented per quad.

entirely of salt marsh or a combination of mostly water and salt marsh with less than 40.47 ha (100 ac) of upland; or "edge blocks," with fewer than 40.47 ha (100 ac) within Maryland. From the beginning of the MARA project, there was no expectation that the species target would be met in these blocks, as these habitats do not support a high diversity of amphibian and reptile species. When the 1,110 blocks that met the species target are considered relative to the 1,253 land-dominated blocks in the grid, the species target was met in 89% of the blocks. One hundred and fifty atlas blocks lacked species data entirely (appendix A); 141 of these were water blocks (e.g., Chesapeake Bay), salt marsh blocks (southern Dorchester and Somerset counties), or edge blocks (along boundaries of adjacent states or Washington, DC). Several of the remaining 9 blocks were within a military base with restricted access.

The number of effort-hours that volunteers spent in actively searching for Maryland's herpetofauna varied across the state. The effort target of 25 or more search-hours was achieved in most quads. In 75 of the total 260 quads, fewer than 25 search-hours were reported (appendix B). When the 17 quads that are dominated by open water or salt marsh or include less than 40.47 ha (100 ac) of upland in Maryland are removed from the analysis, 58 quads (24%) did not meet the effort target.

No effort target was assigned for blocks, but for blocks that met the species target, an average of 12.9 search-hours were reported. An average of 18.6 species were documented

in these blocks. Of the blocks that did not meet the coverage target but had at least one species documented, an average of 2.5 search-hours were reported, with an average of 5.2 species per block.

County-Level Results

An average of 49.0 species were documented in each of Maryland's 23 counties (including Baltimore City, grouped within Baltimore County throughout), as defined by political boundary (table 8). Volunteers documented the most species in Anne Arundel and Charles counties: 55 species in each. The most amphibian species were reported in Allegany County: 26 species; the most reptile species were documented in Charles County: 33 species (see appendix C). The fewest number of species were reported in Talbot County, with 40 species.

The total number of sightings submitted for each county, defined by assigned quad (not by political boundary), ranged from 529 to 3,394 records, with a median of 1,368 records per county (table 8). The hours of active searching in each county ranged from 149 to 1,786, with a median of 532 hours per county. The difference between search-hours and total-hours varied considerably across the counties, generally due to the different strategies for data collection employed by the MARA county coordinators. The coordinators' strategy often corresponded to the number of volunteers in their

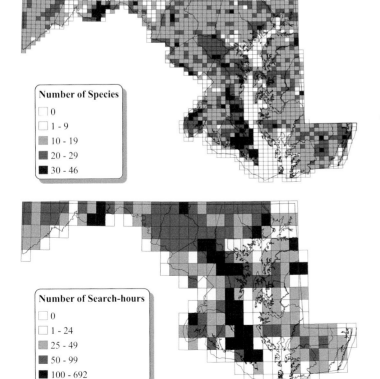

Total number of species documented in each block during the MARA project. Blocks with 10 or more species met the coverage target.

Search-hours recorded in each quad during the MARA project. Quads with 25 or more search-hours met the coverage target.

Table 8. Total sightings, hours, and species by county reported during the Maryland Amphibian and Reptile Atlas project

County	Sightings[a]	Search-Hours[a]	Total-Hours[a]	Total Species[b]	Amphibian Species[b]	Reptile Species[b]
Allegany	1,317	731	1,052	52	26	26
Anne Arundel	2,456	1,047	2,221	55	23	32
Baltimore	1,561	637	1,331	53	23	30
Calvert	1,971	1,786	3,579	53	21	32
Caroline	973	470	724	45	22	23
Carroll	1,115	588	1,321	41	20	21
Cecil	987	282	530	51	24	27
Charles	1,555	620	1,078	55	22	33
Dorchester	2,017	488	802	46	18	28
Frederick	2,231	957	1,546	54	25	29
Garrett	1,966	693	1,523	42	23	19
Harford	1,331	801	1,982	49	24	25
Howard	791	532	1,487	49	24	25
Kent	818	331	577	47	22	25
Montgomery	1,339	469	814	51	23	28
Prince George's	1,583	396	1,135	52	24	28
Queen Anne's	1,368	459	668	48	22	26
Somerset	529	149	359	44	17	27
St. Mary's	3,394	1,492	2,821	54	22	32
Talbot	915	177	325	40	18	22
Washington	1,706	821	1,199	49	23	26
Wicomico	1,254	389	918	47	19	28
Worcester	1,733	441	1,186	50	18	32

[a]Sightings and effort data per county calculated by assigned atlas quads.

[b]Number of species per county calculated by the political boundary.

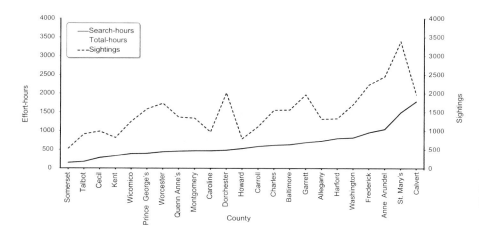

Total effort-hours and sightings recorded in each county during the MARA project.

county. In general, the counties with the highest number of volunteers had the highest number of total-hours. For example, in some counties, such as Worcester and Howard counties, groups of volunteers collectively surveyed particular blocks; in other counties, such as Queen Anne's County, single individuals conducted independent surveys in various blocks. In counties with coordinators that employed the former strategy, the total-hours were much greater than the search-hours. However, regardless of the data collection strategy, the number of sightings increased with effort.

The most effort-hours (search- and total-hours) were reported from Anne Arundel, St. Mary's, and Calvert coun-ties. More than 1,000 hours of active searching were reported in each of these counties. The greatest number of effort-hours was documented in Calvert County, with 1,786 search-hours and 3,579 total-hours (table 8). A collaborative effort between the environmental education program for the Calvert County Public School System (CHESPAX) and the MARA county coordinator for Calvert County most likely contributed to the impressive number of total-hours reported. CHESPAX developed a seventh grade science curriculum based on the MARA project, which was used in a number of public schools in the county; groups of students actively collected data that were later submitted to the project.

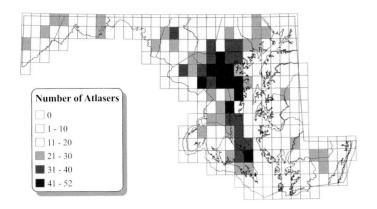

Distribution of MARA volunteers who were registered in the MARA database.

Volunteer Results

More than 1,400 active atlasers and incidental observers were registered in the MARA database. Nearly 600 additional volunteers also contributed to the MARA project by granting access to land for surveys, assisting in herpetofaunal surveys, and participating in volunteer outreach. The number of records submitted from each volunteer varied significantly. Overall, an average of 23.7 total sightings were submitted per person. The average number of species documented per volunteer was five species, and 54% of the volunteers submitted only one record. A core group of 203 volunteers submitted 90% of the total distributional records, and 55 MARA volunteers submitted more than 100 records each. An even smaller group (16 people), representing only 1% of volunteers, submitted more than 500 records each. The most data submitted by any one volunteer was an impressive 3,364 records.

Three-hundred and fifty volunteers reported effort-hours for when they actively searched for herpetofauna. The median number of hours of active searching reported per person was 6 hours. The highest number of search-hours recorded by a single volunteer was 1,281 hours. The average number of quads visited per person was 3 quads, with a maximum of 139 quads visited by a single volunteer. The average number of atlas blocks for which volunteers submitted data, including incidental reports, was 5 blocks. However, one volunteer submitted data for 576 blocks.

Data Verification

Photographs, audio recordings, and remains (e.g., snake sheds, turtle shells), submitted as vouchers, accompanied 19,597 records—more than half of the total 34,910 records submitted. The types of voucher evidence most often submitted were photographs and anuran call recordings. Species identification was confirmed for nearly 94% of the records accompanied by verification evidence. Generally, volunteers accurately identified species. Overall, the verification committee detected species identification error in only 3% of the records. The 10 species most commonly misidentified, calculated as total number of misidentified records, were Northern Two-lined Salamander, American Toad, Fowler's Toad, Cope's Gray Treefrog, Gray Treefrog, Spring Peeper, Southern Leopard Frog, American Bullfrog, Common Five-lined Skink, and Eastern Ratsnake. However, when considering the proportion of number of records misidentified for each species relative to total number of records submitted for that species, the results vary. The species with the highest proportion of misidentified records included Seal Salamander (22%), Broad-headed Skink (22%), Allegheny Mountain Dusky Salamander (11%), Spring Salamander (9%), Eastern Musk Turtle (8%), Common Five-lined Skink (7%), Gray Treefrog (7%), Red-eared Slider (6%), Cope's Gray Treefrog (6%), and North American Racer (6%). The verification committee was unable to make a confident species identification for only 3% of the records submitted with voucher evidence, usually due to poor quality of the voucher.

Overall, the MARA dataset yielded a robust, reliable representation of the distribution of Maryland's herpetofauna. This was a result of stringent verification procedures and a low misidentification rate, offset by a redundancy of records (a single species documented by multiple records for a given quad or block).

Taxonomic Results

Overall

In total, MARA project volunteers documented 85 of Maryland's 89 species of amphibians and reptiles (see appendix D). Species richness was highest on the Atlantic Coastal Plain, especially on the Western Shore, where the number of volunteers and incidental atlasers helped to boost the species totals. Other areas of high richness were found on the Piedmont and Ridge and Valley provinces, including the intensively surveyed Green Ridge State Forest and adjacent C&O Canal National Historical Park along the Potomac

River. Solomons Island Quad in St. Mary's County had the highest number of documented species of any quad, with 50 species. Solomons Island SE, also in St. Mary's County, had the highest number of species documented in a single block, with 46 species.

The number of species with essentially statewide distributions was similar to the number of species with restricted ranges. Nineteen species were each located in 22 or all 23 counties, while 17 species were found in only one or two counties. Thirty-seven species were located in 13 or fewer counties, and 48 species were documented in 14 or more counties.

At the geographic scale of quads, 12 species were found in only one or two quads each (fewer than 1% of all quads), 29 species were each found in fewer than 10% of all quads, and 34 species were each found in 15% or fewer quads. At the other end of the spectrum, the most widely distributed species, those observed in 85% or more of all quads, were Spring Peeper (90%), Green Frog (89%), Eastern Ratsnake (87%), American Bullfrog (85%), and Snapping Turtle (85%). These were followed by Common Watersnake (84%), Eastern Box Turtle (84%), Painted Turtle (83%), Common Gartersnake (80%), and Eastern Red-backed Salamander (77%).

The results of reviewing species' distributions at the more detailed geographic scale of block level were a little different from those for quad level. Six species were located in only one or two blocks. Half of Maryland's herpetofauna, 42 species, were each found in fewer than 10% of all blocks, and 67% of species were each located in fewer than 20% of all blocks. Spring Peeper was still the most widely distributed species, found in 80% of all blocks. The percentages drop off more steeply for blocks than for quads, however, with only two other species above 60%: Green Frog (75%) and American Bullfrog (64%). The top 10 most widely distributed species at the block level were virtually the same as at the quad level, with one exception: American Toad replaced Common Watersnake.

Although amphibians topped the list for being found in the most quads and blocks, reptiles were encountered by the most people. In fact, the top five species recorded by the most volunteers were all reptiles: Eastern Ratsnake (381), Eastern Box Turtle (376), Snapping Turtle (310), Common Gartersnake (292), and Common Watersnake (271). These were followed by Green Frog (242) and American Toad (222). At the opposite extreme, two species, Wehrle's Salamander and Scarletsnake, were each found by only one person, while an additional 10 species were each recorded by only two or three volunteers. These species are among the rarest and/or most difficult to survey, most likely due to their fossorial or cryptic nature.

Amphibians

Volunteers recorded observations for 39 of Maryland's 42 species of amphibians (appendix D). These records constitute 59% of the total number of sightings submitted (table 9). A total of 20,606 amphibian sightings were submitted, including 12,121 unique block records. Of these amphibian block records, 81% were confirmed. Overall, 95% of the amphibian records resulted from encounters with adults and juveniles rather than larvae or eggs. Only 3% of amphibian records were based on remains. Anuran recordings accompanied 51% of the frog and toad records.

Observed amphibian richness varied across the state, with areas of high richness scattered throughout the physiographic provinces. In general, amphibian richness was highest in the central region of the Ridge and Valley, in the eastern Appalachian Plateaus, and on the Western Shore of the Atlantic Coastal Plain. The high richness in the western regions of Maryland was driven by salamanders; the high-richness areas on the Western Shore were most influenced by anurans. Richness was mostly consistent throughout the Piedmont and Blue Ridge provinces. The fewest amphibians were found on the mid- and lower Eastern Shore, with the exceptions of a few isolated areas in Talbot, Wicomico, and Worcester counties.

Salamanders
During the MARA project, 21 of Maryland's 22 species of salamanders were documented (appendix D). A total of 4,722

Table 9. Total number of sightings and quad and block records per taxonomic group recorded during the Maryland Amphibian and Reptile Atlas project

Taxa	Sightings (% confirmed)	Quads with ≥1 species	Quad Records (% confirmed)	Blocks with ≥1 species	Block Records (% confirmed)
AMPHIBIANS	20,606 (54)	—	3,432 (91)	—	12,121 (81)
Salamanders	4,722 (68)	229	1,179 (92)	1,025	3,306 (85)
Frogs and Toads	15,884 (50)	243	2,253 (91)	1,231	8,815 (79)
REPTILES	14,211 (54)	—	3,692 (87)	—	9,454 (72)
Turtles	6,042 (50)	241	1,371 (86)	1,137	3,740 (70)
Lizards	1,027 (49)	160	311 (79)	447	668 (65)
Snakes	7,142 (58)	241	2,010 (89)	1,188	5,046 (75)

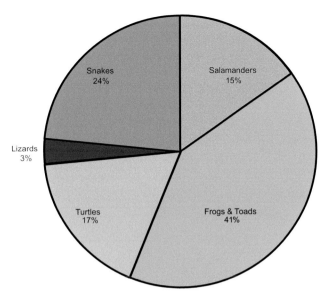

Contribution of each taxonomic group to 21,574 unique block records documented during the atlas project.

land salamander not documented during the atlas period. This species was found in Maryland in 1972. David S. Lee (1972b) reported three individuals from the Youghiogheny River near Sang Run in Garrett County. This observation remains the only known occurrence of the species in Maryland. It is unclear whether this was an established population or an anomalous occurrence. The species has been introduced through releases in multiple locations in the New England region (Powell et al. 2016). The MDNR designates the Mudpuppy as endangered extirpated in Maryland (MDNR 2016b).

Distributional data collected during the MARA project indicate that salamander richness was highest in the western panhandle of Maryland, in Allegany and Garrett counties. The heavily forested landscape, numerous streams, varied topography, and cooler climate of the region are ideal for salamanders. During the project, some salamander species, including the Hellbender, Green Salamander, Seal Salamander, Allegheny Mountain Dusky Salamander, Valley and Ridge Salamander, and Wehrle's Salamander, were documented in this region only. Conversely, salamander richness was lowest on the lower Eastern Shore on the Atlantic Coastal Plain. This is not surprising, as the habitat, climate, and land use in this region do not necessarily favor diverse communities of salamanders. Abundant tidal marsh and dry, sandy soil uplands lead to fewer specialized species. However, the Eastern Tiger Salamander and Southern Two-lined Salamander were con-

salamander sightings were submitted, with 3,306 of those sightings being unique block records (table 9). Volunteers documented one or more salamander species in 229 quads (88%), and species identification was confirmed for 92% of these. One or more salamander species were documented in 3,306 blocks (71%), with species identification confirmed in 85% of the records.

The Mudpuppy (*Necturus maculosus*) was the only Mary-

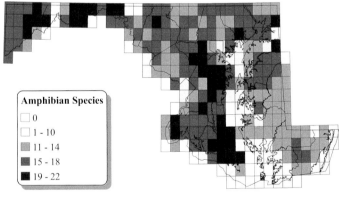

Amphibian species richness in Maryland. Richness is defined by the total number of amphibian species documented per quad.

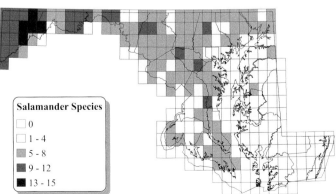

Salamander richness in Maryland. Richness is defined by the total number of salamander species documented per quad.

firmed only on the Eastern Shore during the atlas project. Thus, even though richness on the Eastern Shore was not as high as in other areas of Maryland, the region is home to species found nowhere else in the state. Salamander richness through the Piedmont and Blue Ridge provinces was nearly uniform, with isolated areas of high richness in Baltimore, Frederick, and Howard counties. Some species were found throughout the state, including Spotted Salamander and Eastern Red-backed Salamander. The Eastern Red-backed Salamander was the only salamander species confirmed in all 23 counties.

Occurrence records for salamander species that are of conservation concern were collected during the MARA project. The Jefferson Salamander was documented in five counties—an expansion from the historical distribution of three counties documented by Herbert Harris (1975). The state-endangered Eastern Tiger Salamander was confirmed in two counties on the Eastern Shore, a slight decline from the historical distribution. The Hellbender, which is also listed as endangered in Maryland and was noted by Harris (1975) as becoming "increasingly rare," was documented only in Garrett County during the atlas project. The state-endangered Green Salamander was also documented only in Garrett County, similar to its historical distribution in Maryland. Historically, Wehrle's Salamander was noted as a species that may occur in the state (H. S. Harris 1975), although there were no documented sightings until 1978 (Thompson and Gates 1978). During the atlas project, Wehrle's Salamander was documented in Allegany and Garrett counties. Within Maryland, this species is listed by the state as in need of conservation (MDNR 2016b).

Frogs and Toads

Twenty anuran species are native to Maryland; 18 of these species were documented during the MARA project (appendix D). Volunteers submitted 15,884 sightings for frogs and toads. Unique block records of anuran species totaled 8,815, of which 79% were confirmed (table 9). Anurans were documented in 243 (93%) atlas quads. Species identifications were confirmed in 91% of these.

The Mountain Chorus Frog was not documented during the Maryland Amphibian and Reptile Atlas project. Photograph by Wayne G Hildebrand

The only anuran species that was expected (based on historical distributions) but was not documented during the atlas project was the Mountain Chorus Frog (*Pseudacris brachyphona*), despite concentrated efforts to relocate this species at its historical localities. Historically, this species was restricted to western Maryland (J. A. Fowler 1947c), and it is currently listed as endangered in the state (MDNR 2016b). Historical records documented 11 localities for the species in Allegany and Garrett counties (H. S. Harris 1975). Intense survey efforts at historical localities and other sites from 1996 to 1999 documented only one breeding colony of 17 adults (Forester et al. 2003).

Near the end of the data collection period for the MARA project, the Mid-Atlantic Coast Leopard Frog (*Lithobates kauffeldi*) was recognized as a taxonomic species (Feinberg et al. 2014). The species is closely related and morphologically similar to the Southern Leopard Frog. Since the conclusion of the atlas project, the Mid-Atlantic Coast Leopard Frog has been confirmed by genetic analysis in Cecil, Dorchester, Wicomico, and Worcester counties (Schlesinger et al. 2017).

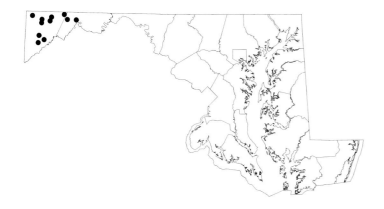

Historical range of the Mountain Chorus Frog in Maryland. Adapted from H. S. Harris (1975)

Observed anuran richness was highest on the Atlantic Coastal Plain. Species richness was documented as consistently high throughout the Western Shore and much of the Eastern Shore. Mild climate, diversity of habitat, and loose, unconsolidated soils in these areas support diverse communities of Maryland's anurans. Some species, such as the Eastern Narrow-mouthed Toad and Carpenter Frog, are found only in this region of Maryland. Conversely, the Appalachian Plateaus Province, where cooler climate, higher elevations, and rocky substrate represent generally unfavorable habitats for most of Maryland's frog and toad species, had the lowest richness in the state. Anuran richness was moderately high throughout much of the Piedmont, Blue Ridge, and Ridge and Valley, although many areas of higher richness were documented on each of the provinces. Many of Maryland's frogs and toads were broadly distributed in the state; however, Spring Peeper, American Bullfrog, Green Frog, and Pickerel Frog were the only species confirmed in all 23 counties.

Frog and toad species of conservation concern were documented during the MARA project. Eastern Narrow-mouthed Toad, a state-endangered species, was confirmed in St. Mary's, Dorchester, Wicomico, and Worcester counties; some of these localities were previously unknown for the species. Distributional records for the state-endangered Barking Treefrog were confirmed in two counties. Barking Treefrog was not included in Herbert Harris's (1975) publication of herpetofaunal distributions in Maryland. The Carpenter Frog, a species of conservation concern in Maryland, was documented in six counties during the atlas project.

Reptiles

Forty-six of Maryland's 47 species of reptiles were documented during the MARA project (appendix D). A total of 14,211 reptile sightings were submitted. Of those, 9,454 were unique block records and 72% were confirmed (table 9). Remains were observed as evidence for 21% of the reptile records; the evidence for 28% of snake records and 71% of sea turtle records was remains.

Observed reptile richness varied across Maryland. Areas of high richness were found throughout much of the state. In general, richness was highest on the Atlantic Coastal Plain, with the most species observed on the Western Shore and lower Eastern Shore. The mild climate, diverse aquatic and freshwater habitats, and loose, unconsolidated soils are favored by many of the state's reptile species. One concentrated area of diversity was found in the central region of the Ridge and Valley Province in eastern Allegany County, where survey efforts were focused in Green Ridge State Forest and along the Potomac River. The fewest reptile species were found on the Appalachian Plateaus.

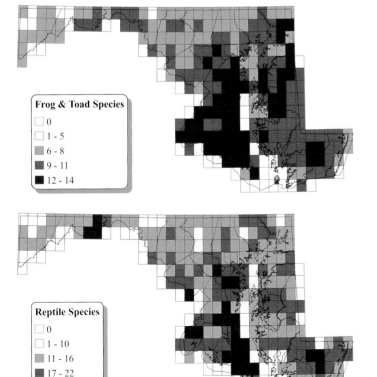

Anuran richness in Maryland. Richness is defined by the total number of frog and toad species documented per quad.

Reptile species richness in Maryland. Richness is defined by the total number of reptile species documented per quad.

Turtles

Of the 18 expected species of turtles in Maryland, 17 were documented during the MARA project (appendix D). Volunteers submitted 6,042 sightings of turtles; 3,740 were unique block records (table 9). Across Maryland, turtles were documented in 241 (93%) of the quads and confirmed in 86% of these. Turtle species were documented in 79% of the blocks, with 70% of the block records confirmed.

The Hawksbill Sea Turtle (*Eretmochelys imbricata*) was not documented during the data collection period. While this species was reported by Herbert Harris (1975), there are no known documented records of occurrence in Maryland waters. Harris (1975) described a single preserved specimen in the NHSM collection that was labeled as "occasionally occurring in the Chesapeake Bay and the Maryland Atlantic Coast." Other authors alluded to the possibility of this species occurring in the Maryland area (F. J. Schwartz 1967; J. D. Hardy 1972; Klimkiewicz 1972). However, without verified locality data, it is impossible to state with confidence whether the Hawksbill Sea Turtle ever occurred in Maryland waters.

Observed species richness for turtles was highest on the Atlantic Coastal Plain and lowest on the Appalachian Plateaus. The Atlantic Coastal Plain's mild climate, diversity of aquatic habitats, and slow-moving waterways are favored by many turtle species in the state. The Diamond-backed Terrapin and all of Maryland's documented sea turtle species were confirmed in this region. The Piedmont and Blue Ridge provinces are also favorable regions for the state's turtle species, with species richness similar to that of the Atlantic Coastal Plain. On the Piedmont, species such as the Bog Turtle and Northern Map Turtle were confirmed during the atlas project. Three turtle species were confirmed in all 23 counties: Snapping Turtle, Painted Turtle, and Eastern Box Turtle.

During the MARA project, data were collected on Maryland's rare, threatened, and endangered turtle species. All of Maryland's sea turtles, with the exception of the Hawksbill Sea Turtle, were confirmed during the atlas period. These species are all federally and state listed as threatened or endangered. The Bog Turtle, a federal- and state-listed species, was confirmed in four counties during the project. Likewise, the state-endangered Northern Map Turtle was confirmed in two counties. The Spiny Softshell, a species in need of conservation, was confirmed in two counties, although the observation from Frederick County was the result of a captive release.

Lizards

All six expected lizard species in Maryland were documented during the MARA project (appendix D). For Maryland's lizards, 1,027 sightings were submitted, of which 668 were unique block records (table 9). Of all the taxonomic orders in Maryland, lizards were documented in the fewest number of quads, 160 (62%), during the atlas project. Seventy-nine percent of these quads included confirmed records. Volunteers documented lizard species in 31% of atlas blocks; 65% of the block records were confirmed.

In general, lizard species richness was high on the Atlantic Coastal Plain. The lower Western Shore had the broadest area of high lizard richness in the state. Five of the state's six species were documented in Anne Arundel, Charles, Calvert, and St. Mary's counties. On the Eastern Shore, more species of lizards were documented in Dorchester, Somerset, Wicomico, and Worcester counties than in other Eastern Shore counties, with four species documented in each. In general, lizard richness was low throughout the Piedmont, especially in the northern counties. In Western Maryland, lizard richness increased on the Ridge and Valley, particularly along the Potomac River, where some Atlantic Coastal Plain species occurred, including the Six-lined Racerunner. The Coal Skink, documented on the Ridge and Valley in Allegany County, contributed to the lizard richness observed in Western Maryland. No lizards were documented in Garrett County on the Appalachian Plateaus during the MARA project. The Common Five-lined Skink and Eastern Fence Lizard were documented in more counties during the project than any other lizard species—in 21 and 20 counties, respectively (appendix D).

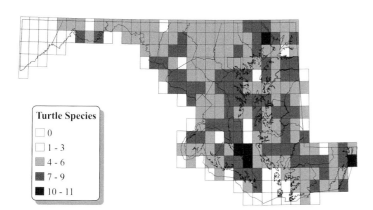

Turtle species richness in Maryland. Richness is defined by the total number of turtle species documented per quad.

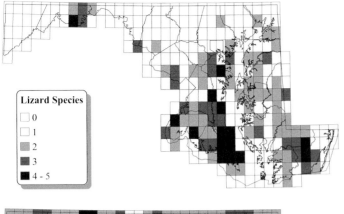

Lizard species richness in Maryland. Richness is defined by the total number of lizard species documented per quad.

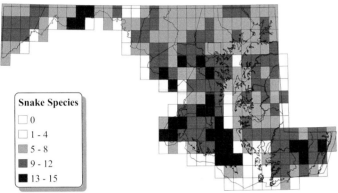

Snake species richness in Maryland. Richness is defined by total number of snake species documented per quad.

Maryland's only state-endangered lizard, the Coal Skink, was documented during the MARA project. Historically, the Coal Skink was known from several localities in Garrett County (H. S. Harris 1975). Museum specimens labeled "Allegany County" also exist, but these lack specific locality data (McCauley 1945; Lemay and Marsiglia 1952; H. S. Harris 1975; Thompson 1984b). During the project, Coal Skink was confirmed in Allegany County in two blocks; it was not documented in Garrett County.

Snakes

All 23 species of snakes native to Maryland, and both Smooth Earthsnake subspecies, were documented during the MARA project (appendix D). Volunteers submitted 7,142 snake sightings, of which 5,046 were unique block records (table 9). Volunteers documented one or more snake species in 241 (93%) atlas quads. Eighty-nine percent of the quads contained confirmed records. Snake species were documented in 82% of atlas blocks, and 75% of block records were confirmed.

Observed species richness for snakes was relatively high across the state, especially on the Atlantic Coastal Plain, specifically on the Western Shore, and in the southern counties of the Piedmont region. Areas of high richness were found in the central region of the Ridge and Valley, whereas fewer snake species were found throughout much of the western region of the Ridge and Valley and northwestern Appala-

chian Plateaus. But, in general, snake richness was moderately high throughout Maryland because of the widespread distributions of many species. Five species of snakes—Eastern Ratsnake, Ring-necked Snake, Common Watersnake, Dekay's Brownsnake, and Common Gartersnake—were confirmed in all 23 counties in Maryland.

During the MARA project, the state-endangered Rainbow Snake was confirmed in Charles County, and the Mountain Earthsnake, a state-endangered subspecies of Smooth Earthsnake, was confirmed in Garrett County. The Plain-bellied Watersnake, a species of conservation concern, was confirmed in Dorchester, Wicomico, and Worcester counties on the lower Eastern Shore. The distributions of these species were all similar to their known historical distributions (H. S. Harris 1975). However, two species of conservation concern showed changes from their historical distributions. The Scarletsnake was confirmed in a new county, Howard, but not in any of the five counties in its historical distribution (H. S. Harris 1975). The Timber Rattlesnake was confirmed in Allegany, Garrett, Washington, and Frederick counties, similar to the historical distribution of the species with the exception of Baltimore County (H. S. Harris 1975).

Non-native Species

During the atlas period, 84 sightings of 30 non-native species of amphibians and reptiles (table 10) were confirmed in

Table 10. Non-native taxa documented during the Maryland Amphibian and Reptile Atlas project

Common Name	Scientific Name	Sightings
AMPHIBIANS		
Salamanders		
Northwestern Salamander	*Ambystoma gracile*	1
Siren	*Siren* sp.	2
Frogs & Toads		
Oriental Fire-Bellied Toad	*Bombina orientalis*	1
Squirrel Treefrog	*Hyla squirella*	1
Asian Painted Frog	*Kaloula pulchra*	1
Cuban Treefrog	*Osteopilus septentrionalis*	1
REPTILES		
Crocodilians		
American Alligator	*Alligator mississippiensis*	3
Turtles		
Russian Tortoise	*Agrionemys horsfieldii*	2
Florida Softshell	*Apalone ferox*	5
African Spurred Tortoise	*Centrochelys sulcata*	6
Mississippi Map Turtle	*Graptemys pseudogeographica kohnii*	5
Northern False Map Turtle	*Graptemys pseudogeographica pseudogeographica*	5
Texas Map Turtle	*Graptemys versa*	1
Alligator Snapping Turtle	*Macrochelys* sp.	1
Burmese Mountain Tortoise	*Manouria emys*	1
Chinese Softshell	*Pelodiscus sinensis*	1
River Cooter	*Pseudemys concinna*	5
Yellow-bellied Slider	*Trachemys scripta scripta*	12
Lizards		
Green Anole	*Anolis carolinensis*	3
Brown Anole	*Anolis sagrei*	3
Wood Slave	*Hemidactylus mabouia*	1
Mediterranean Gecko	*Hemidactylus turcicus*	11
Savannah Monitor	*Varanus exanthematicus*	1
Snakes		
Boa Constrictor	*Boa constrictor*	3
Southern Black Racer	*Coluber constrictor priapus*	1
California Kingsnake	*Lampropeltis californiae*	1
Honduran Milksnake	*Lampropeltis abnorma*	2
Western Ratsnake	*Pantherophis obsoletus*	1
Burmese Python	*Python molurus bivittatus*	1
Ball Python	*Python regius*	2

decades. Often when these animals grow too large or are no longer wanted, their owners release them in local parks or natural areas, even though doing so is illegal in Maryland. For example, in July 2010, a 2-m (6.5-ft) Boa Constrictor (*Boa constrictor*) was captured at Scientist Cliffs in Calvert County and turned over to a local nature center (A. Brown 2010). Most species of released or escaped pets probably would not survive Maryland's winter climate. In December 2011, a 3.6-m (12-ft) Burmese Python (*Python molurus bivittatus*) was found dead on the side of Hammonds Ferry Road on the Baltimore County side of the Patapsco River.

Some of the non-native species from elsewhere in North America probably were inadvertently introduced to the state through the interstate transport of plants and other products. One such species was the Northwestern Salamander (*Ambystoma gracile*). The animal was discovered in late November 2012 in a Christmas tree purchased in Calvert County (S. A. Smith 2013). The tree originated from Oregon, a state in which the Northwestern Salamander is native, and was transported by truck to North Carolina and then to Maryland. Interstate transport of the Northwestern Salamander has been documented under similar circumstances in other states (Rochford et al. 2015). A Green Anole (*Anolis carolinensis*) was found at a retail nursery in Queen Anne's County in June 2012, most likely traveling north on a plant from the southeastern United States. A Cuban Treefrog (*Osteopilus septentrionalis*) was found at a home in Dorchester County in May 2014, a day after the homeowner's parents arrived in a camper from Florida.

Fortunately, most non-native species do not become established in Maryland. However, there are exceptions, such as the Red-eared Slider and the Mediterranean Gecko (*Hemidactylus turcicus*). The Red-eared Slider was well established in the state by the mid-1970s (H. S. Harris 1975), and a population of Mediterranean Geckos existed in Baltimore City as early as 1974 (Norden and Norden 1989). Although Mediterranean Geckos were not detected in this locality during the MARA period, this species was confirmed in 11 atlas blocks in nine counties. It is unknown whether these new localities represent established populations, but this is plausible given that the species is often associated with buildings that provide warmth from Maryland's harsh winters. Additional species that might now have established populations in Maryland include the Northern False Map Turtle (*Graptemys pseudogeographica pseudogeographica*) and the Yellow-bellied Slider (*Trachemys scripta scripta*). Five Northern False Map Turtles were confirmed in four counties, including one female laying eggs in Howard County in June 2010. The Yellow-bellied Slider was confirmed in 12 atlas blocks in seven counties. If the Yellow-bellied Slider has become established, the long-term population sustainability remains in question, as breeding with Red-eared Sliders may eliminate

Maryland, representing 52 quads and 71 blocks. These numbers exclude the introduced Red-eared Slider, which has become "naturalized" and a well-established breeding species within Maryland. The non-native species found by MARA volunteers included 19 North American species that did not historically occur in the state and 11 species not native to North America.

Many of the non-native species were probably escaped or released pets. Reptile expos, where non-native species of amphibians and reptiles are offered for sale, have been common events in Maryland and neighboring states over the past few

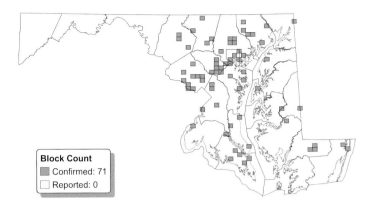

Blocks in which one or more non-native species were documented during the Maryland Amphibian and Reptile Atlas project.

Block Count
Confirmed: 71
Reported: 0

genetic distinctness. Finally, a population of sirens (*Siren* sp.) was discovered in Lake Artemesia in Prince George's County in June 2014 (S. A. Smith 2014). Follow-up surveys by the MDNR documented additional individuals. The origin of this population is debatable; it may have become established as a result of fishermen using these animals as purchased bait and then releasing them when unused. Herbert Harris (1975) reported a siren species from a single Calvert County specimen, although the validity of this record has been questioned (R. W. Miller 1980).

Species Accounts

This section contains accounts for each amphibian and reptile species documented during the MARA project. Accounts are arranged phylogenetically based on higher-level taxonomy, beginning with salamanders. Within each taxonomic group, species are arranged alphabetically, first by taxonomic family and then by scientific name. Each account includes a description of the species followed by its overall geographic and Maryland range, habitat, life history, and ecology. At the conclusion of each account, atlas results are presented and compared with the distribution mapped by Herbert Harris (1975), and conservation implications for the species are discussed. Total body size measurement ranges and maximum lengths are as provided in Powell et al.'s (2016) field guide, and overall geographic ranges are described from range maps in the same field guide. Each species is illustrated with a photograph, usually taken by a MARA volunteer. The majority of photographs are of Maryland specimens or of specimens from surrounding states.

Species occurrences are mapped at the quad and block scales for most species. Some of Maryland's herpetofauna are subject to excessive human impacts, such as overcollection, disturbance of their limited habitats, or overt destruction. For these species, MARA data are presented only at the quad scale. The species occurrence statuses used during data collection in the MARA project (see "Designing and Implementing the Atlas Project") were reduced to two statuses in the species accounts: "confirmed" and "reported." Confirmed records are those with photographic or auditory verification evidence that were reviewed and agreed upon by the verification committee. Reported records are those that were submitted without such verification evidence. Records deemed questionable by the steering committee were excluded from the data summaries and distributional maps. These included some extralimital observations (e.g., Smooth Greensnake reported from the Atlantic Coastal Plain) and observations that more likely represented a guess by the observer rather than character-based identification (e.g., depredated turtle nests, toad eggs).

Although the taxonomy of many amphibians and reptiles of Maryland includes subspecies, data were collected and submitted to the MARA project at the species level for all taxa except Eastern Smooth Earthsnake, Mountain Earthsnake, and Red-eared Slider, for two reasons. First, for most taxa, only one subspecies is known in Maryland, and second, when multiple subspecies do exist, broad intergra-

dation zones make identification to the subspecies level impractical. The species-level taxonomy is maintained in the species accounts, with the exception of Red-eared Slider. The Smooth Earthsnake subspecies are discussed but combined under a single account. Scientific and common (English) species names follow Crother (2017).

Common Five-lined Skink, one of the six native species of lizards documented in Maryland during the Maryland Amphibian and Reptile Atlas project. Photograph by George M. Jett

The Smooth Greensnake, one of two green-bodied snakes in Maryland, is found from the western region of the Piedmont Province through the Appalachian Plateaus Province. Photograph by Scott McDaniel

Amphibians

Photograph by Matthew Kirby

Jefferson Salamander
Ambystoma jeffersonianum

The Jefferson Salamander is one of four salamanders in the Family Ambystomatidae (Mole Salamanders) found in Maryland. This species is a brownish gray to dark gray salamander with a light gray belly. In comparison to other *Ambystoma*, its body is more slender, and it can be distinguished from all other salamanders of similar size by its extremely long toes, particularly on the hind limbs. Small bluish flecks occur along the sides of the body, limbs, and tail. The intensity of these bluish markings is variable and is reduced in older individuals. The tail is substantially laterally compressed. Males in breeding condition have a conspicuously swollen cloaca. Adult Jefferson Salamanders grow to 10.7–18 cm in total length.

The Jefferson Salamander ranges from western New England and southeastern Canada through southern New York, northern New Jersey, and most of Pennsylvania, southwestward to Indiana, Kentucky, West Virginia, and Virginia. Throughout the northern part of its range, the Jefferson Salamander forms a unisexual complex resulting from hybrid origin with the Blue Spotted Salamander (*Ambystoma laterale*) and other *Ambystoma* species (Petranka 1998). In Maryland, the Jefferson Salamander occurs from extreme northwestern Montgomery and central Frederick counties, west to eastern Garrett County. It was first reported in Maryland by Brady (1937) in his list of the amphibians and reptiles of Plummers Island, Montgomery County; however, no specimen existed (Netting 1946; Stine and Simmons 1952). Netting (1946) reported that the first specimen was collected in 1937 from Allegany County. Through subsequent years,

a handful of additional sites were discovered, and H. S. Harris (1975) reported seven known sites. In 1977, a status and distributional survey conducted for Jefferson Salamanders documented 54 breeding sites (Thompson et al. 1980). Since that time, the author (Thompson) and others have located additional breeding populations. Despite much searching, the author has failed to locate any Jefferson Salamander sites in central or western Garrett County.

Few people ever see a Jefferson Salamander. Its cryptic coloration coupled with its fossorial life history make it a hard creature to observe. Adults spend most of their lives underground in deciduous or mixed deciduous forests (Thompson et al. 1980; Petranka 1998). Populations are patchily distributed across the landscape and are restricted to forested areas where suitable breeding pools occur (Bishop 1943; Petranka 1998). The ideal breeding site is a woodland vernal pool, but various fishless, human-made habitats such as neglected farm ponds, abandoned canals, and any other wet spot that may hold water for several months have been used in Maryland (Thompson et al. 1980).

Within their range, Jefferson Salamanders are the first amphibian of the season to commence breeding activity (Brodman 1995). In Maryland, breeding activity may begin as early as the end of January through the first week of February. In an average year in western Maryland, the first wave of breeding activity begins in mid-February, when adults leave their woodland locations and migrate to breeding pools. Males outnumber females in any given breeding year (J. T. Collins 1965; Douglas 1979; Downs 1989). Available information suggests that males breed annually, but females skip one or more years before returning to breed (Downs 1989). Female Jefferson Salamanders almost exclusively attach their eggs to aquatic vegetation or small sticks that may be in the breeding pool (Piersol 1910). The typical egg mass is sausage shaped rather than round, and the gelatinous covering is always clear. Several masses containing an average of 15–20 eggs, but ranging from 4 to 60 eggs, are laid per female (Bishop 1941; Seibert and Brandon 1960; C. K. Smith 1983). The eggs hatch into gilled aquatic larvae in approximately 30–45 days, depending on water temperature (Bishop 1941). These larvae are predators and grow rapidly, feeding on small zooplankton at first, then graduating to practically any other living thing they can engulf (C. K. Smith and Petranka 1987). The larvae metamorphose in 2–4 months (Bishop 1941). Petranka (1998) reports that at least 3 years are needed to reach sexual maturity, which may be an underestimate, especially for females in more northern populations.

After breeding, the adults return to their woodland locations. Using radiotelemetry, Faccio (2003) found that adults moved 30–219 m from the edge of a breeding pool in Vermont. His research also showed that adults chose small mammal burrows almost exclusively as their terrestrial refuges. Deep vertical tunnels and highly branched horizontal tunnels were both utilized.

Jefferson Salamanders were recorded in 22 quads and 36 blocks during the atlas period. The limits of their documented range in Maryland are not dramatically different from the known range published by H. S. Harris (1975). However, since that publication, Jefferson Salamanders have been documented in eastern Garrett County and have been found to be more prevalent in Allegany County and western Washington County, as also reflected in the atlas results.

Although not often encountered, the Jefferson Salamander appears to be secure in the state. A number of breeding pools and associated terrestrial habitats on pub-

licly owned land have been designated as "Ecologically Sensitive Areas" and thus receive special management protection. A few sites on private land have been given special protection, as well. Some of the largest populations are known to occur within sections of the C&O Canal National Historical Park. Throughout the canal's length, various sections mimic the seasonal fluctuations of a vernal pool. The threat posed by amphibian diseases casts a dark cloud over the future of many species, especially those that concentrate en masse at breeding sites. To what extent this issue will affect Jefferson Salamander populations can only be speculated on at this time.

ED THOMPSON

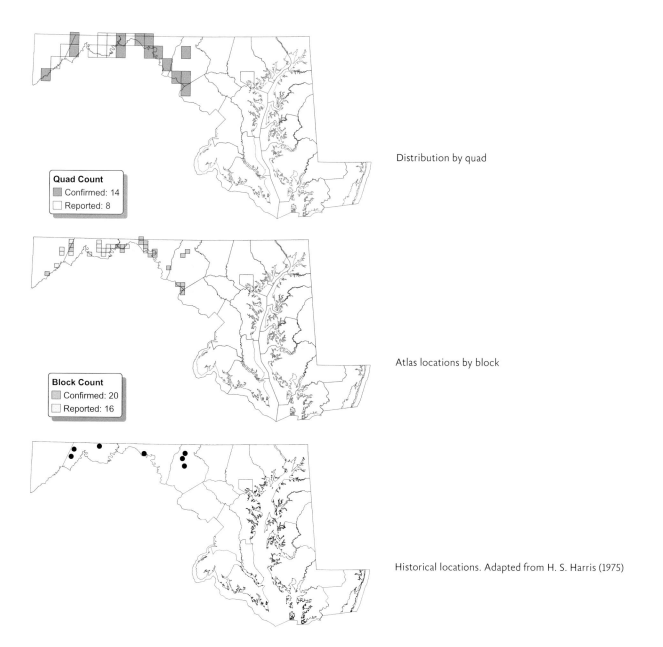

Distribution by quad

Quad Count
Confirmed: 14
Reported: 8

Atlas locations by block

Block Count
Confirmed: 20
Reported: 16

Historical locations. Adapted from H. S. Harris (1975)

Photograph by Don C. Forester

Spotted Salamander
Ambystoma maculatum

The Spotted Salamander is the most recognizable of the Mole Salamanders (Ambystomatidae) in Maryland. Its generic name, *Ambystoma*, is thought to represent a *lapsus calami* (slip of the pen) for *Amblystoma*, a composite of two Greek words: *amblys*, meaning "blunt," and *stoma*, meaning "mouth" (Beltz 2006). The specific epithet, *maculatum*, is Latin for "spotted," a reference to the presence of prominent dorsal spots (Jaeger 1972).

This species is characterized by two irregular rows of yellow spots running from behind the eyes along the dorsum to the tip of the tail. In some populations, the spots on the neck and head may be orange, or they may be mostly absent. The ground color ranges from black to steel gray on the dorsum and uniform gray on the venter. The Spotted Salamander is a robust ambystomatid, with adults ranging from 11.2 cm to 19.7 cm in total length. Breeding females may be slightly larger than breeding males, due to the fact that females take longer to reach sexual maturity (Flageole and Leclair 1992). Juveniles resemble adults but have punctate spots that grow disproportionately larger with age.

Spotted Salamanders occupy much of eastern North America from the Canadian Maritime Provinces in the Northeast, west to the Great Lakes region. From southern Illinois, the range extends westward to include parts of Missouri, the eastern edge of Oklahoma and Texas, and down to the Gulf Coastal Plain of Louisiana. In the south, the species is absent from Florida through southeastern Georgia and most of North Carolina, but has a statewide distribution in South Carolina. In Maryland, Spotted Salamanders are distributed across the state from the Appalachian Plateaus (J. A. Fowler 1944; Thompson and Gates 1982) to the Atlantic Coastal Plain, but are largely absent from the lower Eastern Shore (H. S. Harris 1975).

Spotted Salamanders have a biphasic life history, and their preferred habitat is deciduous forest adjacent to vernal pools (Petranka 1998) and fishless impoundments (Bahret 1996). Although the larvae lack chemical protection against vertebrate predators (Kats et al. 1988), on the Atlantic Coastal Plain larvae often occur in the backwaters of streams containing predatory fish (Semlitsch 1988). Once they leave the water, juveniles and adults are fossorial, burrowing beneath the forest substrate in search of earthworms, grubs, millipedes, and insects (Mitchell and Gibbons 2010), usually within 250 m of the pond margin (Petranka 1998). During the winter, they frequently occupy small mammal burrows, especially those of deer mice and shrews (Madison 1997). Spotted Salamander phenology is variable and dependent on temperature and precipitation (Sexton et al. 1990). Individuals are most reliably encountered during the breeding season when adults migrate to and from their breeding sites. They may occasionally be encountered during the nonbreeding season by rolling fallen logs on the forest floor.

The onset of reproduction varies across the species' range and is influenced significantly by latitude and altitude (Petranka 1998). In Maryland, reproduction commences in mid-February to mid-March and coincides with late winter or early spring rains, when air temperatures reach 3-10 °C (D. M. Hillis 1977). Spotted Salamanders exhibit philopatry, returning to their natal pond in successive years (Shoop 1968), but there is evidence of measurable genetic exchange between adjacent populations (Zamudio and Wieczorek 2007). During breeding migrations, Spotted Salamanders respond to olfactory cues emanating from their natal pond (McGregor and Teska 1989), but they do not respond to the auditory cues of syntopic anurans (Forester and La Pasha 1982). The duration of the breeding season ranges from 3 to 60 or more days (R. N. Harris 1980). Males typically precede females to the breeding site (Shoop 1974; D. M. Hillis 1977), and during a typical mating session, the sex ratio is skewed heavily toward males (Sexton et al. 1986).

During courtship, males and females engage in a stereotypic nuptial dance during which the male deposits a mushroom-shaped spermatophore that the female picks up with her cloaca (Arnold 1976). In a single night of courtship, a male may deposit more than 40 spermatophores (Arnold 1977). Following spermatophore pickup, the pair separates, and the female oviposits 2-4 ovoid egg masses, each containing up to 250 individual eggs (Bishop 1941). Egg masses are attached to underwater sticks and vegetation or are deposited directly on the pond bottom (Petranka 1998). The outer jelly membranes of an egg mass are usually either clear or milky white (L. M. Hardy and Lucas 1991). However, some egg masses contain a mutualistic green alga (*Oophila amblystomatis*), which is present in the female's tissue and transferred to her eggs as they pass through the oviducts (Kerney et al. 2011). The algae provide oxygen to the developing embryos and receive carbon dioxide for photosynthesis as a by-product of the embryos' respiration (Bachmann et al. 1986). Development is temperature dependent; in Maryland, eggs hatch in 38-52 days and reach metamorphosis in approximately 60 days (Worthington 1968, 1969). Premetamorphic survival is variable, often less than 2% (Petranka 1998). Males reach maturity in 2-3 years and females in 3-5 years (Wilbur 1977b).

During the MARA project, Spotted Salamanders were detected in 156 quads and 417 blocks across the state, similar to the overall distribution reported by H. S. Harris (1975). The species remained undetected in Wicomico, Worcester, and Somerset counties on Maryland's lower Eastern Shore. However, new records were reported for Dorchester and Caroline counties. Mitochondrial DNA from two of three populations on Maryland's Eastern Shore are fixed for a single haplotype (Forester 1992). Fixed mitochondrial haplotypes are indicative of a recent historical invasion (C. A. Phillips 1994), suggesting that the species may still be expanding its range on the Delmarva Peninsula.

Although the Spotted Salamander is not presently threatened in Maryland, it remains vulnerable. Increased urbanization and suburban sprawl are contributing to the destruction of breeding sites and the adjacent upland forest necessary to support the species. As local populations become more isolated, gene flow is reduced and genetic variability decreased. Conservation of vernal pools and other freshwater wetlands in a network of forested corridors is necessary for the continued survival of this species and other amphibians throughout Maryland.

DON C. FORESTER

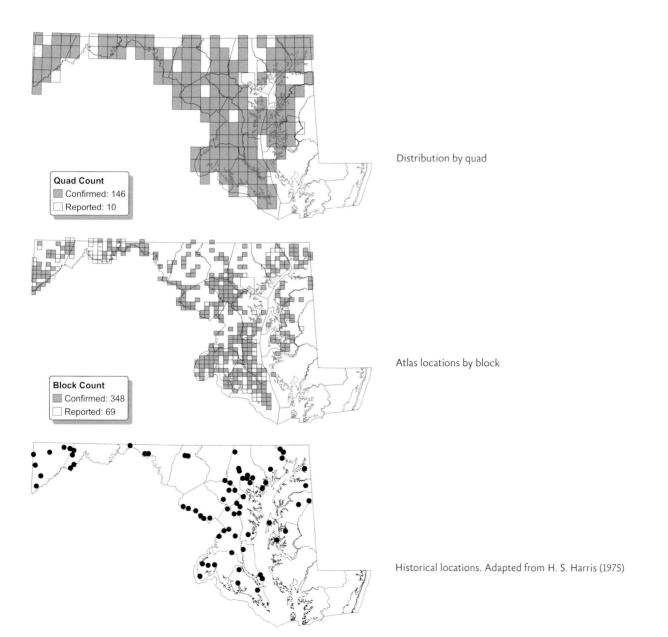

Distribution by quad

Quad Count
Confirmed: 146
Reported: 10

Atlas locations by block

Block Count
Confirmed: 348
Reported: 69

Historical locations. Adapted from H. S. Harris (1975)

Photograph by Michael Kirby

Marbled Salamander

Ambystoma opacum

The smallest of Maryland's *Ambystoma* salamanders, the Marbled Salamander is stout bodied with a black background color marked with white or light gray crossbands, roughly hourglass shaped, on the head, back, and tail (Petranka 1998). These crossbands are variable, and some may be broken, absent, or run together (Petranka 1998). Males typically have brighter crossbands than females, although in some, this difference may be difficult to discern. Adults measure 9–10.7 cm in total length. Recently metamorphosed juveniles are dark brown to black with light flecking that coalesces into crossbands by a few months of age (Petranka 1998).

The Marbled Salamander is fairly common throughout its range. It occurs throughout much of the eastern deciduous forests from southern New England to northern Florida and west to eastern Missouri and Texas. It is absent from much of the southern Appalachians. In Maryland, Marbled Salamanders occur statewide except for the Appalachian Plateaus (H. S. Harris 1975).

In Maryland, the Marbled Salamander appears to be found primarily in soil types dominated by moderately coarse-textured soils and does not generally occur in areas with medium-textured soil types (F. P. Miller 1967). These coarse-textured soils are most likely conducive to the fossorial behavior of this salamander. Forested habitats with temporary ponds or vernal pools are essential for this salamander species. Forests can be deciduous, mixed hardwoods, or pine stands in floodplains or uplands (Mitchell and Gibbons 2010). Within forests, Marbled Salamanders may be found under logs, rocks, and other surface debris within the vicinity of vernal pools, most often in the spring

and fall (Hulse et al. 2001). Throughout much of the year, adults and juveniles are fossorial, spending most of the time underground, but they may be active on the ground surface after rains in the summer and fall (Petranka 1998).

Breeding season for the Marbled Salamander is in late summer and early fall. On rainy nights starting in September, adults migrate to the dry basins of temporary wetlands (Mitchell and Gibbons 2010). Males generally arrive at the breeding areas a week or two before the females (Krenz and Scott 1994). In Maryland, breeding occurs primarily in October (Lee 1973d). At a study site in Anne Arundel County, half of all the Marbled Salamanders moving to breeding ponds did so during mid- to late September (Molines 2008). Unlike the other *Ambystoma* species in Maryland, females nest in the dry beds of temporary ponds and vernal pools in advance of winter and spring precipitation (Petranka 1998). Following courtship, eggs are laid in nests constructed under logs, vegetation, or leaf litter in a spot likely to be flooded by winter precipitation (Hulse et al. 2001). Females typically remain within their nest to guard the eggs until the nest is flooded by rising water and the eggs hatch, although nest abandonment does occur (Petranka 1998). For females that guard their eggs through hatching, the brooding period may last a few weeks to several months, depending on the seasonal pattern of pond filling (Petranka 1998). In New Jersey, Hassinger et al. (1970) estimated a 3-month brooding period. The larval period lasts about 135 days, depending on water temperature (Hulse et al. 2001). Metamorphosis occurred at a site in Prince George's County in mid- to late May (Worthington 1968); it typically occurs from mid-June to early July in Pennsylvania (Hulse et al. 2001). Young can be found under logs and other debris in the vicinity of ponds shortly after they metamorphose.

During the atlas period, Marbled Salamanders were documented in 90 quads and 203 blocks. Records were scattered throughout Maryland, with their distribution primarily on the Atlantic Coastal Plain. The most records occurred in southern Maryland and on the central Eastern Shore. None of these salamanders were found on the Appalachian Plateaus, and they were rare on most of the Blue Ridge and Piedmont. The distribution of the Marbled Salamander was similar to that documented by H. S. Harris (1975), except on the Piedmont. Historically, in Baltimore and Harford counties, this species was found in more locations. The absence of records from these areas during the atlas period may be a function of limited access or survey effort. In Dorchester and Howard counties, more Marbled Salamander locations were documented during the atlas period than were historically known (H. S. Harris 1975).

Conservation of forests with temporary ponds and vernal pools is essential for the continued existence of the Marbled Salamander in Maryland. Fortunately, many of these wet woods are not ideal for development or agricultural production. In addition, many of the larger nontidal wetlands that serve as breeding habitat for this species are protected by state wetlands law. The Marbled Salamander in Maryland seems to be secure currently, but increased protection for smaller nontidal wetlands and larger protected wetland buffers would help ensure their security in the future.

GLENN D. THERRES

Quad Count
Confirmed: 82
Reported: 8

Distribution by quad

Block Count
Confirmed: 170
Reported: 33

Atlas locations by block

Historical locations. Adapted from H. S. Harris (1975)

Photograph by Matthew Kirby

Eastern Tiger Salamander

Ambystoma tigrinum

The Eastern Tiger Salamander is the largest terrestrial salamander in eastern North America. The common name "tiger" refers to the irregular vertical blotches or bars of yellow to olive gray on a dark brown to black background that cover its robust back and sides. In juveniles and young adults, these can appear more as spots, which extend further down the sides than on Spotted Salamanders (Petranka 1998; White and White 2007; Mitchell and Gibbons 2010). Tiger Salamanders have a large, round, dorsally flattened head with relatively small eyes and have stout limbs and a long, knife-edged tail. Adult Eastern Tiger Salamanders typically measure 18-21 cm in total length and reach a maximum total length of 33 cm. Adult females are the same size or slightly larger than males, although the male's tail is proportionally longer and taller. Males also can be distinguished by a swollen cloaca during the breeding season (Stine 1984; Petranka 1998).

The Eastern Tiger Salamander is patchily distributed across the eastern United States, from Long Island, New York, west through Minnesota and south to northern Florida and eastern Texas. It is mostly absent from the Appalachian Highlands and occurs in scattered, disjunct populations on the Atlantic Coastal Plain and lower Mississippi Delta. In Maryland, Eastern Tiger Salamanders have been reported only from the Atlantic Coastal Plain. They were first discovered in Maryland in 1933 in Dorchester County (Netting 1938; Stine et al. 1954) and were subsequently found in Caroline County in 1937, Kent and Queen Anne's counties in 1952 (Stine et al. 1954; Stine 1984), and Charles County in 1953 (Stine 1984). Unverified reports include two

from the 1950s for Somerset County (Reed 1957c; Stine 1984) and one from 1964 for Worcester County (Stine 1984). The only known Western Shore population, located in Charles County, was extirpated in 1963 when construction of a golf course near La Plata destroyed the breeding pond (H. S. Harris 1975; Stine 1984). A single adult Eastern Tiger Salamander was found in Anne Arundel County in 1962, but this was believed to have been an introduction, possibly transported in the root ball of horticultural plantings from Kent County (Stine 1984).

This salamander is primarily fossorial and is rarely encountered except during the breeding season. Its habitat is highly variable range-wide, but optimal habitats include sandy or otherwise friable soils associated with suitable breeding ponds (Petranka 1998). Tiger Salamander habitats on the Delmarva Peninsula include moist, often sandy deciduous, coniferous, or mixed woodlands associated with breeding pools, such as Delmarva bays, fishless human-made ponds, and borrow pits (Arndt 1989; White and White 2007). Simpson (2009) reported that hydroperiods in six Maryland breeding ponds were moderate to long, with some drying as early as mid-June and others still inundated in August. Longer hydroperiods are critical for successful development because larval periods can be long and are highly variable, depending on site conditions, timing of egg laying, and availability of food resources (Petranka 1998). In years with low rainfall, most or all breeding ponds have no successful metamorphosis (Petranka 1998; S. A. Smith, unpubl. data).

Breeding in Maryland can occur from late November to early April (Stine 1984; S. A. Smith, unpubl. data) and is triggered by warm, heavy rains with a few nights above freezing. Breeding often occurs during the so-called winter thaw in late January or early February. Males enter breeding ponds a few days to weeks before females arrive. After a nuptial dance by the male that includes snout-rubbing the female's body and grasping her side, the male deposits 8-37 pyramidal-shaped, gelatinous spermatophores on the pond bottom (Petranka 1998). If the female is receptive, she will collect a spermatophore with her cloaca, thus fertilizing her eggs (Petranka 1998). Tiger Salamanders exhibit the greatest variation in clutch size of any salamander, attaching several gelatinous egg masses containing 5-165 embryos onto live and dead vegetation or other support structures in the pond, often near the bottom (Stine 1984; Petranka 1998). Stine (1984) reported a range of 17-144 embryos per egg mass with an average of 45 embryos for a Maryland pond. Individual females lay 5-8 egg masses (J. D. Anderson et al. 1971). The egg masses of Eastern Tiger Salamanders can be distinguished from those of Spotted Salamanders as they are flimsy and tear apart when lifted out of water, while Spotted Salamander egg masses are solid and hold their form out of water. The larval period lasts 2.5-5 months. Sexual maturity is reached in as little as 2 years (Petranka 1998), though it can take up to 8 years (Kathy Clark, pers. comm.). Tiger Salamander longevity is 16–25 years (Petranka 1998).

During the atlas project, Eastern Tiger Salamanders were confirmed in only two quads and four blocks in two counties, Caroline and Kent. The MARA results show considerably fewer occurrences than were shown by H. S. Harris (1975), who considered the entire Atlantic Coastal Plain of Maryland as the species' probable range. Localities on Harris's map covered eight counties, including historical records and unverified reports. The MARA results are more similar to the map published by Stine (1984), which showed only seven extant breeding ponds in Caroline and Kent counties. Surveys from 1997 to 2015 by Bryan DuBois, Nathan Nazdrowicz and Jim White, and S. A. Smith (all unpubl. data) and by Otto (2006) and Simpson (2009) reported breeding at 21 total ponds in Caroline and Kent counties. An adult was also found in Dorchester County adjacent to a breeding pond located in Sussex County, Delaware. Not all of these documented sites were surveyed specifically for Eastern Tiger Salamanders during the atlas project, but if included would have increased the totals to three quads and seven blocks.

The Eastern Tiger Salamander is listed as a state-endangered species in Maryland (MDNR 2016b), but many of the known breeding ponds are clustered on public and private conservation lands, so the potential for successful management efforts is high. Protection of Delmarva bays, vernal pools, and other breeding ponds and the forested matrix that surrounds and connects them is critical to long-term population viability for this salamander species.

SCOTT A. SMITH

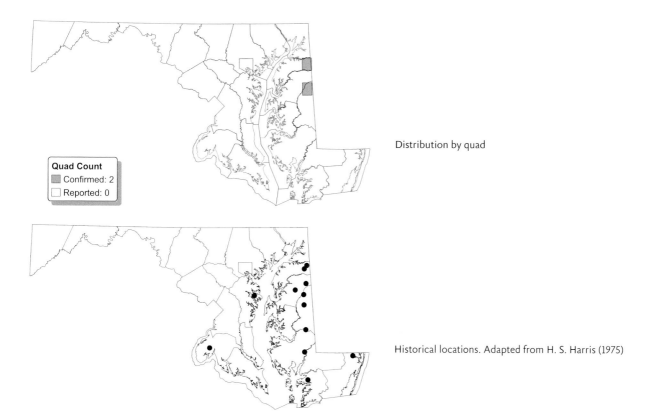

Distribution by quad

Quad Count
Confirmed: 2
Reported: 0

Historical locations. Adapted from H. S. Harris (1975)

Photograph by Robert T. Ferguson II

Hellbender
Cryptobranchus alleganiensis

The Hellbender is fully aquatic and is an unusual-looking creature. The overall coloration is variable, but it is usually some shade of brown, olive brown, or gray brown. Dark, irregular spots or blotches are often present throughout the dorsal surface. The head is large and flattened, with small, lidless eyes. A wrinkled fold of skin resembling the curled edge of a lasagna noodle runs along the side between the front and hind limbs. The legs are short and squat, and the toes are thick and blunt. The tail is laterally compressed, with a high dorsal keel. During the breeding season, adult males develop a swollen cloaca. Hellbenders are slippery and very hard to hold; if handled roughly, they exude prodigious amounts of slime. The Hellbender is the largest salamander in North America. Stout and robust, adults may be 28-56.8 cm in total length.

Hellbenders range from extreme southern New York south through Pennsylvania, southeastern Ohio, West Virginia, Kentucky, and western North Carolina to northern Georgia and Alabama. Disjunct populations occur in east-central Missouri and the Ozark region of southern Missouri and northeast Arkansas. In Maryland, Hellbenders were historically known from the Susquehanna drainage in northeastern Maryland (H. W. Fowler 1915) and the Ohio drainage in far western Maryland (H. S. Harris 1975). They are now considered extirpated in Maryland's Susquehanna region. Gates et al. (1985a) surveyed 44 sites in the Susquehanna River system and other streams emptying into the northern Chesapeake Bay and found no evidence of Hellbenders. Currently, the species is known only from a few tributaries

in the Youghiogheny River watershed in Garrett County. McCauley and East (1940) and J. A. Fowler (1947a) provided early documentation of Hellbenders in the Youghiogheny River watershed. Anecdotal reports of Hellbenders exist for the Potomac River and tributaries, but Gates et al. (1985a) sampled 124 sites within the Potomac River system in both Maryland and West Virginia and found no evidence of Hellbender occupation.

Hellbenders inhabit medium-sized streams to large rivers, preferring clean, cold water with abundant cover in the form of large flat rocks, ledges, and other hiding places. Other than during the breeding season, they are rarely encountered out in the open during daylight hours. Most of their activity occurs after dark. Hellbenders are predators and seem to prefer crayfish as their main source of food, but they will eat just about any creature they can overpower and swallow (Nickerson and Mays 1973; Petranka 1998).

In Maryland, their breeding season begins in August and lasts at least through early September. Researchers in West Virginia and Pennsylvania have reported similar timing (Green and Pauley 1987; Hulse et al. 2001). Males excavate a nest chamber, usually under a large rock, and one or more females are attracted to the chamber (Alexander 1927; Bishop 1941). Males often compete for quality nest sites and will actively fight one another (Alexander 1927; Horchler 2010). As the female lays her eggs, the male exudes a cloud of milt containing the sperm, fertilizing the eggs externally. An egg and its gelatinous covering are about the size of a marble, and the 200-400 eggs are connected in string-like fashion. After egg-laying, the female leaves or is driven away, and only the male Hellbender remains with the eggs (Bishop 1941). The male stays with the eggs throughout the incubation period, actively guarding them from various predators, including other Hellbenders. Eggs hatch in 68-75 days, depending on water temperature (Bishop 1941). It has recently been discovered that the larvae stay in the nest with the male at least a month after hatching (Jeff Briggler, pers. comm.). Hellbenders become sexually mature when 5-8 years old (Bishop 1941; Dundee and Dundee 1965); males normally mature at a younger age and smaller size than females. Hellbenders are relatively long-lived animals, surviving at least 25-30 years in the wild (Petranka 1998).

Adult Hellbenders seem to have relatively small home ranges, and because individuals defend only their immediate turf, they can occur in relatively high densities. R. E. Hillis and Bellis (1971) estimated a population of 152 adult Hellbenders in a 220 m stretch of stream in Pennsylvania. There are probably very few streams that support densities such as this today. When one imagines all of the large rivers, creeks, and streams inhabited by this species before European colonization, it is easy to speculate that this salamander was one of the most common large aquatic animals in the mountainous regions of the eastern United States. In contrast, many populations today are isolated, declining, or extirpated (Unger and Williams 2012).

During the atlas period, Hellbenders were documented in only one atlas block in Garrett County, at a known site in the Youghiogheny River watershed. However, a multistate project that analyzed water samples for the presence of Hellbender DNA detected their presence at additional new sites in the Youghiogheny River watershed, which suggests that these salamanders are still present in parts of their historical range in western Maryland.

In Maryland, the Hellbender is listed as endangered (MDNR 2016b). The MDNR has been monitoring one population since 1997, and the data indicate a slow, steady decline to the point that this population is in danger of extirpation. Deforestation and pollution have taken a heavy toll on Hellbender populations. Miles of once-occupied habitat have been compromised by the damming of major rivers. Perhaps the worst insult for Hellbenders is the extreme siltation that has occurred in many waterways. Some populations have been reduced because of years of collecting. Being aquatic and secretive, this giant has undergone a steady, practically invisible decline in much of its range for many years. Being long lived and slow to replace themselves, old adults may be found in a given stream long after the population has passed a threshold toward extirpation.

ED THOMPSON

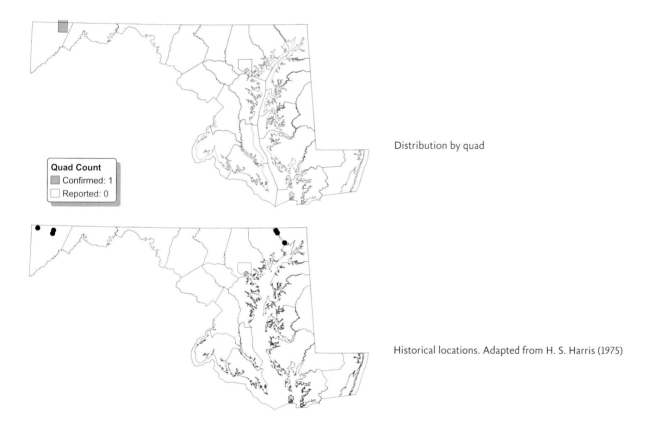

Quad Count
Confirmed: 1
Reported: 0

Distribution by quad

Historical locations. Adapted from H. S. Harris (1975)

Photograph by Ed Thompson

Green Salamander

Aneides aeneus

The Green Salamander is unlike any other Maryland salamander and is easily distinguished by its green coloration, flattened head and body, and enlarged, squared-off toe tips. The background color is dark brown tending toward black, with yellow-green to bright green lichen-like markings all over the body, head, tail, and legs. The ventral surface is light colored and unmarked. The amount of green can vary among individuals; older adults often have less. Adults may attain a total length of 14.8 cm, but most individuals average between 8.3 cm and 12.5 cm. Adult males and females are similar in size, but secondary sexual characteristics are often exhibited, especially during the breeding season (Petranka 1998). Sexually mature males tend to have a wider head due to enlarged jaw muscles, have a mental gland on the chin, and exhibit a swollen cloaca.

There are six species within the genus *Aneides* in North America; the Green Salamander is the only one that occurs east of the Mississippi River. Its range extends from southwestern Pennsylvania through the Appalachian Mountains to northern Alabama and extreme northeastern Mississippi. Disjunct populations also occur in the southern Blue Ridge Mountains of North Carolina, South Carolina, and extreme northeastern Georgia, and along the Ohio River in Ohio, Indiana, and Kentucky. This salamander is uncommon to rare in much of its range. In Maryland, it is restricted to a 70-km^2 area along one river drainage in northwestern Garrett County. Green Salamanders were first discovered in Maryland in 1966 (H. S. Harris and Lyons 1968a). Only two locations were known until 1980, when a status and distributional survey for this elusive animal was conducted (Thompson 1984a). Rock outcrops throughout Garrett County and western Allegany County were sampled during this survey and additional sites were discovered, but the range was found to be very restricted.

The flattened body, prehensile tail, and unique toes of Green Salamanders serve them well for climbing and negotiating tight places. Throughout their range, they are primarily associated with cliff faces and rock outcrops of sandstone, granite, limestone, or schist (Netting and Richmond 1932; Walker and Goodpaster 1941; Gordon and Smith 1949; Bruce 1968). The rock type does not seem to matter, but the structure or microhabitat of the rocks is important. Cracks and crevices that penetrate deep into the rocky habitat are critical. These crevices can be neither too dry nor too wet. In Maryland, nearby West Virginia, and Pennsylvania, Green Salamanders appear to be restricted to sandstone of the Pottsville formation (Netting and Richmond 1932; Thompson and Taylor 1985; Hulse et al. 2001). Additionally, there are reports throughout the literature associating Green Salamanders with arboreal tendencies (Bishop 1928; Pope 1928; Brimley 1941; J. A. Fowler 1947b; Gordon 1952; A. Schwartz 1954). Waldron and Humphries (2005) studied this phenomenon and discovered that Green Salamanders in South Carolina spent more time in trees than anyone had thought. They postulated that the reason these salamanders are hard to find in the rock habitat during the summer is that a significant portion of the population is spending time in nearby trees. Larger trees and hardwoods are preferred, according to their research. It is easy to imagine that this arboreal lifestyle may have been a more important part of this salamander's life history in the old growth forest conditions under which the species evolved. The author (Thompson) has observed Green Salamanders on tree trunks within one meter of the rock habitat in Maryland.

Green Salamanders become active in Maryland in April and start courtship and breeding activity in May. Suitable egg-laying sites, usually a crevice within the rock habitat, are first occupied by males, who compete for these locations (Cupp 1991). Most egg laying in Maryland is completed by mid-June (E. Thompson, pers. obs.). Canterbury and Pauley (1994) reported similar timing for egg laying at several sites in West Virginia. The female stays with the eggs until hatching, which in Maryland primarily occurs in the first two weeks of September. The brooding female takes an active role in protecting the eggs from pathogens and predators, and orphaned clutches invariably fail to survive (Snyder 1973). Green Salamanders undergo direct development: the young hatch fully developed from the egg and resemble miniature adults. Despite their small size, Green Salamanders take a number of years to become sexually mature. Petranka (1998), acknowledging that there is little specific information on this topic, speculates that it takes 2–3 years to reach sexual maturity. However, Waldron and Pauley (2007) report that Green Salamanders grow slowly for plethodontids and may take 7–8 years to reach sexual maturity. This information certainly has implications for their conservation.

During the atlas period, Green Salamanders were documented in two quads and five blocks. This distribution was similar to that described by H. S. Harris (1975), but these salamanders were not documented in several known localities discovered by MDNR surveys and reported by Thompson (1984a). Reasons for this discrepancy are most likely related to survey effort, given that many populations are isolated and difficult to access.

The Green Salamander is listed as a state endangered species in Maryland (MDNR 2016b) because of its restricted range and observed declines at some localities. Monitoring has occurred at various sites in the state since the 1980s, and results from these efforts are conflicting: the species continues to do well at some sites, while other populations appear to have declined. About half of the known sites in Maryland occur on public land, where the habitat can be protected from drastic land-use changes. The long-term future of Green Salamanders in Maryland may depend on the fate of sites on private land and how this animal adapts to climate change.

ED THOMPSON

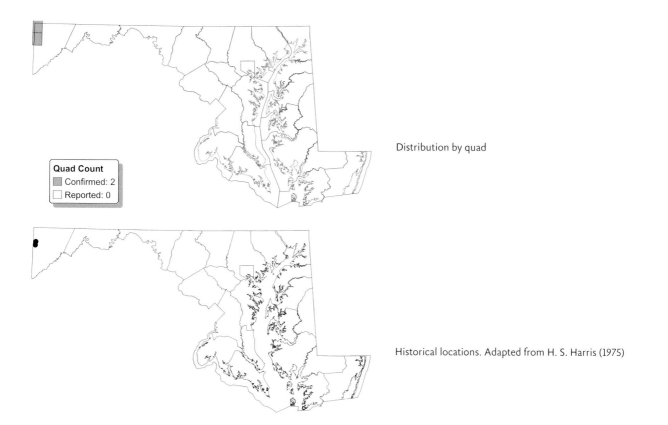

Quad Count
Confirmed: 2
Reported: 0

Distribution by quad

Historical locations. Adapted from H. S. Harris (1975)

Photograph by Lance H. Benedict

Northern Dusky Salamander
Desmognathus fuscus

The Northern Dusky Salamander is a denizen of freshwater streams throughout Maryland. The species is highly variable in coloration, with a dark ground color and slightly lighter dorsum. The dorsum may be bordered by a dark wavy or broken line, marked with variable dark spots or streaks, or punctuated with faint chestnut brown oval spots. The dorsal surface of the base of the tail is typically a lighter brown flanked with dark scallops. The venter is typically pale and lightly mottled with brown. The posterior one-third of the tail exhibits a knife-edged keel, and a prominent white line extends from the eye to the posterior corner of the jaw. Adults range in total length from 6.4 cm to 11.5 cm, with males being slightly larger. Juveniles have five to eight paired or alternating dorsal light brown spots between the forelimbs and hind limbs, which coalesce with age.

The Northern Dusky Salamander is distributed throughout eastern North America from the Canadian Maritime Provinces and Quebec in the north, down through New England and the mid-Atlantic states to the Piedmont of South Carolina in the east. It is absent from the Atlantic Coastal Plain of North and South Carolina. Its western range extends to the shores of Lake Ontario and Lake Erie and along the western foothills of the Appalachians to Kentucky and Tennessee. In Maryland, H. S. Harris (1975) reported Northern Dusky Salamanders occurring statewide with the exception of Dorchester, Somerset, and Worcester counties. In general, the species is uncommon throughout the Atlantic

Coastal Plain in Maryland, especially on the Eastern Shore, due to a lack of suitable habitat.

The preferred habitat of Northern Dusky Salamanders includes shallow, rocky streams associated with wooded valleys (Hom 1988), from near sea level to 1,200 m (Petranka 1998). They are often densely packed in headwater streams, where they seek refuge in seeps and springheads. They may be found beneath rocks, moss, and debris in these habitats, as well as near the edge of first- and second-order streams in all but the coldest months (Hamilton 1943; Organ 1961; Orser and Shure 1975; Snodgrass et al. 2007; Plenderleith 2009). Juveniles and adults forage among cobble bars within the stream channel and seldom move more than a few meters from the stream margin (Bishop 1943; Hom 1988; Plenderleith 2009). Northern Dusky Salamanders are relatively sedentary. Hom (1988) reported that adult salamanders in a Tennessee population remained within the same 15-m stream segment over a two-year period, and Plenderleith (2009) reported an average upstream movement of 7 m over two years in central Maryland. Barthalmus and Bellis (1972) estimated the home range of a Pennsylvania population at less than 3 m^2. Despite their sedentary behavior, Northern Dusky Salamanders apparently have a well-developed spatial map, and they can return to a home range following displacement of 30 m (Barthalmus and Bellis 1969; Barthalmus and Savidge 1974). Females can locate their nests following natural and experimental displacement (Dennis 1962; Forester et al. 2008).

The reproductive ecology of the Northern Dusky Salamander is complex. There are two courtship periods (fall and spring), and egg deposition occurs in late spring to early summer (Petranka 1998). Females excavate a nesting cavity in terrestrial situations beneath rocks, moss, or other surface debris adjacent to streams or seepages. Clutch size ranges from 24 to 50 or more eggs and is dependent on the body size of the female (Danstedt 1975; Hom 1987). Following oviposition, the female coils about her eggs and remains with them for 46-61 days until hatching (Juterbock 1986; Hom 1987). The presence of the female contributes to the survivorship of the clutch, and unattended eggs experience catastrophic consequences (Hom 1987). After hatching, Northern Dusky Salamanders crawl into the water and spend 9-12 months as aquatic larvae before metamorphosing into terrestrial juveniles (Danstedt 1975). In two low-elevation populations in Maryland, males reached sexual maturity at 2 years of age and females at 3 years (Danstedt 1975, 1979), while in higher-elevation populations in Virginia, maturity was delayed until 3.5 years for males and 4.5 years for females. In these populations, females produced their first clutch in their fifth year (Organ 1961).

The diet of Northern Dusky Salamanders consists primarily of terrestrial and semiterrestrial invertebrates, with the inclusion of increasingly larger prey as the animal grows (Burton 1976; Sites 1978; Krzysik 1979). Adults have also been reported to engage in opportunistic cannibalism (Wilder 1913; Hamilton 1943). During the brooding period, females cease active foraging and feed opportunistically (Montague and Poinski 1978; Krzysik 1980). Brooding females practice oophagy (egg eating) on dead and diseased eggs (Bishop 1941; R. L. Jones 1986).

During the MARA project, Northern Dusky Salamanders were found in 318 blocks within 110 quads across the state. The distributional pattern approximated that reported by H. S. Harris (1975) and emphasized the spotty distribution of this species below the Fall Line.

Northern Dusky Salamanders are sensitive to stream pollution (Gore 1983) and urbanization (Orser and Shure 1972). More recently, Bank et al. (2006) implicated the stocking of fishes, fungal pathogens, substrate embeddedness, and widespread atmospheric pollution in the decline of the species in Acadia National Park in Maine. In Maryland, the species continues to maintain robust populations throughout the Piedmont west to the Appalachian Plateaus. However, the species' preferred nesting habitats (seepages and the headwaters of first-order streams) are increasingly affected by deforestation, development, and urbanization (Snodgrass et al. 2007).

DON C. FORESTER

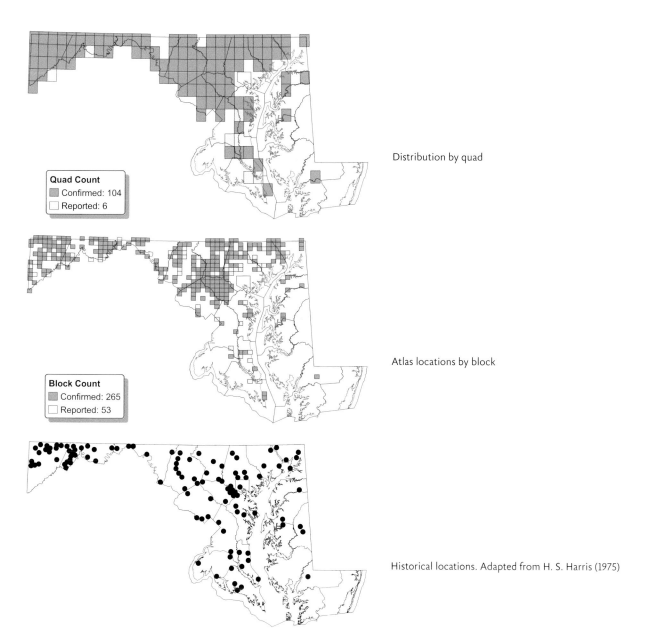

Distribution by quad

Quad Count
Confirmed: 104
Reported: 6

Atlas locations by block

Block Count
Confirmed: 265
Reported: 53

Historical locations. Adapted from H. S. Harris (1975)

Photograph by Michael Kirby

Seal Salamander
Desmognathus monticola

The Seal Salamander is a heavy-bodied stream salamander found in the mountains of western Maryland. Although similar to the Northern Dusky Salamander, it is distinct upon close evaluation. The Seal Salamander is the most "bug-eyed" of the dusky salamanders found in Maryland. It has a laterally compressed tail that is dorsally keeled and makes up approximately half of total length. Dorsal coloration varies from pale gray to brown, and older adults may be quite dark. Most individuals have a mottled dorsal pattern, often described as wormy, reticulated, or netlike, though this pattern may not be present in older or aberrant specimens. The dorsal portion of the tail often has distinctly light, paired spots. The venter is generally a cream to gray color and without markings, but small, pale white or blue flecks may extend from the flanks. Blue flecks may also be present on the dorsum. The species does not exhibit sexual dimorphism in size, and adults typically measure 7.5-12.5 cm in total length, with a record of 14.9 cm. Juveniles often have four to five distinct pairs of staggered, chestnut blotches on the dorsum between the limbs, which fade with age.

The Seal Salamander ranges from southwestern Pennsylvania, southwest to extreme western Florida. The range is generally centered on the Appalachian Mountains and foothills, but isolated populations occur in Piedmont and Gulf Coastal Plain regions of the southern United States. Historically, the Seal Salamander was believed to be confined to the Appalachian Plateaus in Maryland (Franz 1972). Franz (1972) documented Seal Salamanders on the Allegheny Front, in the upper reaches of streams on the eastern slopes of Dans and Piney Mountains. Seal Salamanders were not detected in the lower-elevation stretches of these streams or in any streams originating on the Ridge and Valley Province, and Franz (1972) suggested that the LaVale and Cumberland valleys were a physical barrier to dispersal. H. S. Harris (1975) reported the Seal Salamander occurring "just off the plateau in a few streams," but whether he was referring to the same localities mapped by Franz (1972) is not known.

In Maryland, the Seal Salamander inhabits mountainous seepages, springs, rills, and streams in forested areas. These habitats typically have stony substrates, though in some areas of the Chestnut Ridge in Pennsylvania, the author (Ruhe) has observed Seal Salamanders in sandy-bottomed springs within clumps of moss and woody debris. Seal Salamanders typically do not venture far from running water and are most common within and adjacent to aquatic habitats (i.e., stream splash zone). They are typically found under rocks and, less commonly, logs, moss, and other debris during daylight hours. At night, individuals can be found exposed on the surface in and adjacent to streams. Petranka (1998) reports that adults can climb to heights of 2 m above the ground. While not as common as Allegheny Mountain Dusky Salamanders or Northern Dusky Salamanders, Seal Salamanders may be found in relatively high densities in aquatic habitats in mature, mountainous forests. The species is active throughout the year, but these salamanders retreat below cover objects in flowing water during the coldest months.

Organ (1961) reported that mating occurs in the spring and fall, with egg deposition in June and July. Fifteen to 40 eggs are laid among and under rocks, under roots and logs, and occasionally in human-made refuse such as broken clay pots or between vinyl sheets (B. Ruhe, pers. obs.). These nest sites are typically located not in streams but in adjacent areas that are moist but without flowing or standing water. Females brood-guard eggs and remain with egg clutches until hatching. According to Petranka (1998), eggs hatch in late summer to early fall and juveniles have a larval period of 8-11 months. Females have biennial reproductive cycles. Petranka (1998) also reports that adults can live as long as 11 years in the wild and do not become sexually mature until 3-5 years of age.

As reported by Hulse et al. (2001), gut content analysis showed that Seal Salamanders in Pennsylvania ate mostly insects, but a significant minority (15%) had also eaten salamanders (*Desmognathus* and *Eurycea*). Hulse et al. (2001) also reported that nearly 46% of preserved specimens examined showed tail regeneration. As Seal Salamanders are highly territorial, Hulse et al. speculated that the majority of this regeneration is due to intraspecific aggression.

The MARA project recorded the presence of Seal Salamanders in 42 blocks within 18 quads. All observations were from Allegany and Garrett counties, on the Appalachian Plateaus and along the Allegheny Front. MARA locations generally mirrored those reported by H. S. Harris (1975), though atlas records were lacking from northwestern Allegany County, an area where the species was previously abundant. Whether this absence is an artifact of survey effort or indicates a range contraction is unknown.

No conservation efforts specifically targeting the Seal Salamander have been conducted in Maryland. The species has undoubtedly benefited from the creation of state forests and conservation lands in Garrett County and, to a lesser extent, in Allegany County. Acid mine drainage, forest fragmentation, and surface mining most likely affect Seal Salamander habitat. Unfortunately, the species is often used as fishing bait, and in one study (Jensen and Waters 1999), 67% of all salamanders found for sale in bait shops were Seal Salamanders.

BRANDON M. RUHE

Distribution by quad

Quad Count
■ Confirmed: 15
□ Reported: 3

Atlas locations by block

Block Count
■ Confirmed: 27
□ Reported: 15

Historical locations. Adapted from H. S. Harris (1975)

Photograph by Michael Kirby

Allegheny Mountain Dusky Salamander
Desmognathus ochrophaeus

The Allegheny Mountain Dusky Salamander is the smallest of the three *Desmognathus* salamanders that are found in Maryland. Like other *Desmognathus*, this species has a light-colored diagonal line extending from behind the eye to the back of the mouth and hind limbs that are larger than the forelimbs. The dorsum has a broad and relatively straight-edged, reddish to light brown stripe that continues onto the tail and is marked with a middorsal row of chevrons to irregularly round dark spots. Patterns in older individuals fade to a uniform dark brown or black, with a head that may be slightly lighter in color than the body. The ventral color is a uniform light to dark gray. The tail is rounded dorsally, a trait that distinguishes this species from the Northern Dusky Salamander and Seal Salamander. Males have a sinuate jaw line when viewed from the side (Petranka 1998). Adult Allegheny Mountain Dusky Salamanders range from 7 cm to 10 cm in total length. Juveniles have a solid red to yellowish straight-edged dorsal stripe, which resembles the overall pattern of Two-lined Salamanders or Eastern Red-backed Salamanders.

The Allegheny Mountain Dusky Salamander is restricted to the Adirondack and Appalachian Mountain regions of eastern North America, extending from extreme southeastern Quebec and northern New York to northern New Jersey and northeastern Ohio, and south through western Pennsylvania to northern Tennessee. In Maryland, it is restricted to Garrett County and western Allegany County (H. S. Harris 1975). Franz (1972) suggested that this species

was restricted to the Appalachian Plateaus in Maryland, but R. W. Miller (1980) reported specimens collected from Piclic Run in Green Ridge State Forest, approximately 30 km east of the closest known locality on the Appalachian Plateaus.

Allegheny Mountain Dusky Salamanders inhabit seepages, springs, and streams, as well as adjacent forested areas. These salamanders stray farther from the water's edge than other *Desmognathus* species of Maryland (Hulse et al. 2001). In summer months, they may be found in moist woodlands under logs, rocks, and other debris 100 m or more from the nearest water (Hulse et al. 2001). In the winter, they retreat to springs and small streams and may remain active at the surface where water does not freeze. During the coldest months, they move to underground passages or into cracks and crevices of shale banks. Allegheny Mountain Dusky Salamanders can occur in high numbers along streams; densities of 0.6–1.2 salamanders/m^2 have been reported (Hall 1977; L. P. Orr 1989).

Breeding primarily takes place in the spring but may also occur in late summer and fall. Females reproduce annually; oviposition occurs in March through September, with peak oviposition in May to July (Petranka 1998). Nest sites are located underground in hollowed out depressions in seepage banks or under embedded logs or rocks in spring seeps. Females lay 8–24 eggs singly in grapelike clusters. Average clutch size from a population in Pennsylvania was 16 eggs (Hall 1977). Females remain with the eggs through hatching, which peaks in October but depends on timing of oviposition and hydroperiod. The larval period is usually very short, lasting 2–3 weeks, but may last up to 8 months. Sexual maturity is reached in 3–4 years, and these salamanders have lived up to 20 years in captivity (Niemiller 2011).

The home range of the Allegheny Mountain Dusky Salamander is reported at less than 1 m^2 (Holomuzki 1982). These salamanders are territorial and will aggressively defend their territories from conspecifics and Eastern Red-backed Salamanders. They can successfully displace Eastern Red-backed Salamanders from their cover objects (E. M. Smith and Pough 1994). In the presence of Northern Dusky Salamanders, Allegheny Mountain Dusky Salamanders tend to move farther from the water (Petranka 1998). They feed on various invertebrates, including snails, earthworms, isopods, and larval and adult insects (Petranka 1998). Birds and small snakes are reported predators, and Cupp (1994) found that these salamanders will avoid cover objects marked with the scent from Ring-necked Snakes.

During the MARA project, Allegheny Mountain Dusky Salamanders were found in 19 quads and 90 blocks, all on the Appalachian Plateaus. Although the distributions were generally similar, the MARA results found Allegheny Mountain Dusky Salamanders more ubiquitous than reported by H. S. Harris (1975), and these salamanders most likely occur in every block comprising the Appalachian Plateaus. Specimens reported by R. W. Miller (1980) from Green Ridge State Forest in Allegany County warrant scrutiny, as this report appears to be out of range for Maryland and adjacent Pennsylvania counties (Brandon Ruhe, pers. comm.).

The Allegheny Mountain Dusky Salamander is a common species of the Appalachian Plateaus in Maryland and is presently in little need of conservation. Maintaining forested buffers along springs and small streams and preventing degradation of water quality by acid mine drainage and other pollutants are important conservation strategies for this species.

NATHAN H. NAZDROWICZ

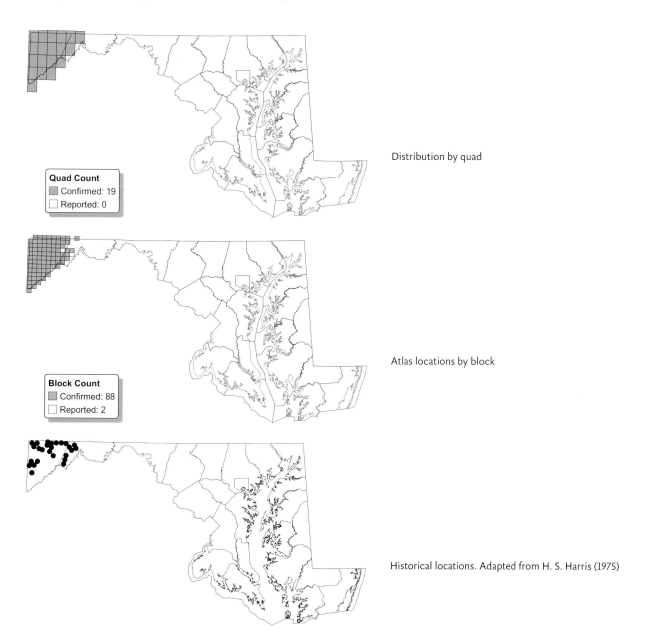

Quad Count
Confirmed: 19
Reported: 0

Distribution by quad

Block Count
Confirmed: 88
Reported: 2

Atlas locations by block

Historical locations. Adapted from H. S. Harris (1975)

Northern Two-lined Salamander. Photograph by Lance H. Benedict

Northern Two-lined Salamander
Eurycea bislineata
Southern Two-lined Salamander
Eurycea cirrigera

The Northern and Southern Two-lined Salamanders are part of a cryptic species complex that also includes the Blue Ridge Two-lined Salamander (*Eurycea wilderae*) (Jacobs 1987). The two species are nearly identical but are easily distinguished from other salamanders by an overall yellowish color marked by two dark brown to black lines that extend from the eyes to the end of the tail, although these stripes may become broken on the tail. Scattered brown to black spots are often present along the back between the two dorsolateral stripes. Occasionally, these spots form a third, broken stripe along the midline (Hulse et al. 2001). The venter is a uniform pale to bright yellow. The body is slender, with four thin, well-developed limbs of equal size. The tail is laterally compressed and makes up roughly half of the total length (White and White 2007). Average adult size is similar in males and females, with total length of 6.4–9.5 cm.

The Northern Two-lined Salamander is distributed from southern Canada and New England through the mid-Atlantic states. The Southern Two-lined Salamander ranges from the southern Delmarva Peninsula west to Indiana and south through the southeastern United States, excluding peninsular Florida. The contact zone between the two species occurs in north-central Virginia, southern West Virginia, and eastern Ohio (Pauley and Watson 2005). Histori-

cally, all Two-lined Salamanders in Maryland were assumed to be Northern Two-lined Salamanders (H. S. Harris 1975). Although the species was common throughout most of the state, records were lacking from most of the southern Eastern Shore, except for Wicomico County (H. S. Harris 1975). Discovery of populations in Worcester County and on the Eastern Shore of Virginia in 2007 and 2008 prompted an investigation of species identification for the Delmarva Peninsula (Nathan Nazdrowicz, pers. comm.). Nathan Nazdrowicz and Paul Sattler (unpubl. data) used protein electrophoresis (Guttman and Karlin 1986; Jacobs 1987) to determine that Southern Two-lined Salamanders were present in these southern counties of Delmarva. Worcester County remains the only known county of occurrence for the Southern Two-lined Salamander in Maryland, as the historical localities from Wicomico County have not been rediscovered, and no specimens exist for Dorchester or Somerset counties. The range of the Northern Two-lined Salamander comprises the remainder of the state.

Two-lined Salamanders are primarily stream dwelling and are associated with seeps, springs, brooks, and floodplain wetlands. They can be found beneath rocks, logs, woody debris, and leaf packs in the stream channel and along the wetted margin or around spring seeps (Petranka 1998; White and White 2007). While Two-lined Salamanders are most commonly found among cover in and around streams and wetland habitat, adults may occasionally be found upwards of 50 m from water. These salamanders may also be observed in open terrain and crossing roads during periods of high humidity and heavy rain at night (Hulse et al. 2001).

The full life cycle occurs in these habitats, with adults and larvae found throughout the year; winter activity is restricted to headwater areas, springs, and seeps. Courtship may begin in the fall (Hulse et al. 2001), but the breeding season typically extends from January to April (Bishop 1943), and even into May to July throughout the full geographic range of the species (Hulse et al. 2001). Eggs have been found from January through April in southeastern Virginia (J. T. Wood and McCutcheon 1954), in May in northern Virginia (Beane et al. 2010), and in March through early April in Maryland (H. S. Harris 1975). Clutches of up to 100 eggs are laid in a single layer attached to the underside of rocks, woody debris, and other cover in the stream channel (Petranka 1998; Hulse et al. 2001). Eggs are guarded by the female until they hatch into aquatic larvae. The larval period is variable, ranging from 1 year to 3 years, depending on temperature and latitude (Hulse et al. 2001).

Larvae and adults feed year-round on a variety of invertebrates (Petranka 1998). Southern Two-lined Salamander larvae occasionally prey on vertebrates such as heterospecific salamander larvae (Pauley and Watson 2005). As with other stream-dwelling salamanders, predators of the Two-lined Salamander include various fish, snakes, birds, shrews, and other salamanders (*Desmognathus* and *Gyrinophilus*) (Petranka 1998; Sever 2005). The Two-lined Salamander most often flees as an antipredator mechanism, but sometimes it assumes a coiled defensive posture to protect the head and positions the tail to be autotomized.

The Northern Two-lined Salamander was the second most commonly encountered salamander species during the MARA project. There were 836 sightings, recorded in 146 quads and 556 blocks. In contrast, the Southern Two-lined Salamander was documented only in Worcester County in two quads and four blocks. The distribution of the Two-lined Salamander complex remained consistent with historical records (H. S. Harris 1975), with the exception of no localities reported from Wicomico County. Two of the localities for the Southern Two-lined Salamander were from populations previously analyzed, including the northernmost record submitted to MARA (N. Nazdrowicz and P. Sattler, unpubl. data). Although the Northern Two-

lined Salamander was found throughout most of Maryland, it was most abundant in the higher-gradient stream systems of the mountain and Piedmont regions of the state and was less prevalent on the Atlantic Coastal Plain. Habitat differences may account for this disparity in occurrences.

Land conversion, habitat destruction, loss of headwater streams, impaired water quality, increased contaminants, and sedimentation (Petranka 1998; Southerland et al. 2004; Pauley and Watson 2005)—all have negative implications for all life stages of Two-lined Salamander species. Continued monitoring of freshwater streams, widespread watershed assessment, and species population analyses are useful tools for evaluating the current and future distribution of both species. Additional genetic analyses may be warranted to better define the range of the two species, particularly if additional localities are discovered on the southern Eastern Shore.

RACHEL GAUZA GRONERT

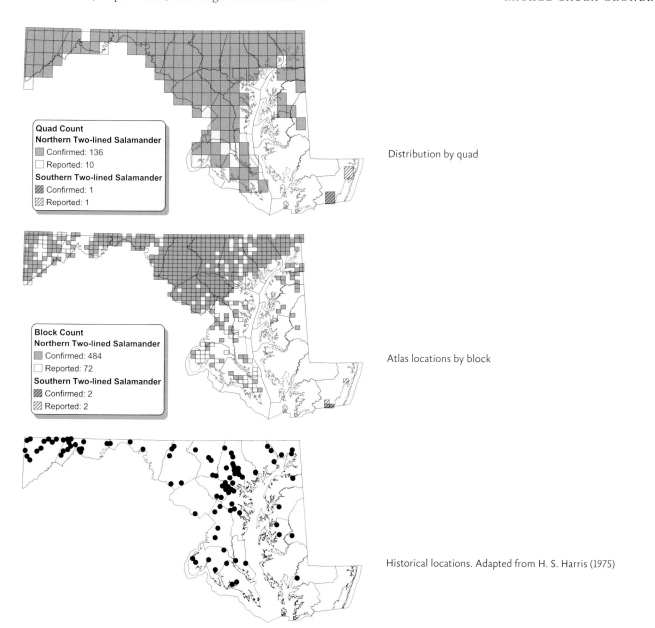

Quad Count
Northern Two-lined Salamander
Confirmed: 136
Reported: 10
Southern Two-lined Salamander
Confirmed: 1
Reported: 1

Distribution by quad

Block Count
Northern Two-lined Salamander
Confirmed: 484
Reported: 72
Southern Two-lined Salamander
Confirmed: 2
Reported: 2

Atlas locations by block

Historical locations. Adapted from H. S. Harris (1975)

Photograph by Andrew Adams

Long-tailed Salamander
Eurycea longicauda

The aptly named Long-tailed Salamander sports a tail that is nearly two-thirds its total length and is the longest tail in relation to body size of any salamander species in Maryland. The tail is laterally compressed and has dark, vertical herringbone-shaped marks running its length along the sides. The fragile, ribbon-like tail is an extension of a slender body, which ranges in color from light yellow to deep orange. Black spots adorn the back and sides. Dorsally, spots are scattered or may form a middorsal row, and a distinctive line of spots is present dorsolaterally from the eye continuing to the tail. The ventral surface is largely unmarked and is pale yellow to cream. The head and eyes are prominent, and the limbs are well-developed. The total length for the Long-tailed Salamander is 9.2-15.9 cm in mature adults. Males and females are similar in appearance, although females are slightly longer.

The distribution of the Long-tailed Salamander extends along the Appalachian Mountains from New York to northern Alabama and the Ohio River Valley, and continues west into the Ozark and Boston Mountains in Arkansas, Kansas, Missouri, and Oklahoma. In Maryland, Long-tailed Salamanders have been documented predominantly above the Fall Line from the Piedmont to the Appalachian Plateaus provinces. H. S. Harris (1975) reported that this species does not occur on the Atlantic Coastal Plain. However, R. W. Miller (1979) documented a population on the Atlantic Coastal Plain in Anne Arundel County, approximately 5.0 km from the zone of contact with the Piedmont.

A lungless salamander, this species requires moist and cool microhabitats. Adults and juveniles are most frequently found near the margins of springs, seeps, streams, and spring-fed ponds (Petranka 1998; Hulse et al. 2001), often associated with karst systems such as those with limestone and shale substrates (Beane et al. 2010). Their occurrence in these habitats varies seasonally (Hulse et al. 2001; White and White 2007; Nazdrowicz 2015). Long-tailed Salamanders can also be found near or within caves (Franz 1967a; N. Taylor and Mays 2006), mine shafts (Mohr 1944), and springhouses, human-made stone structures built over spring seeps (Nazdrowicz 2015). Adults and juveniles can be found by turning over cover such as logs and rocks near suitable habitats during spring and summer months. Long-tailed Salamanders may also be observed among vegetation and along vertical surfaces such as rock walls after dusk and on rainy or humid nights (Nazdrowicz 2015).

The annual activity cycle of the Long-tailed Salamander is characterized by a migration between terrestrial and subterranean habitats. Adult Long-tailed Salamanders are terrestrially active from April through August, though immature individuals may remain active through November (Nazdrowicz 2015). Underground habitats associated with spring systems are utilized by the species for the remaining eight months of the year (Mohr 1944; Nazdrowicz 2015). Mating and oviposition occur underground in stream passages and rock fissures within springs (Petranka 1998). Adults migrate to subterranean habitats in August and September, where they congregate to mate, sometimes in high abundances, before moving deeper within the spring to lay eggs and overwinter (Nazdrowicz 2015). Approximately 60-120 eggs are deposited singly underwater, scattered along the edges of submerged rocks, roots, and woody debris, primarily in November (Mohr 1944; Nazdrowicz 2015). Franz (1964) observed a cluster of eggs suspended above water in a cave in Maryland, but Nazdrowicz (2015) suggests that these eggs may have been exposed due to a drop in water level rather than oviposited out of water. The incubation period lasts 7-10 weeks (Mohr 1943; Nazdrowicz 2015), and recently hatched larvae have been found from November through March (Petranka 1998). The larval period is typically less than one year, with metamorphosis occurring in June or July (Franz and Harris 1965; Franz 1967a; Nazdrowicz 2015). Sexual maturity is reached in 1-2 years (Petranka 1998).

During the MARA project, Long-tailed Salamanders were recorded in 68 quads and 145 blocks. Encounters were largely limited to west of the Fall Line and Interstate 95, which is an approximate division between the Atlantic Coastal Plain and Piedmont. The exceptions were confirmed sightings in two blocks east of I-95 and one block bisected by I-95. Overall, sightings of Long-tailed Salamanders during the atlas project closely aligned with those reported by H. S. Harris (1975).

Long-tailed Salamanders may be affected by changes in environmental conditions, including land conversion, habitat destruction, loss or degradation of headwater streams, and impaired water quality (Southerland et al. 2004). Cool, relatively high-quality streams and springs, as well as human-made springhouses, have been associated with reproducing populations of Long-tailed Salamanders. Preserving the quality and integrity of springs and headwater streams is an important conservation measure for Long-tailed Salamanders. As such, mapping and inventory of springs, seeps, and intermittent and headwater streams would benefit these salamanders and other species that occupy these sensitive habitats. Nazdrowicz (2015) notes also that habitat quantity and connectivity may be important components influencing Long-tailed Salamander distributions and that maintenance of forested corridors in proximity to and between springheads would be beneficial. Springhouses could also be maintained and restored to reduce light and promote humidity. Lastly, avoidance and minimization of impacts related to coal mining, strip mining (Petranka 1998), and hydraulic fracturing, with strict enforcement of best management practices, should be considered for this and other salamander species in western Maryland.

RACHEL GAUZA GRONERT

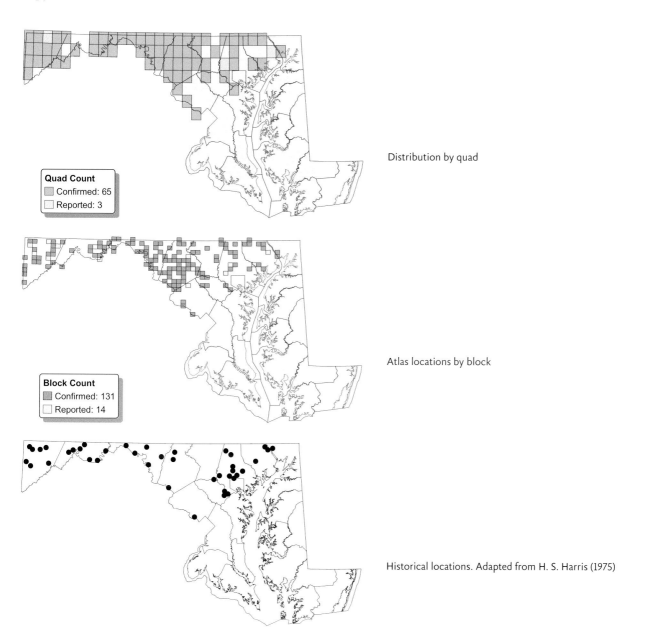

Distribution by quad

Quad Count
Confirmed: 65
Reported: 3

Atlas locations by block

Block Count
Confirmed: 131
Reported: 14

Historical locations. Adapted from H. S. Harris (1975)

Photograph by Robert T. Ferguson II

Spring Salamander
Gyrinophilus porphyriticus

The Spring Salamander was once known as the Purple Salamander because of the purplish brown color of older adults. Younger adults are typically brownish pink or orange, overlain with diffuse dark markings that at times form a reticulated pattern. A light-colored line bordered by dark gray or blackish lines runs from the eye to the nostrils along a raised ridge called the canthus rostralis, which functions much like a gunsight to zero in on prey (Dodd 2003). Adults are large and robust, ranging in size from 12.1 cm to 19 cm in total length.

The Spring Salamander is distributed along the Appalachian Mountains and adjacent foothills from Maine and southern Quebec to northern Georgia, central Alabama, and extreme northeastern Mississippi. In Maryland, the species ranges from central Frederick County along the western edge of the Piedmont west through the Appalachian Plateaus. Historical records are scant from much of the Ridge and Valley. As reported by H. S. Harris (1975), the species showed a disjunct distribution, occupying the Blue Ridge region in the east and the Appalachian Plateaus and adjacent edge of the Ridge and Valley in the west. Harris excluded the central portion of Ridge and Valley Province from the range because extensive surveys throughout this region by Richard Franz in the early 1960s yielded no localities (Platania 1976). However, Platania (1976) documented a population of Spring Salamanders in a small rocky stream that drained to Fifteenmile Creek within Green Ridge State Forest, filling a portion of the gap in distribution suggested by Harris (1975).

Adult Spring Salamanders are most commonly found in springs and seepages, associated caves, and small head-water streams (Bruce 1972b; Petranka 1998; D. M. Green et al. 2013). In western Maryland, Platania (1976) found this species in a small, clear, rocky upland stream. Adults are primarily aquatic, though they can also be found in wet depressions under logs, stones, and leaves in adjacent forest (Conant and Collins 1998). Spring Salamanders are primarily active at night and remain hidden during the day in crevices and under cover objects. They can sometimes be found by turning over rocks and logs or sifting through leaf litter in their habitat. They can also be found moving across roads on rainy nights (Petranka 1998).

Mating takes place in the spring or fall throughout their range, with egg laying generally occurring in the summer months, most likely within subterranean recesses of springs or the stream bed (Petranka 1998). Clutch size ranges from 16 to 106 eggs and is related to body size and elevation; females attend eggs attached in single layers to the undersides of rocks and other objects (Petranka 1998). Eggs hatch in late summer and early fall (Bruce 1980). Larvae hide in gravel beds, beneath stones and logs, or in subterranean recesses (Petranka 1998). The larval period of the Spring Salamander is one of the longest among plethodontid salamanders. It is believed that most larvae take at least 4 years to metamorphose (Petranka 1998).

The diet of larval and adult salamanders consists mainly of aquatic invertebrates, snails, and worms (Hamilton 1932). However, a study in New Hampshire found that adult Spring Salamanders ate a higher percentage of terrestrial prey (Burton 1976). Adults in the southern portion of the range have been found to have a high percentage of salamanders in their diet (Bruce 1972b).

During the MARA project, Spring Salamanders were recorded in 47 blocks within 22 quads. All but two records were from the Blue Ridge west to the West Virginia state line. The two records east of the Blue Ridge included an adult from the Walkersville watershed on the Piedmont of Frederick County and an adult from a spring along the Fall Line in Howard County. During the atlas project, the species was also found on the Ridge and Valley Province in the Town Creek and Fifteenmile Creek drainages, including records within Green Ridge State Forest. However, this species remained elusive throughout large portions of the Ridge and Valley, particularly in Washington County and western Allegany County. Unlike H. S. Harris (1975), the atlas project documented no Spring Salamanders from the Marsh Creek and Antietam Creek watersheds in southeastern Washington County or from the northern Allegheny Front in western Allegany County. Given that Spring Salamanders can be difficult to find because of their tendency to remain within subterranean recesses in springs and small streams, the absence of records in these areas during the atlas period is possibly attributable to survey effort. Gaps in the range documented by Harris (1975) persist, however, suggesting that this species may be absent from some portions of the Ridge and Valley Province. It is noteworthy that Harris showed only two records of Spring Salamander in Garrett County, but during the atlas project, the species was found throughout the county.

Overall, the Spring Salamander population in Maryland appears stable, particularly within Garrett County and Green Ridge State Forest in Allegany County. Laws passed to protect wetlands and streams have benefited the species by protecting its habitat. However, loss of natural upland habitat adjacent to wetlands and streams, increases in impervious surfaces resulting from development, and acid mine drainage in western Maryland could result in reduced water quality that cause population declines.

DAVID R. SMITH

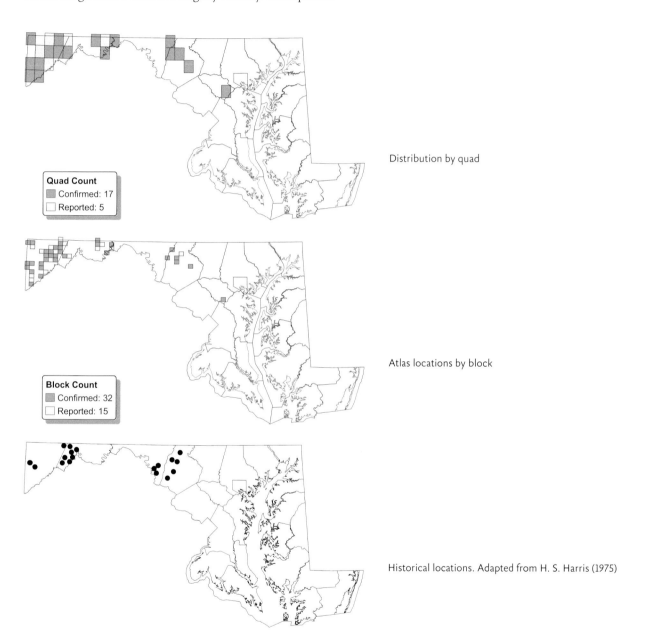

Distribution by quad

Quad Count
Confirmed: 17
Reported: 5

Atlas locations by block

Block Count
Confirmed: 32
Reported: 15

Historical locations. Adapted from H. S. Harris (1975)

Photograph by Ed Thompson

Four-toed Salamander
Hemidactylium scutatum

The Four-toed Salamander is fairly easy to identify once captured. Its underside is white with large, scattered black spots. The dorsal side is usually a rusty brown on the head, body, and tail, changing to a grayish coloration laterally. As its name implies, this salamander has four toes on its hind feet instead of the usual five in most salamanders. A noticeable constriction is present at the base of the tail, from where the tail will easily autotomize when the animal is attacked by a predator or handled (White and White 2007). Adults measure 5.1-9 cm in total length. Adult males are smaller and more slender than females, although they have relatively longer tails, and the snout is more squarely truncated (Bishop 1943).

The Four-toed Salamander occurs from Nova Scotia south to the Gulf of Mexico and west to Minnesota, Missouri, and southeastern Oklahoma, although populations become discontinuous in the western portion of the range. It is absent from much of the Atlantic Coastal Plain from southeastern Virginia through Florida but occurs sporadically along the Gulf Coast from the panhandle of Florida to Louisiana. In Maryland, Four-toed Salamanders have been found in all physiographic provinces; however, no records were documented by H. S. Harris (1975) for Carroll, Frederick, or Somerset counties. The first record of this species for the Baltimore area was in 1942 (Mansueti and Simmons 1943), and for Talbot County in 1969 (Grogan 1974a). R. W. Miller (1980) discovered the first Carroll County individual in 1980, and the first record for Somerset County was as late as 2008 (Nazdrowicz 2009), two years before the atlas project.

Forested habitats are essential for the Four-toed Salamander. Adults and juveniles inhabit forests surrounding fishless wetlands such as swamps, bogs, marshes, and vernal pools, which serve as breeding sites (Petranka 1998). This species prefers hardwood forests in the southeastern portion of its range (Mitchell and Gibbons 2010), but in Virginia it has been found in coniferous forests as well (J. T. Wood 1955). Within these forested habitats, Four-toed Salamanders are most often found under cover objects on the forest floor, including logs, rocks, bark, and other natural cover (Petranka 1998; Hulse et al. 2001), and they can occasionally be found in rotting logs (Mansueti and Simmons 1943).

Mating among Four-toed Salamanders occurs in the fall. Females migrate to breeding sites to lay eggs during late winter and early spring rains, usually during a 2- to 6-week period. Migrations occur later on the Appalachian Plateaus than the Atlantic Coastal Plain (Petranka 1998). In Virginia, eggs have been documented as early as February (J. T. Wood 1955), and J. E. Cooper (1953) found eggs in late March in Maryland. Females lay 4-80 eggs (Mitchell and Gibbons 2010), usually just above the water line, in crude cavities under moss or other debris. Sphagnum or other mosses that form hummocks along the margins or small islets in wetlands seem to be the preferred nesting sites (Petranka 1998). However, not all breeding sites contain moss clumps; sites such as sedges and rushes, liverworts, loose bark, mounds of pine needles, and rotten wood may also be utilized (J. T. Wood 1955). Four-toed Salamanders may nest communally (Goodwin and Wood 1953; J. T. Wood 1955; R. N. Harris and Gill 1980). In Virginia, as many as 868 eggs were found in one communal nest (J. T. Wood 1953). Females remain with the eggs until hatching or may abandon the nest prior to hatching; for communal nests, usually only the first ovipositing female remains with the eggs (Petranka 1998). This nest attention increases embryonic survival (R. N. Harris and Gill 1980). Eggs hatch about 1-2 months after oviposition, with the incubation period dependent on local site conditions (Petranka 1998). After hatching, the larvae wriggle the short distance to water. The larval period is exceptionally brief, lasting only 3-6 weeks (Petranka 1998). In a Virginia population, adult size was reached in about 2 years (J. T. Wood 1953).

During the atlas period, Four-toed Salamanders were found in scattered locations throughout the state. This species was found in 63 quads and 92 blocks, with a similar overall distribution to that reported by H. S. Harris (1975). More locations of this salamander were found on the Piedmont during the atlas period than were reported by Harris; however, no Four-toed Salamanders were found in Montgomery County. Several locations were documented in Carroll County and two in Frederick County, compared with none reported by Harris (1975).

With little change in distribution over time, it appears that the Four-toed Salamander is secure in Maryland for the foreseeable future. Loss of wetland breeding sites in forested areas is the main threat (Mitchell and Gibbons 2010). Fortunately, Maryland's nontidal wetland laws provide protection to much of this species' breeding habitat, including a 7.62-m (25-ft) surrounding buffer.

GLENN D. THERRES

Distribution by quad

Quad Count
Confirmed: 57
Reported: 6

Atlas locations by block

Block Count
Confirmed: 82
Reported: 10

Historical locations. Adapted from H. S. Harris (1975)

Photograph by Robert T. Ferguson II

Eastern Red-backed Salamander
Plethodon cinereus

The tiny Eastern Red-backed Salamander may be the most ubiquitous vertebrate of the great eastern forests of North America. The species is polymorphic, and two common color morphs of this slender salamander exist in Maryland: a striped morph, which has a red to orange dorsal stripe continuing from the head onto the tail, and an unstriped or "lead-back" morph, which is uniformly brown, slate gray, or black. Occasionally, in striped individuals, the dorsal stripes will be white, silver, tan, yellow, or brown. Striped and lead-back individuals generally have many white, silver, or gold flecks along the flanks and limbs. These flecks are often found throughout the dorsum on the lead-back morph. The venter is the same in both Maryland morphs: a mottled black and white that is often called a "salt-and-pepper" pattern. Erythrism (red), albinism (pinkish due to lack of melanin), and leucism (abnormally light color, typically yellowish or white) have also been reported in Maryland (Dyrkacz 1981; J. M. Moore and Ouellet 2014). The prevalence of color morphs varies from location to location, and researchers have not yet made any convincing correlations between predators, habitat type, climate, or any other environmental factors and the presence of color morphs (J. M. Moore and Ouellet 2015). The number of costal grooves can be variable among individuals and ranges from 17 to 22, with 19 typically observed (Conant 1975). The species does not exhibit sexual dimorphism in size, and adults typically measure 5.7-10 cm in total length.

The Eastern Red-backed Salamander ranges from Cape Breton Island, Nova Scotia, west through southern Canada to Minnesota, and southeast to North Carolina. While the species is quite common throughout most of its range, it becomes uncommon along the periphery and is patchily distributed in warm, xeric habitats. H. S. Harris (1975) reported the Eastern Red-backed Salamander from all Maryland counties and physiographic provinces. He also described the absence of observations from a significant portion of the Ridge and Valley Province of Allegany and Washington counties.

This salamander is a denizen of the eastern forests, preferring mesic soils with appreciable duff layers under deciduous, mixed deciduous-coniferous, or coniferous canopies, but may also occur in gardens or other disturbed areas adjacent to forests. The species is most often found under cover objects, including rocks, logs, debris, leaf litter, and even human refuse. Petranka (1998) reports that Eastern Red-backed Salamander densities are greatest in forests with well-drained soils, but low to absent in permanently wet, rocky, or highly acidic soils. This species is relatively well-known to the public due to its high densities, wide distribution, and presence close to dwellings and woodlots in suburban and urban areas. This is the "backyard" salamander often found in gardens and while raking leaves.

Eastern Red-backed Salamanders are typically surface-active from late winter to fall, but they can be found during warm spells throughout the winter in the mid-Atlantic region (Petranka 1998). During the coldest and driest months of the year, these salamanders move to subsurface retreats. In Maryland, breeding occurs from September to April (Sayler 1966). Females breed biennially and typically oviposit in June (Sayler 1966). Six to nine large eggs are deposited in a small cluster suspended from the ceiling of subsurface burrows or cavities in or under embedded logs or rocks (Petranka 1998). Females brood-guard the eggs, which undergo direct development, and hatching occurs approximately 6 weeks after oviposition (Sayler 1966; Petranka 1998). Sexual maturity is reached at approximately 2 years of age, at a length of 3.2-3.9 cm, snout to vent (Petranka 1998).

This salamander species is important ecologically, as it comprises a significant portion of the vertebrate biomass in forest communities. Burton and Likens (1975b) determined that the biomass of all Eastern Red-backed Salamanders at a study site in New Hampshire exceeded that of all birds during the nesting season in the same area and was comparable to that of all small mammals in the same area. Amazingly, their estimates did not take into account the large percentage of individuals that were not surface-active (Highton 2005). Thus, the Eastern Red-backed Salamander is a very important low-level consumer, mobilizing energy from tiny invertebrate food items for use by large invertebrate and vertebrate consumers (Burton and Likens 1975a). These sala-manders are opportunistic carnivores and will consume a wide array of invertebrates and occasionally conspecifics.

During the MARA project, Eastern Red-backed Salamanders were found in 762 blocks within 201 quads. The species was documented in all counties and all physiographic provinces, with an absence from most of the Ridge and Valley Province in western Washington County and much of Allegany County, as similarly reported by H. S. Harris (1975). This region is the stronghold of the congeneric Valley and Ridge Salamander. Atlas records for the Eastern Red-backed Salamander were most commonly obtained from the Piedmont and eastern and western portions of the Appalachian

Mountains, with a widespread but patchy distribution on the Atlantic Coastal Plain.

The Eastern Red-backed Salamander is considered common throughout Maryland and eastern North America. As such, no conservation measures are in place specifically for this species. Despite this abundance, deMaynadier and Hunter (1998) found that these salamanders were adversely affected in areas with forest clear cutting and within edge habitats near cut-over areas. Actions such as selective forestry practices and preservation of wooded upland habitats in Maryland will most likely benefit the Eastern Red-backed Salamander.

BRANDON M. RUHE

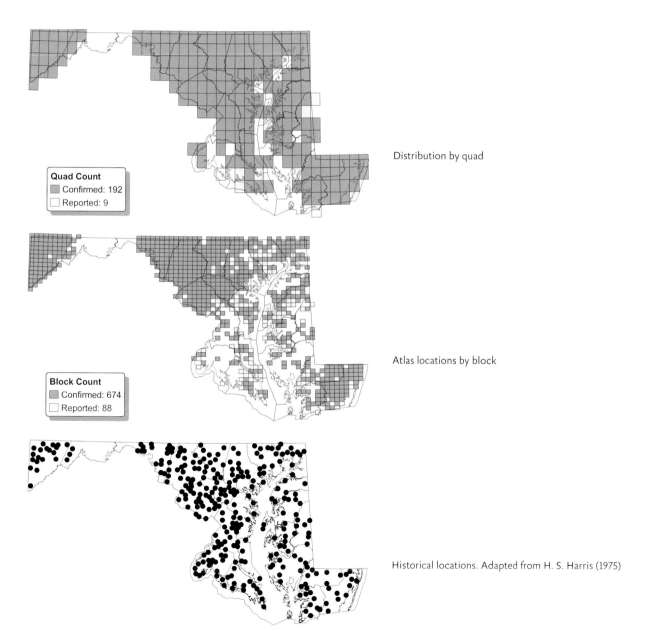

Quad Count
Confirmed: 192
Reported: 9

Distribution by quad

Block Count
Confirmed: 674
Reported: 88

Atlas locations by block

Historical locations. Adapted from H. S. Harris (1975)

Photograph by Andrew Adams

Northern Slimy Salamander
Plethodon glutinosus

The common name of this species, Northern Slimy Salamander, is a misnomer. "Northern Sticky Salamander" would more accurately describe this large, woodland salamander. In fact, the specific epithet, *glutinosus*, is Latin for "gluey or full of glue." *Plethodon* is derived from the Greek *plethore*, meaning "fullness or full of," and *odon*, meaning "teeth." While Northern Slimy Salamanders do have a plethora of teeth, they are more famous for their adhesive skin secretions.

The Northern Slimy Salamander is bluish black or black with numerous white, yellow, or brassy dorsal spots and white lateral flecking. The tail is relatively long and paler in color than the trunk (B. T. Miller and Niemiller 2011). The unspotted venter is generally lighter in color than the dorsum (Powell et al. 2016), and white spotting may be present on the chin and throat. Adult males have a large, circular mental gland on the underside of the chin (Bishop 1941; Highton 1956). Adult Slimy Salamanders attain a total length of 12.1-17.2 cm.

The Northern Slimy Salamander is one of 13 morphologically similar but genetically distinct species of slimy salamanders extensively distributed throughout the eastern and southeastern United States. Collectively, they comprise the *Plethodon glutinosus* species complex (Highton et al. 1989). The Northern Slimy Salamander is distributed from central New York and southern Illinois south to west-central Alabama and central Georgia. A disjunct population is found in New Hampshire. This species is largely absent from the Atlantic Coastal Plain. In Maryland, Slimy Salamanders are

distributed west of the Fall Line through the Piedmont, Blue Ridge, Ridge and Valley, and Appalachian plateaus, but are absent from the Great Valley in Washington County (J. A. Fowler 1944; Marsiglia 1950; H. S. Harris 1975). Hybrid populations of Northern Slimy Salamander and White-spotted Slimy Salamander (*Plethodon cylindraceus*), as recognized by genetic analysis, have been reported north of the Potomac River in Washington County (Highton and Peabody 2000).

Northern Slimy Salamanders inhabit mature hardwood and mixed forests, moist ravines, shaded hillsides, rock crevices, cave entrances, mines, and shale banks (Petranka 1998; Dodd et al. 2001; Mitchell and Gibbons 2010). Moist microhabitat under rotting logs, fallen branches, leaf litter, and rocks is ideal for these salamanders. While not adept at excavating their own burrows, Northern Slimy Salamanders will commandeer and widen existing root tracks, earthworm passages, and other natural openings for their own use (Wells 2007). They have also been documented as using hollow portions of standing trees and crevices under bark as refugia (W. H. Smith et al. 2011). In Maryland, Cunningham Falls State Park in Frederick County is an example of model habitat for this species (Prince 1957). One study estimated a density of 5,600-8,400 Northern Slimy Salamanders per hectare (Semlitsch 1980).

Females reproduce biennially (Highton 1962a; Semlitsch 1980). Courtship activity occurs in September to October, with males using chemical, tactile, and visual cues to locate and entice females (Highton 1962a; Petranka 1998). Fertilization is internal, and eggs undergo direct development. Oviposition occurs in May to June (Highton 1962a; H. S. Harris 1975). Clutch size in Slimy Salamanders is quite variable. In Frederick County, clutches containing 16-34 eggs were documented (Highton 1962a). Eggs are typically suspended from the ceiling of a small cavity. Few nests of the Northern Slimy Salamander have been found in nature, but females most likely use subsurface burrows and cavities, logs, crevices in bluffs, or caves (Petranka 1998; Cunningham and Smith 2010; Mitchell and Gibbons 2010). Females attend the eggs through development, which lasts 2-3 months, and guard the young until they disperse from the nest approximately 2-3 weeks later. Juveniles do not appear on the forest floor until 7-8 months of age (Semlitsch 1980). In Maryland, Northern Slimy Salamanders reach sexual maturity at 4 years of age. Longevity in the wild is estimated to exceed 6 years (Beamer and Lannoo 2005a), but 20 years has been documented in captivity (Snider and Bowler 1992).

Northern Slimy Salamanders are opportunistic, generalist predators of various forest-floor invertebrates (Mitchell and Gibbons 2010). Although these salamanders may forage on the forest floor at night or during warm, rainy periods, they more commonly employ a sit-and-wait strategy. Generally, Slimy Salamanders remain in their burrows and ambush prey that pass by the burrow entrance. They are active at the surface most of the year but do not normally stray far from their burrows, to which they exhibit site fidelity (Petranka 1998). Northern Slimy Salamanders will defend their burrows and territories from interlopers and engage in competitive interactions with other Slimy Salamanders and congeners (Thurow 1976; Bailey 1992; Marvin 1998; Wells 2007; Cunningham et al. 2009). They readily retreat to their burrow to escape predation (Petranka 1998). If escape fails and an attack ensues, they ooze a milky-white, sticky secretion from their skin. This secretion is unpalatable to predators and serves as a defense mechanism by adhering to the face, feathers, or fur of predators (Brodie et al. 1979; Dodd 2004).

During the atlas project, Northern Slimy Salamanders were documented in 55 quads and 140 blocks. The majority of records were from Garrett County. Overall, the distribution was similar to that described by H. S. Harris (1975), although no Northern Slimy Salamanders were recorded from Montgomery County. Atlas efforts confirmed the species in

Howard County, from which Harris (1975) mapped a single, unsubstantiated report.

Across much of the species range, the Northern Slimy Salamander is considered stable (IUCN 2014); however, troubling and unexplained declines have been documented in Great Smoky Mountains National Park in Tennessee (Caruso and Lips 2013). In Maryland, the species is considered secure. Although no localities in Montgomery County were documented during the atlas project, this may reflect a lack of targeted efforts. Threats to Northern Slimy Salamanders include habitat loss and degradation, climate change, invasive species, and emerging disease. Protection of mature, forested upland habitats will greatly contribute to the long-term presence of the Northern Slimy Salamander in Maryland.

HEATHER R. CUNNINGHAM

Distribution by quad

Quad Count
Confirmed: 52
Reported: 3

Atlas locations by block

Block Count
Confirmed: 120
Reported: 20

Historical locations. Adapted from H. S. Harris (1975)

● Specimen Examined
◑ Specimen Reported

Photograph by Michael Kirby

Valley and Ridge Salamander
Plethodon hoffmani

Within Maryland, the Valley and Ridge Salamander, as the name implies, occurs primarily on the Ridge and Valley Province of western Maryland. This small woodland salamander did not receive species status until the early 1970s, having been recognized previously as a distinct population of the Ravine Salamander (*Plethodon richmondi*) (Highton 1962b, 1972). The Valley and Ridge Salamander is similar in appearance to the "lead-back" morph of the Eastern Red-backed Salamander. The dorsal color is brown to black with scattered small, white specks and abundant brassy flecking (Highton 1972, 1986). Larger, white spots may occur on the sides. The belly is predominantly dark with white spotting, which contrasts with a predominantly white chin (Highton 1972, 1986; Conant and Collins 1998; Petranka 1998). Adults typically range from 9.1 cm to 11.2 cm in total length and have an average of 20–21 costal grooves.

The Valley and Ridge Salamander ranges throughout the Ridge and Valley Province from central Pennsylvania southwest to the New River in Virginia and West Virginia (Highton 1986). Its range also encroaches westward onto the Appalachian Plateaus and eastward to the Blue Ridge Province in the southern part of its range in Botetourt County, Virginia (Highton 1986). In Maryland, Valley and Ridge Salamanders have previously been documented on the Ridge and Valley and eastern edge of the Appalachian Plateaus from Hancock in western Washington County to Big Savage Mountain in eastern Garrett County (H. S. Harris 1975). The first Maryland specimen of this species was collected

from Allegany County in 1937, but was considered the Ravine Salamander at that time (Stine and Fowler 1956).

Valley and Ridge Salamanders prefer mature, deciduous forest on well-drained soils (Petranka 1998; Highton 1999). In West Virginia, habitat was described as hillside slopes of mixed deciduous forest with flat stones (N. B. Green and Pauley 1987). Conant and Collins (1998) described the species as living beneath flat stones on forested hill slopes up to 4,600 feet in elevation. Valley and Ridge Salamanders are active at the surface primarily in spring and fall; they remain underground during hot, dry conditions in summer and during the coldest winter months (Fraser 1976a; Petranka 1998). According to Fraser (1976a), only a small proportion of Valley and Ridge Salamanders in a population will occur above ground even under ideal conditions. These salamanders remain under rocks and logs during the day and forage for small invertebrates at night (Petranka 1998). Valley and Ridge Salamanders eat an assortment of small invertebrates, including worms, snails, millipedes, centipedes, spiders, and insects (Fraser 1976a, 1976b; Petranka 1998).

Valley and Ridge Salamanders typically mate in the spring (Petranka 1998), with egg laying in Maryland occurring in late May and early June (Angle 1969). Angle (1969) postulated that breeding probably takes place on a biennial cycle similar to that reported by Sayler (1966) for the Eastern Red-backed Salamander. Courtship in the Valley and Ridge Salamander has not been specifically described, but Petranka (1998) suggests it is likely to include a tail-straddle walk similar to that of other *Plethodon* species. Egg laying is assumed to occur underground (Petranka 1998; Beamer and Lannoo 2005b), with clutch size ranging from 3 to 8 eggs, based on dissections of mature females collected in Maryland and Pennsylvania (Angle 1969). Incubation lasts about 2 months, and eggs undergo direct development. Hatching occurs in late August and early September (Angle 1969). Females reach sexual maturity in their third spring, while males appear to be sexually mature by the second fall (Angle 1969).

During the atlas project, Valley and Ridge Salamanders were recorded in only 31 blocks within 15 quads in Allegany and Garrett counties. They were found from Sideling Hill along the border of Allegany and Washington County in the east to just west of Backbone Mountain in Garrett County in the west. These locations generally fall within the historical range of the species as reported by H. S. Harris (1975), including a few records on the Appalachian Plateaus. However, targeted searches for this species in the Hancock area produced no Valley and Ridge Salamanders (Wayne Hildebrand, pers. comm.), even though Harris (1975) reported the species extending east to approximately Hancock, Maryland.

Valley and Ridge Salamanders continue to occupy suitable sites within the drier habitats of the Ridge and Valley

Province in Maryland. Large portions of this range occur on state-owned lands (e.g., Green Ridge State Forest) that afford some habitat protection. Logging of forested habitat could lead to local extirpations of the species, but populations would be expected to rebound following managed timber harvests (Petranka 1998). More permanent extirpations would be expected if forest were to be converted to agricultural or urban uses (Petranka 1998). Habitat conversion in the Hancock area over past decades may explain the absence of the species from these historical sites.

DAVID R. SMITH

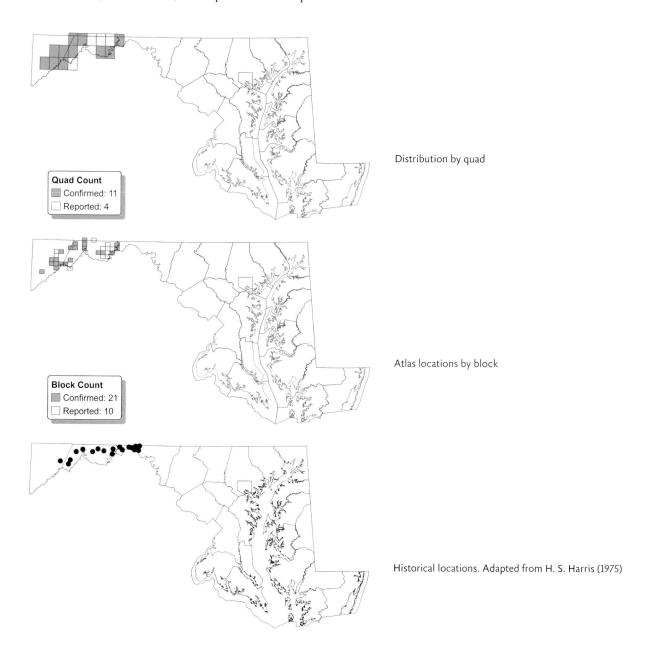

Distribution by quad

Quad Count
Confirmed: 11
Reported: 4

Atlas locations by block

Block Count
Confirmed: 21
Reported: 10

Historical locations. Adapted from H. S. Harris (1975)

Photograph by Ed Thompson

Wehrle's Salamander
Plethodon wehrlei

Wehrle's Salamander is named for Richard W. Wehrle, a naturalist from Indiana, Pennsylvania, who collected the individuals from which the species was first described. This overall dark gray to black salamander exhibits a good bit of variation throughout its geographic range. In Maryland, these salamanders typically have a dark gray to bluish black ground color on the dorsal surface, with white to cream-colored spots and flecks confined to their sides. This spotting usually extends from the vicinity of the hind legs to the side of the jaws. The density of these whitish markings is moderate but may be reduced or absent on some individuals. The belly is slate gray, with a lighter, often whitish throat. The hind feet are more noticeably webbed than in other *Plethodon* species of the region. The eyes are very prominent. This species has a more slender build than the Northern Slimy Salamander, with which it may be confused. A medium-sized plethodontid, adults range from 10 cm to 13.3 cm in total length.

Wehrle's Salamander is mostly restricted to the Appalachian Plateaus region from extreme southwestern New York through west-central Pennsylvania, through most of West Virginia, south to west-central Virginia, and just into adjacent North Carolina. An isolated population occurs in southeastern Kentucky and adjacent northeastern Tennessee. In Maryland, Wehrle's Salamanders occur only on the Appalachian Plateaus Province of Garrett County and extreme western Allegany County. They were first discovered in Garrett County in 1977 (Thompson and Chapman 1978). This was not a surprising find, as H. S. Harris (1975) included this species as a possible addition to the Maryland herpe-

tofauna, given that the known range at that time extended around Garrett County from Pennsylvania to West Virginia. Over the years, at least nine sites have been documented for this salamander in western Maryland (Thompson 1984b; E. Thompson, unpubl. data).

This salamander is a forest dweller. As with most woodland salamanders, the older the forest is, the better it is for them. For Wehrle's Salamander, mature forests with an abundance of rock ledges or flat rocks appear to be ideal habitat (M. Brooks 1945; Hall and Stafford 1972). In West Virginia, N. B. Green and Pauley (1987) reported that these salamanders are found in mature forests at practically any elevation on the Appalachian Plateaus up to 1,463 m, basically the highest elevations in that state. At high-elevation sites, they are found under rocks and logs in mixed red spruce and yellow birch forests (Petranka 1998). They can also be found in the twilight zone of caves (N. B. Green and Pauley 1987) and in deep rock crevices (Conant and Collins 1998). Indeed, all the documented sites in Maryland, with two exceptions, have been found while searching the crevices of rock outcrops with a flashlight.

The various aspects of the reproductive cycle have been little studied in this secretive species. Most breeding occurred in March and April in a West Virginia population (Pauley and England 1969). However, Hall and Stafford (1972) reported breeding in September and October in northern Pennsylvania. Eggs have rarely been found. The only report of a nesting record is for a female found brooding a clutch of 6 eggs in mid-August in a small cavity within a Virginia cave (J. A. Fowler 1952). As in other species of the genus *Plethodon*, eggs undergo direct development. Hall and Stafford (1972) found that males begin maturing sexually during their third year and first breed in their fourth year. Females do not breed until their fourth or fifth year of life.

During the atlas period, Wehrle's Salamander was found in only two blocks within two quads, one from a known site in western Allegany County and one from a new site in Garrett County. It is the author's (Thompson) opinion that this paucity of records during the MARA study was due to the rarity and very secretive nature of the species rather than a population decline. Documenting the sites known prior to the atlas project took many years of fieldwork.

Wehrle's Salamander is a rarely encountered, secretive species. It is listed as in need of conservation in Maryland (MDNR 2016b). Four of the known locations occur in protected habitat on state forest and state park lands. Other areas of potential habitat also occur on public land. Some variation exists in the habitat of the known locations. This fact, coupled with the known locations being rather scattered throughout the Appalachian Plateaus of Maryland, might mean that this salamander is more prevalent than records indicate. A good amount of potential habitat exists on the ridges of the Appalachian Plateaus. The species' preference for rocky ridge tops has its down side, however. The construction of wind power facilities on mountain ridges is a serious threat to Wehrle's Salamanders, as this land-use practice causes deforestation. Range-wide, deforestation is perhaps the biggest threat to woodland salamanders such as Wehrle's (Mitchell et al. 1999).

Further fieldwork is needed to document new sites for Wehrle's Salamander because additional populations are likely to exist in Maryland. Targeting potential habitat on the state's publicly owned lands is particularly important. As it is, several of the known sites on protected lands should support this secretive amphibian for a long time to come.

ED THOMPSON

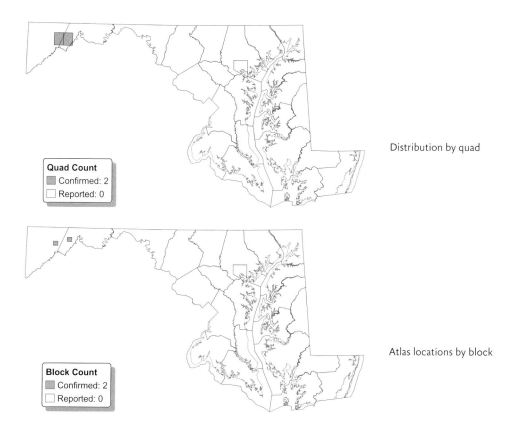

Quad Count
Confirmed: 2
Reported: 0

Distribution by quad

Block Count
Confirmed: 2
Reported: 0

Atlas locations by block

Photograph by Nathan H. Nazdrowicz

Mud Salamander
Pseudotriton montanus

The Mud Salamander may be one of the most elusive salamanders in Maryland. It was formerly referred to as the Rare Red Salamander (Mansueti 1941a) and Baird's Red Salamander (J. E. Cooper 1953). The Mud Salamander has a reddish body with small, round, black spots somewhat evenly distributed across the dorsal and lateral surfaces. Adults darken with age, and the dorsal spots become less conspicuous (Petranka 1998). The irises are brown and difficult to distinguish from the pupils, giving the entire eye a dark appearance. This is a medium-sized, moderately stout salamander with an adult total length of 7.5-16.5 cm.

The Mud Salamander is distributed along the East Coast from southern New Jersey to northern Florida and west to Louisiana, with a discontinuous range in the western lowlands of the Appalachians in West Virginia, Tennessee, Kentucky, and Ohio. Despite their wide geographic range, these salamanders are uncommon throughout the southeastern United States (Mitchell and Gibbons 2010). In the southern portion of their range, they occur on the Atlantic Coastal Plain and Piedmont; unlike the type specimen, however, which was collected near South Mountain in south-central Pennsylvania (McCoy 1992), they are found predominantly on the Atlantic Coastal Plain in the northern portion of their range (Petranka 1998). Maryland is at the northern periphery of this species' range (Petranka 1998), with only a few records from New Jersey (Conant 1957), Pennsylvania (Hulse et al. 2001), and Delaware (Heckscher 1995). The first known specimen of the Mud Salamander in Maryland was collected in 1926 from Branchville, Prince George's County (R. W. Miller 2011). Mansueti (1941a) published the first dis-

tributional note on this species in Maryland, followed by J. A. Fowler (1941), though some of those records were later determined to be the Red Salamander (R. W. Miller 2011). H. S. Harris (1975) documented the Mud Salamander's distribution in Maryland as occurring throughout the Atlantic Coastal Plain, with more populations on the Western Shore. R. W. Miller (2011) questioned the validity of some records that formed the basis of Harris's distributional map, but removal of those records did not change the overall distribution in Maryland. Mud Salamanders have proved extremely rare on the Delmarva Peninsula (White and White 2007), with only nine localities known from Maryland's Eastern Shore (Heckscher 1995) before the start of the atlas project.

Mud Salamanders inhabit muddy or mucky habitats in or along the margins of swamps, springs, floodplain forests, and small headwater streams (Petranka 1998). Saturated areas with sphagnum moss and a decaying leaf layer in hardwood forests are favorable habitats (Mitchell and Gibbons 2010). These salamanders usually hide in burrows or beneath dead leaves, logs, or sphagnum (Mitchell and Gibbons 2010). In addition to finding them under debris or moss, they may be found by raking the top layer of mud and leaves and exposing them in their burrows (Heckscher 1995).

Due to their scarcity and secretive nature, few reports are available on the reproduction of Mud Salamanders. Mating occurs in late summer and fall, and females lay eggs during the fall and early winter (Petranka 1998). Presumably, females choose subsurface cavities in or near water to lay their eggs. J. A. Fowler (1946) found eggs of this species in late December 1946 in a seep at the base of a hillside in Maryland, and Frank Hirst and Arnold Norden found a clutch of well-developed eggs in early May 1994 in Worcester County (Heckscher 1995). Females lay approximately 75-190 eggs, depending on body size (Bruce 1975). Eggs hatch in late winter or spring (Mitchell and Gibbons 2010), and larvae live beneath leaf litter and aquatic vegetation in seepages, muddy streams, or floodplain wetlands (Petranka 1998). The larval period lasts more than a year, with most larvae metamorphosing after 15-17 months (Petranka 1998).

During the MARA project, Mud Salamanders were confirmed in only four quads and six blocks: five in Anne Arundel County and one in Prince George's County. None were found in Baltimore, Calvert, Charles, Montgomery, or St. Mary's counties or on the Eastern Shore, including Worcester County, where one was last found in 1995 (Foley and Smith 1999). Subsequent to the atlas period, an old adult Mud Salamander was found near Indian Head, Charles County, by Seth Berry (pers. comm.) in late December 2015.

These results suggest that the Mud Salamander's distribution in Maryland may have been significantly reduced from its historical range. An MDNR survey of 46 suitable sites for Mud Salamanders in 2006 documented a similar trend (Chalmers 2006). R. W. Miller (2014) questioned the conclusions of this study, believing this species should occur anywhere on Maryland's Atlantic Coastal Plain where suitable habitat occurs. Because this salamander is fossorial and its habitat is difficult to access, it possibly went undetected in some atlas blocks. None of the records during the atlas project were from historical locations, and the species was found at only one of the locations where Mud Salamanders were found during the MDNR survey in 2006. Surveys of suitable habitat in Delaware during 1993-1994 did not document any additional sites (Heckscher 1995), suggesting that this species may be declining there also.

The Mud Salamander warrants further studies to determine its population status in Maryland. Degradation of nontidal wetlands may have contributed to the observed reduction in its distribution. Although this species most likely occurs in more sites in Maryland, the scarcity of records suggests that there may be serious conservation needs to keep the Mud Salamander a viable component of Maryland's biodiversity. Protection of suitable habitat is one such conservation measure that is essential to sustaining this species.

GLENN D. THERRES

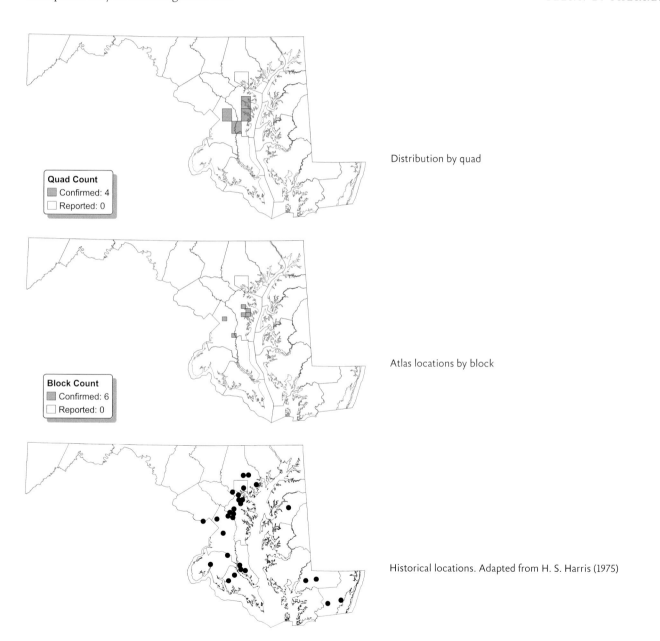

Distribution by quad

Atlas locations by block

Historical locations. Adapted from H. S. Harris (1975)

Photograph by Matthew Kirby

Red Salamander
Pseudotriton ruber

The Red Salamander is one of two large-bodied, reddish salamanders in Maryland. It is similar in size and appearance to the secretive Mud Salamander: both have a red body with black spots. However, the Red Salamander has yellow irises instead of brown. The spots on its body are roundish to irregularly shaped and some may fuse together. Generally, the spots are also more numerous and densely concentrated dorsally than those of the Mud Salamander. Juveniles and young adults can be particularly bright, ranging from orange to coral red (White and White 2007). Adults darken with age, becoming dull orange to purplish brown, and the spots often become less distinct (Petranka 1998). Adult Red Salamanders range from 7 cm to 15.2 cm in total length.

The Red Salamander ranges from southern New York, Pennsylvania, and western Ohio to the Gulf Coast of Mississippi, Alabama, and the panhandle of Florida. It is absent from much of the Atlantic Coastal Plain from the southern Delmarva Peninsula through Florida. In Maryland, the species ranges widely, being found on all physiographic provinces of the state. Historically, Red Salamanders have been documented in all counties with the exception of those of the lower Eastern Shore, including Dorchester, Somerset, Wicomico, and Worcester counties, where they are presumed absent (H. S. Harris 1975). The southernmost record on the Eastern Shore was reported from northern Talbot County, near the Queen Anne's county line.

Throughout the year, Red Salamanders occupy aquatic habitats such as small headwater streams, seeps, and spring-fed bogs, but they may also move to adjacent terrestrial habitats in spring and early summer (Bruce 1978). Adults and juveniles can be found beneath accumulated leaf litter, rocks, logs, or other cover objects in or adjacent to their aquatic habitats (Petranka 1998). On rainy nights, they may be observed crossing roads (White and White 2007).

Like many salamanders that occupy spring seep habitats, Red Salamanders are active throughout the year. Females retreat to subsurface cavities during the fall and early winter to oviposit and guard their eggs. In Maryland, courtship occurs in the spring and fall, and females can store sperm until oviposition (Petranka 1998). Females lay 30-130 eggs, depending on body size, within seepages, springs, or streambanks; the incubation period lasts 2-3 months (Petranka 1998). Larvae remain sheltered within these aquatic habitats beneath rocks, mud, decaying leaf litter, and vegetation, where they feed on worms, insect larvae, and other invertebrates (Petranka 1998). The larval period is prolonged and variable, lasting 1.5-3.5 years, with higher-altitude populations having a longer larval period than Atlantic Coastal Plain populations (Petranka 1998). Sexual maturity is reached at 4-5 years of age (Bruce 1972a).

The Red Salamander has been postulated to be a mimic of the toxic red eft stage of the Eastern Newt as a means of protection from predators (R. R. Howard and Brodie 1973). This form of mimicry, known as Batesian mimicry, suggests that the Red Salamander is nontoxic and gains protection from predators by mimicking the toxic newt. However, other studies have shown that the Red Salamander is at least noxious to some potential predators (Brandon et al. 1979) and that the species is better described as a Müllerian mimic, occurring with the Eastern Newt along a spectrum of unpalatability (Brandon and Huheey 1981).

During the MARA project, Red Salamanders were recorded in 207 blocks within 92 quads. They were widely distributed on the Western Shore of the Atlantic Coastal Plain west through the Piedmont, Blue Ridge, Ridge and Valley, and Appalachian Plateaus provinces. The atlas distribution of the Red Salamander was similar to the historical distribution in Maryland (H. S. Harris 1975), with two exceptions: MARA contributors documented fewer blocks on the outskirts of Washington, DC, and none on the Atlantic Coastal Plain of the Eastern Shore. However, Red Salamanders can be difficult to find because of their tendency to remain buried in mucky substrates within springs and along streams with poor access. Before the atlas period, in 2005, two Red Salamanders were captured during a trapping survey for the Eastern Pinesnake (*Pituophis melanoleucus*) in Idylwild Wildlife Management Area in southeastern Caroline County (Scott Smith, pers. comm.). Suitable habitat remains within this area and elsewhere in Cecil and Kent counties, and the absence of records during the atlas period may be attributable more to insufficient survey effort than a decline in the species (Scott Smith and Nathan Nazdrowicz, pers. comm.).

Overall, the Red Salamander population within Maryland appears stable. Laws passed to protect wetlands and streams have benefited the species by protecting its habitat. Loss of natural upland habitat adjacent to these wetlands and streams, however, as well as increases in impervious surfaces resulting from development, could reduce water quality and cause population declines.

DAVID R. SMITH

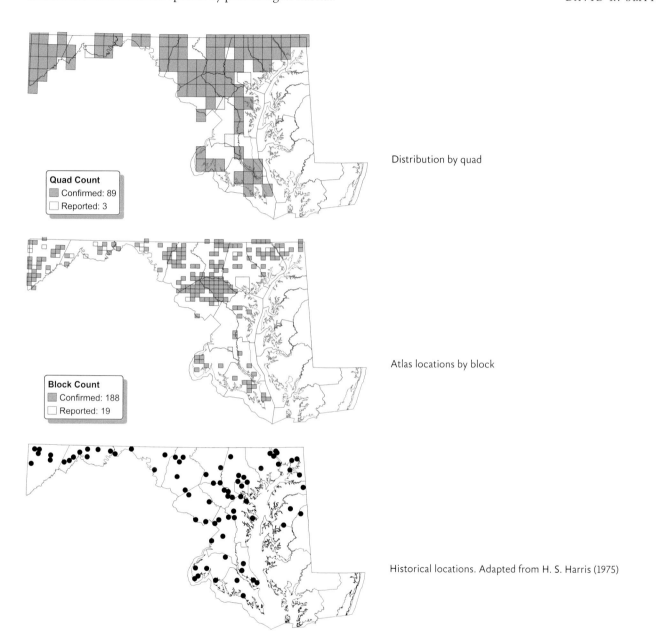

Distribution by quad

Quad Count
Confirmed: 89
Reported: 3

Atlas locations by block

Block Count
Confirmed: 188
Reported: 19

Historical locations. Adapted from H. S. Harris (1975)

Eastern Newt (adult). Photograph by Andrew Adams

Eastern Newt (red eft). Photograph by Wayne G Hildebrand

Eastern Newt
Notophthalmus viridescens

The Eastern Newt is unique among Maryland's salamanders, exhibiting a complex life cycle that involves four distinct life history stages: egg, aquatic larva, terrestrial red eft, and aquatic adult (Petranka 1998). The life cycle is punctuated by two periods of metamorphosis, the first between the larval and red eft (juvenile) stages and the second between the red eft and adult stages (Healy 1970, 1975; Hurlbert 1970a, 1970b). The duration of the life cycle varies regionally and is dependent on elevation and latitude (Hunsinger and Lannoo 2005).

The generic name of the Eastern Newt, *Notophthalmus*, is a compilation of two Greek words: *noto*, meaning "mark," and *ophtalmus*, meaning "eye," in reference to the eye-like, red and black spots on the dorsum. The specific epithet, *viridescens*, is Latin for "slightly green," in reference to the dorsal ground color of adults (Beltz 2006). The venter of the adult Eastern Newt is pale yellow with a stippling of black spots. Despite their size similarities, males typically have larger hind limbs than females, and during the breeding season, the sexes may be differentiated based on male secondary sexual characteristics (cornified toe tips, horny nuptial pads on the inner thighs, a filamentous tail fin, and a swollen cloaca). Adult Eastern Newts of both sexes range from 5.7 cm to 12.2 cm in total length. The skin of the red eft is granular,

bright orange, and adorned with the red and black eyespots present in adults. The skin of all life history stages contains granular glands that secrete tetrodotoxin, but the skin of red efts is 10 times more toxic than that of larvae or adults (Brodie 1968). The red eft's conspicuous orange coloration is aposematic, serving to warn potential predators that it is unpalatable (Brandon and Huheey 1975; Brandon et al. 1979).

The Eastern Newt is the second-most widely distributed salamander in the United States (Hunsinger and Lannoo 2005), ranging over most of eastern North America from the Canadian Maritime Provinces in the Northeast to southern Florida in the southeast, westward across southern Quebec and Ontario, southward through eastern Minnesota to the Gulf Coast of eastern Texas. In Maryland, Eastern Newts are distributed across all physiographic provinces, from the Appalachian Plateaus in the west to the Atlantic Coastal Plain in the east (H. S. Harris 1975; R. W. Miller 1984a; Grogan 1994). H. S. Harris (1975) suggested that the species was absent from the lower Eastern Shore, but subsequent records have been documented for Dorchester, Wicomico, and Worcester counties (R. W. Miller 1984a; Grogan 1985, 1994), and it is also present on the Eastern Shore of Virginia (Mitchell and Reay 1999).

Larval and adult Eastern Newts inhabit permanent and semipermanent ponds, lakes, marshes, and the oxbows of rivers, from near sea level to the Appalachian Highlands (Petranka 1998). Adults have lungs and lack gills. They remain aquatic throughout their reproductive lives but may move into upland habitats to overwinter, during droughts, or to avoid parasites (Hurlbert 1969; Gill 1979). Red efts forage in the leaf litter of the forest floor during the day, especially after summer rains. All life history stages are carnivorous; they consume a diet of size-appropriate invertebrates, as well as amphibian eggs and larvae (Petranka 1998).

Seasonal movements are well documented. Healy (1975) determined that red efts moved an average of 800 m from their natal ponds to upland habitats and maintained mean home ranges of 267 m^2 and 353 m^2 in successive years. Adult Eastern Newts exhibit homing behavior within ponds (R. N. Harris 1981) and between ponds (Gill 1979) following experimental displacement. During orientation and migration, Eastern Newts rely on electromagnetic fields (J. B. Phillips 1987), as well as chemical, visual, and geotactic cues (Hershey and Forester 1980).

In Maryland, the period of courtship and egg deposition is prolonged, occurring from early January to late June (Worthington 1968; Lee 1973d; D. M. Hillis 1976). When a male and female come in contact, the male engages in a lateral display in which he undulates his body and tailfin. If the female is receptive, she nudges the male with her snout, and the male responds by depositing one or more spermatophores (Arnold 1977), which the female picks up with her cloaca. If the female is unreceptive, the male may clasp her with his enlarged hind limbs and increase her receptivity by applying pheromones from genial glands (posterior to his eyes) directly onto her nares (Arnold 1977). Eggs are fertilized internally and clutch size varies from 200 to 375 eggs, depending on the size of the female (Hunsinger and Lannoo 2005). Over a period of days or weeks, eggs are deposited individually on submerged vegetation (Bishop 1941) or, in the absence of vegetation, on the surface of rocks or sticks (Gage 1891). Egg development is temperature dependent, requiring 20-35 days to hatch (Bishop 1941). In Maryland, gilled larvae transform approximately 2 months after hatching (Worthington 1968). Transformation initiates the red eft stage of the life cycle. Using skeletochronology to examine the age structure of Eastern Newts from the Blue Ridge Province of Maryland, Forester and Lykens (1991) determined that

red efts spend 4-7 years on land before entering the aquatic habitat as reproductive adults. Adults in their study were between 4 and 9 years old, but Gill (1985) reported adults as old as 15 years in a Virginia population.

During the MARA project, Eastern Newts were recorded in 89 quads and 198 blocks. The species was not found in Somerset, Wicomico, or Worcester counties, where they may occur in isolated populations not surveyed during the atlas period. Overall, when compared with H. S. Harris's (1975) map, the statewide distribution of the Eastern Newt appears healthy and unchanged. The tendency of adults to successfully occupy diverse aquatic habitats, coupled with the ability of red efts to colonize new locations, indicates that no special conservation efforts are required at this time.

DON C. FORESTER

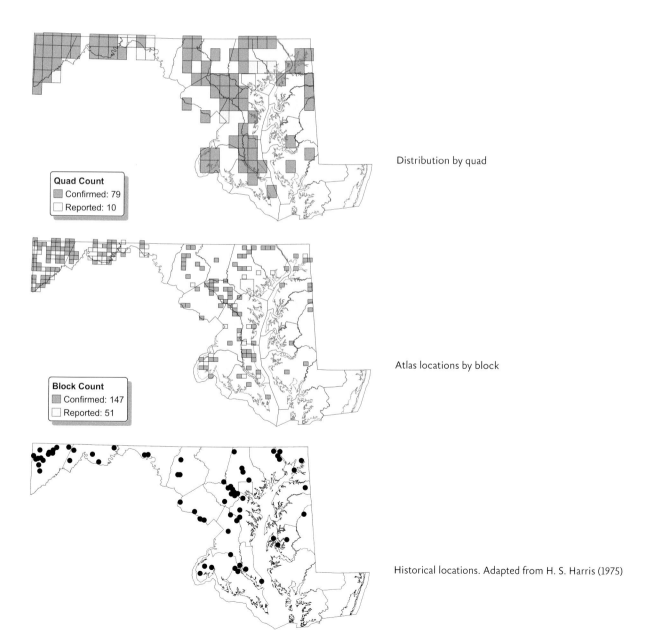

Quad Count
Confirmed: 79
Reported: 10

Distribution by quad

Block Count
Confirmed: 147
Reported: 51

Atlas locations by block

Historical locations. Adapted from H. S. Harris (1975)

Photograph by Ed Thompson

American Toad
Anaxyrus americanus

The American Toad is a terrestrial anuran belonging to the Family Bufonidae (True Toads). The generic name, *Anaxyrus*, is Greek for "king or chief." The specific epithet, *americanus*, is derived from Latin and means "of America" (Jaeger 1972). American Toads are characterized by their stout bodies with shortened limbs, which makes them more suited to walking or hopping than to leaping. The skin is warty and rough, with two prominent, elliptical parotoid glands on the neck, posterior to the eyes. These glands secrete bufotalin, a milky toxin that functions to discourage predators (Müller-Schwarze 2006). A pair of cranial crests project posteriorly from between the eyes. Each crest has a postorbital branch that runs laterally between the eye and the parotoid gland. The branches do not touch the parotoid glands, except sometimes by a short spur. The ground color of the dorsum is variable and may be beige, brown, or brick red. Typically, the dorsum is marked by a light middorsal stripe and a series of dark spots, each containing one or two warts. The venter is cream-colored and mottled anteriorly with dark blotches. During the reproductive season, males may be distinguished from females by a dark throat and prominent nuptial pads on the thumbs. American Toad adults range from 5.1 cm to 9 cm in body length, with a maximum body length of 15.5 cm. Females are often significantly larger than males (Dodd 2013a).

The American Toad is widely distributed across eastern North America, ranging as far north as southern Labrador, west to eastern Manitoba in Canada, and as far south as the Atlantic Coast of southern Virginia, southwest to central Georgia and west to eastern Texas. In Maryland, American Toads occur in all physiographic provinces; however, they have not been documented from the central Eastern Shore in Caroline, Dorchester, Kent, Queen Anne's, and Talbot counties (H. S. Harris 1975; R. W. Miller 1979).

American Toads are nocturnal denizens of rural and suburban landscapes, where they are best described as habitat generalists (Hecnar and M'Closkey 1996; Bonin et al. 1997). They are frequently found in association with old fields (Clawson and Baskett 1982), agricultural landscapes (Knutson et al. 2004), and disturbed habitats such as clear-cuts (Pais et al. 1988). They may also be found in upland deciduous forests, especially along stream corridors and human-made trails (Forester et al. 2006).

American Toads are active throughout the warmer portions of the year, and the duration of their activity period is latitude dependent (Dodd 2013a). Smithberger and Swarth (1993) reported that in Maryland, American Toads are active from late February through October. During the winter months, the toads burrow in the soil below the frost line. They may disperse a linear distance of several kilometers from their breeding sites during the course of a single season (Dodd 2013a). Forester et al. (2006) placed radio transmitters on 16 female American Toads in central Maryland and found that they exhibited linear movements of less than 50 m punctuated by localized activity, during which time they presumably foraged. The authors calculated the average home range size to be 688 m².

These toads are spring breeders throughout their range, but the onset of breeding is temperature dependent (Fairchild 1984; R. D. Howard 1988; Sullivan 1992). In Maryland, reproductive activity begins during mid-March to April, when air temperature rises above 10 °C (Forester and Thompson 1998). Breeding habitat is variable and includes flooded fields, vernal pools, sediment control ponds, and margins of lakes and rivers (Dodd 2013a). Males precede females to the reproductive site, where they establish calling stations. The call is a distinctive trill, varying in length from 4 to 11 seconds (Zweifel 1968) and repeated 0–3 times per minute (R. D. Howard and Young 1998). Males employ three reproductive strategies during the breeding season: active calling, satellite behavior, and terrestrial searching (Forester and Thompson 1998). American Toads occasionally hybridize with Fowler's Toads when the two species occur together (Volpe 1952; Zweifel 1968). Following amplexus, females oviposit two strands of eggs (one from each oviduct). Clutch size is correlated with female body size, ranging from 1,700 to 20,000 eggs (Dodd 2013a). Eggs hatch within 2 weeks into black tadpoles (Wright and Wright 1949). Tadpoles aggregate in the shallows of breeding habitats with members of their own sibling group (Waldman and Adler 1979). Metamorphosis is temperature dependent, with the tadpole stage lasting between 39 and 70 days (Wilbur 1977a; Minton 2001). Age at maturity is 4–5 years (Dorcas and Gibbons 2008).

During the MARA project, American Toads were recorded in 169 quads and 763 blocks. Throughout most of Maryland, the distribution was largely unchanged compared with that reported by H. S. Harris (1975). Lack of records in western Allegany County was likely to be the result of survey effort rather than rarity of this species in the region. Scattered, unconfirmed records were reported from the central Eastern Shore counties, which would represent new locality records for this species. However, due to potential confusion with the Fowler's Toad, which was found to be prevalent in this region during the atlas period, the status of the American Toad in the central Eastern Shore remains in question until verified evidence is documented. This also brings into question the unconfirmed reports from the lower Eastern Shore counties, although a single confirmed record was reported for Worcester County.

The abundance of blocks recorded in the atlas study reveals that this species remains common throughout its historical range. One threat to American Toads may be increased road traffic due to development. Although American Toads can persist in urban areas and were documented in blocks of heavily urbanized areas such as Howard County during the MARA project, breeding pools situated close to roadways can result in significant mortality events as toads migrate to breeding habitats. Urban planning that places stormwater ponds away from roadways may mitigate this problem.

DON C. FORESTER

Distribution by quad

Quad Count
Confirmed: 143
Reported: 26

Atlas locations by block

Block Count
Confirmed: 621
Reported: 142

Historical locations. Adapted from H. S. Harris (1975)

Photograph by Lance H. Benedict

Fowler's Toad
Anaxyrus fowleri

The Fowler's Toad is named in honor of the naturalist Samuel Page Fowler, who described the species in 1843. Like other True Toads (Bufonidae), the Fowler's Toad has dry and warty skin with a prominent parotoid gland behind each eye. The body color varies from brown to gray and is sometimes greenish or brick red, with a narrow, light stripe characteristically down the back. Dark spots on the dorsum typically have three or more small warts. The belly and chest are white and unspotted or with a central dark blotch. The parotoid glands, unlike those of the American Toad, abut the cranial crests, and the femoral warts lack a spiny cornified top. During the breeding season, males can be distinguished from females by a darker throat. Adult Fowler's Toads measure 5.1-7.5 cm in body length, with females typically larger than males (Dodd 2013a).

The Fowler's Toad ranges from eastern New England and the southern Great Lakes region south to the Gulf of Mexico, but is absent from most of Florida and the southeastern Atlantic Coastal Plain. In Maryland, Fowler's Toads are found in all physiographic provinces, although records from the Appalachian Plateaus are unconfirmed (H. S. Harris 1975). They are most abundant on the Atlantic Coastal Plain; populations are more isolated and scattered on the Piedmont, Blue Ridge, and Ridge and Valley provinces (Hildebrand 2005).

Fowler's Toads occur mainly in areas with loose and sandy soils, such as near lake shores, river valleys, coastal dunes, woodlands, and fields (Dodd 2013a). These toads may also be common along roadsides and near homes, around gardens and outdoor lighting, where they feed on attracted insects at night. Fowler's Toads breed in a variety of shallow-water habitats such as flooded fields, ditches, floodplain wetlands, vernal pools, and the margins of ponds.

The breeding season for the Fowler's Toad in Maryland is March through August. The breeding call is a nasal bleat, which is likened to an extended, sheep-like *w-a-a-a-h* and lasts 1-4 seconds (Powell et al. 2016). Males call from the shallow edges of wetlands. Double-stranded egg masses containing up to 8,000 eggs are laid in shallow water, attached to aquatic vegetation (Wright and Wright 1949; Dodd 2013a). Eggs hatch after 2-7 days. The black tadpoles attain a length of about 2.7 cm before transformation at about 21-60 days, depending on water temperature. Both male and female Fowler's Toads reach reproductive maturation at more than 1 year of age. In the wild, Fowler's Toads may live up to 5 years (Kellner and Green 1995).

After their breeding season, adult Fowler's Toads will home to familiar areas over distances up to 1.3 km (Oliver 1955). Orientation occurs through olfactory cues and a sun compass (Landreth and Ferguson 1968; Grubb 1973). Juveniles that have completed metamorphosis establish small home ranges near the shorelines of the breeding habitat (Stille 1952; R. D. Clark 1974). During the summer, Fowler's Toads may become less active when the temperature exceeds 25 °C. Under these conditions, toads will burrow into the ground until conditions improve.

Fowler's Toads feed on a variety of invertebrates, especially beetles and ants. Generally, these toads approach prey at a walk. However, when disturbed or fleeing a predator, Fowler's Toads will hop rapidly (Walton 1988; Walton and Anderson 1988). This switch in locomotion allows the toads to cover up to 37 cm per hop (Rand 1952). The noxious skin secretions of Fowler's Toads repel many potential predators. However, Eastern Hog-nosed Snakes are well-adapted to prey on toads and appear to be immune to their skin secretions.

During the MARA project, Fowler's Toads were documented in 676 blocks within 178 quads. They were distributed almost ubiquitously on the Atlantic Coastal Plain, while records from the Piedmont and Ridge and Valley were mostly associated with rivers, including the Potomac and Monocacy. The MARA results revealed Fowler's Toads were more widely distributed on the Piedmont and Ridge and Valley than depicted by H. S. Harris (1975).

The status of the Fowler's Toad in Maryland appears to be secure, and it remains a commonly observed species, particularly on the Atlantic Coastal Plain. However, roadside calling surveys suggest that this species may be in decline in areas of the Northeast, including Maryland (Weir et al. 2014). Piedmont populations may be at greatest risk due to the isolated nature of their populations and rarity of suitable sandy soils in the landscape.

WAYNE G HILDEBRAND

Distribution by quad

Quad Count
Confirmed: 161
Reported: 17

Atlas locations by block

Block Count
Confirmed: 553
Reported: 123

Historical locations. Adapted from H. S. Harris (1975)

● Specimen Examined
◑ Specimen Reported

Photograph by Nathan H. Nazdrowicz

Eastern Cricket Frog
Acris crepitans

The Eastern Cricket Frog belongs to the Family Hylidae (Treefrogs); however, unlike typical treefrogs, this species is completely terrestrial and lacks the characteristic expanded toe pads used for climbing. The dorsal coloration is variable, with streaks or mottling of black, yellow, green, orange, or red on a background of gray, olive, or brown. The skin is warty, and the belly is white to yellow. A dark, backward-facing triangular spot is present between the eyes, and dark stripes are on the dorsal surface of the thighs. The hind feet are extensively webbed. Male Eastern Cricket Frogs are slightly smaller than females and have a dark yellow vocal sac during the breeding season. The Eastern Cricket Frog is one of Maryland's smallest frogs, with an adult body length of 1.6–3.5 cm.

The Eastern Cricket Frog can be found from southern New York south to the Gulf Coast of Florida but is absent from most of the southeastern Atlantic Coastal Plain from North Carolina through peninsular Florida. The western and northern range limits are the Mississippi and Ohio rivers. The species is absent from the Appalachian Plateaus and parts of the southern Appalachians. In Maryland, the range covers most of the state from the Ridge and Valley through the Atlantic Coastal Plain, although records are lacking from Carroll County (H. S. Harris 1975).

Eastern Cricket Frogs are never far from water that has abundant shoreline and dense emergent vegetation, which is used for cover. They inhabit permanent and seasonal open-canopied wetlands such as Delmarva bays, ponds, ditches, swamps, brackish-water marshes, puddles, and slow-moving streams (White and White 2007). In Maryland,

they are also found in wetlands associated with major river systems such as the Potomac, Susquehanna, and Patuxent rivers. These small frogs often occupy areas along the water's edge and may be observed sitting on floating vegetation and debris or on exposed muddy shores in close proximity to cover (G. R. Smith et al. 2003; Dodd 2013a).

Although Eastern Cricket Frogs tend to remain near wetlands, they will travel between them—generally after periods of rain (Gray et al. 2005). Their strong hind legs allow them to jump up to 0.9 m vertically or horizontally, affording them the ability to cover great distances between suitable habitats (Dorcas and Gibbons 2008). Adults may travel up to 1.3 km between aquatic sites, while juveniles may move more than 91 m from wetlands. Eastern Cricket Frogs do not tolerate freezing. Overwintering occurs in protected terrestrial sites such as under logs or compact dead vegetation, or they may spend winter months buried in the soil (H. Garman 1892; Pope 1944; Walker 1946; Irwin et al. 1999).

Eastern Cricket Frogs breed in Maryland from March through July. The breeding call of the male is a clicking noise like that of a cricket or the sound made when hitting two marbles together: *gick gick gick gick*. The call starts slow but increases in speed, lasting for 20-30 beats (Powell et al. 2016). Eastern Cricket Frogs are frequently heard calling during the day, but also call at night. Small clutches of eggs or single eggs are attached to submerged vegetation or strewn about the bottom of the wetland. A female may lay about 50-400 eggs in a season (Dodd 2013a). The eggs hatch in about 7 days, and transformation from tadpole to frog takes 40-90 days, depending on temperature (Wright and Wright 1949). The tadpole, with its characteristic black-tipped tail, attains a length of 3-5 cm. Sexual maturity occurs at about 1 year for both sexes. Eastern Cricket Frogs are short lived. Gray (1983) documented that few individuals survived through two winters, and average life expectancy is estimated at only 4 months. Complete population turnover may occur in as little as 16 months (Burkett 1984).

Eastern Cricket Frogs are opportunistic feeders. They are known to feed on small invertebrates during the day or night (B. K. Johnson and Christiansen 1976). These tiny frogs are prey to a variety of animals, including fish, bullfrogs, snakes, turtles, birds, and mammals (Dorcas and Gibbons 2008). Generally, Eastern Cricket Frogs rely on their cryptic coloration to avoid predation. When approached, the frogs often jump into the water and later emerge near the shore in the same area.

During the MARA project, Eastern Cricket Frogs were identified in 104 quads and 278 blocks, mostly on the Atlantic Coastal Plain, but with some found on the Piedmont and Ridge and Valley provinces. Piedmont records were not reported from Carroll, Baltimore, Harford, and Cecil counties and were sporadic for Frederick and Montgomery counties. The MARA observations in central and western Maryland were associated with fish hatcheries and areas along the Potomac River. The distribution resulting from the atlas project remained similar to that recorded by H. S. Harris (1975).

Although little change in overall distribution was documented in Maryland, Eastern Cricket Frog populations have experienced catastrophic declines in several northeastern states, including Pennsylvania and New York (Ruhe and Koval 2009; Meshaka and Collins 2010; Kenney and Stearns 2015), where they are listed as state endangered. Roadside chorus surveys also have suggested significant declines, as evidenced by decreased annual occupancy estimates from

2001 to 2011 region-wide in both Maryland and Virginia (Weir et al. 2014). Given these documented declines, particularly in the northern portion of this species' range, populations on the Piedmont and Ridge and Valley in Maryland may benefit from monitoring to detect early signs of decline and the adoption of conservation strategies utilized by other states. For example, in New York, conservation strategies include protection of remaining populations, identification of suitable unoccupied habitat, and management for facilitation of colonization.

WAYNE G HILDEBRAND

Quad Count
Confirmed: 83
Reported: 21

Distribution by quad

Block Count
Confirmed: 194
Reported: 84

Atlas locations by block

Historical locations. Adapted from H. S. Harris (1975)

Photograph by Ed Thompson

Cope's Gray Treefrog
Hyla chrysoscelis
Gray Treefrog
Hyla versicolor

The Gray Treefrog complex is made up of two species that are indistinguishable in appearance. Consequently, both species are treated together in this account. A mistake during the formation of sperm and egg cells in the diploid Cope's Gray Treefrog producing offspring with twice the genetic material (autopolyploidy) has given rise to the tetraploid Gray Treefrog one or more times in the past (Wasserman 1970; Bogart 1980).

Both species vary greatly in color from almost white to greenish, but most are gray with darker blotches on the back. Shades and coloration may change in individuals, depending on the activity of the frog and its environment. A rectangular white marking is present under the eye. The underside of the hind legs is bright yellow to orange in adults, and the tips of the toes are expanded into toe pads typical of treefrogs. The skin of the back is slightly warty, leading to colloquial names such as the Common Tree Toad (Wright and Wright 1949). Both species are of medium size, ranging from 3.2 cm to 6 cm in body length, and females are slightly larger than males. Cope's Gray Treefrogs can be distinguished from Gray Treefrogs based on chromosome number determined by microscopic examination of cells or on analysis of male advertisement calls (Gayou 1984; Matson 1990; Keller and Gerhardt 2001).

Generally, members of the Gray Treefrog complex are found throughout most of eastern North America excluding northern Maine and southern Florida. In the eastern half of this overall range, the Cope's Gray Treefrog is absent from the northeastern United States and has a southerly distribution from southern Ohio east to southern New Jersey and south through northern Florida. The tetraploid Gray Treefrog is present in the Northeast and largely absent from the Southeast. Sympatry occurs in portions of Virginia and Maryland and through most of the western range from Minnesota south to eastern Texas. In Maryland, Cope's Gray Treefrogs are common in eastern and southern Maryland, and Gray Treefrogs are common in central and western Maryland (Otto et al. 2007b). In general, Cope's Gray Treefrogs are found in relatively low elevations with warm, wet environments associated with the shores of the Chesapeake Bay, while Gray Treefrogs occupy the opposite conditions more typical of the central and western portions of the state (Otto et al. 2007b). The species are sympatric in portions of the eastern Piedmont, Prince George's and Anne Arundel counties on the Western Shore, and Caroline, Kent, and Queen Anne's counties on the Eastern Shore.

Members of the Gray Treefrog complex inhabit forests and spend most of their time in trees or on human-made structures such as fences or buildings. During the breeding season, they migrate to small ponds, wetlands, or temporary pools created by heavy rains (White and White 2007). In trees, these frogs are often found in cavities and will readily enter and occupy small pipes placed in trees.

In Maryland, members of the Gray Treefrog complex can be found at breeding ponds from April through early August (Fellers 1979). Individual males may call from perches in trees or other structures in or near the breeding pond during any period of warm temperatures or heavy rain. The call of both species is a short trill, with the Gray Treefrog having a slower call rate than the Cope's Gray Treefrog. In areas where the two species use the same ponds for breeding, differences in advertisement calls are accentuated, and mate choice plays a role in maintaining species boundaries (Gerhardt 1994). Gene flow between the two species rarely, if ever, occurs, and hybrids are sterile or do not survive to reproduce (C. F. Johnson 1963; Gerhardt 1982, 2005). Females deposit eggs individually or in small clusters of 4–40 eggs at the water surface, sometimes attached to stems and leaves of aquatic vegetation or debris. The developmental period lasts 45–60 days, depending on water temperature (Wright 1914). Exposure of tadpoles to invertebrate predators results in development of deeper and more brightly colored tailfins that aid in escaping predators (Van Buskirk and McCollum 2000). Sexes of both species mature at 2 years of age.

During the atlas period, Cope's Gray Treefrogs were

recorded in 207 quads and 693 blocks and Gray Treefrogs in 133 quads and 404 blocks. Both species occurred in 87 blocks, while records from 122 blocks were of unidentified members of the Gray Treefrog complex. Members of this complex occurred in almost all blocks in central and eastern Maryland but were rare in Allegany and Garrett counties. There appears to have been little change in the distribution of the Grey Treefrog complex in Maryland since H. S. Harris's (1975) report, although our understanding of the species' ranges is now greatly improved. During the atlas period, Cope's Gray Treefrog was documented for the first time in southwestern Garrett County.

The distribution of the Gray Treefrog complex has not changed with Maryland's increased urbanization over the past few decades, and both constituent species are known to use human-created habitats for breeding, but these species require forested habitat to complete their life cycle and are susceptible to declines resulting from heavy conversion of forest to human land use (Simon et al. 2009). Preservation of forest in proximity to relatively pollutant-free breeding ponds should allow the persistence of these species in Maryland (Brand and Snodgrass 2010).

JOEL W. SNODGRASS

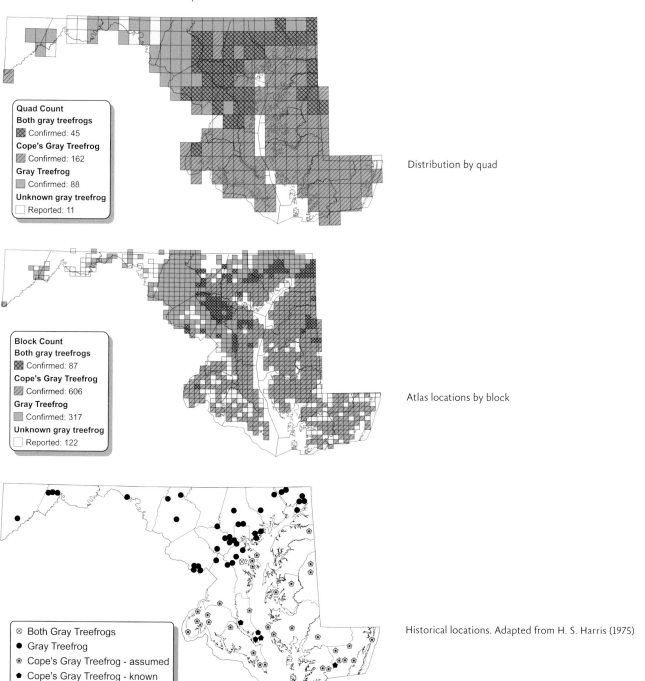

Quad Count
Both gray treefrogs
Confirmed: 45
Cope's Gray Treefrog
Confirmed: 162
Gray Treefrog
Confirmed: 88
Unknown gray treefrog
Reported: 11

Distribution by quad

Block Count
Both gray treefrogs
Confirmed: 87
Cope's Gray Treefrog
Confirmed: 606
Gray Treefrog
Confirmed: 317
Unknown gray treefrog
Reported: 122

Atlas locations by block

⊗ Both Gray Treefrogs
● Gray Treefrog
⊛ Cope's Gray Treefrog - assumed
⬟ Cope's Gray Treefrog - known

Historical locations. Adapted from H. S. Harris (1975)

Photograph by Lance H. Benedict

Green Treefrog
Hyla cinerea

True to its name, the Green Treefrog is typically bright green in color. Dorsal coloration may vary depending on temperature, light, and stress and may also be olive, or brownish to gray, or yellowish (Dodd 2013a). Green Treefrogs become lighter on bright backgrounds and at higher temperatures (King et al. 1994). Small, golden dorsal flecks are present on some individuals. A white or yellowish lateral stripe, occasionally edged in black, extending from the lip to the groin on both sides of the body, is usually present (Conant and Collins 1998), although some individuals lack lateral stripes altogether (Dodd 2013a). At one time, the Green Treefrog was considered two different species based on the presence or absence of this stripe (E. R. Dunn 1937; Wright and Wright 1949; Reed 1956i). The belly is an immaculate white. Adult Green Treefrogs range from 3.2 cm to 5.7 cm in body length.

The Green Treefrog occurs from extreme southern New Jersey southward on the Atlantic Coastal Plain through Florida, throughout the lowlands of the southeastern United States, and westward through the Gulf States to Texas, southeastern Oklahoma, and east-central Arkansas. Its range extends up the Mississippi River as far north as southern Illinois and extreme southwestern Kentucky. In some areas of the Southeast, Green Treefrogs appear to be expanding their range well onto the Piedmont (Dorcas and Gibbons 2008). Maryland is at the northern edge of this species' range. In Maryland, Green Treefrogs occur on both sides of the Chesapeake Bay on the Atlantic Coastal Plain and sections of the eastern Piedmont (H. S. Harris 1975). Harris (1975) reported no records of this species from Caro-

line County. Subsequently, Green Treefrogs were reported in that county in a North American Amphibian Monitoring Program (NAAMP) survey (Hildebrand 2005). A questionable report of a Green Treefrog was also recorded in another NAAMP survey in Howard County (Hildebrand 2005).

Green Treefrogs are found in and around a variety of aquatic habitats, including swamps, ponds, lakes, ditches, and rivers (Dorcas and Gibbons 2008). They prefer wetlands with substantial amounts of emergent vegetation such as cattails (White and White 2007; Dodd 2013a). Unlike most other frogs, Green Treefrogs readily enter brackish water (Conant and Collins 1998) and are commonly associated with tidal areas in Maryland. They also tend to be more willing than other treefrogs to breed in waters inhabited by fish (Dorcas and Gibbons 2008). When not at or near these aquatic habitats during the breeding season, these frogs live in forested areas adjacent to their breeding sites (Dodd 2013a). They also can be found on the sides of nearby houses and other lighted buildings. Green Treefrogs are usually active from mid-spring until late summer. They usually remain hidden during the daytime, perched in trees, or under bark or loose boards on buildings (Dorcas and Gibbons 2008). They become active at dusk to feed or engage in breeding activity.

During the breeding season, Green Treefrogs can be located by listening for loud choruses on warm rainy or humid nights. Calling, which sounds like a nasal *bink bink bink*, begins in earnest at dusk and continues until after midnight (Dodd 2013a). Males call from emergent vegetation such as cattails, reeds, trees, and grasses to attract females (Hildebrand 2005). A chorus usually begins with a few males initiating a calling bout, followed by more and more males joining in until all males are calling and the wetland reverberates in a deafening cacophony of noise (Dodd 2013a). Though Green Treefrogs may call from April through August in Maryland, peak calling is during May and June (1973d). Amplexus is initiated when a female approaches and touches a calling male, and within 4-5 hours, egg deposition occurs (Dodd 2013a). The eggs are attached to floating vegetation in several clusters of about 500 or more eggs (Dorcas and Gibbons 2008). The average clutch size recorded in a study in northwestern Florida was 1,214 eggs (Gunzburger 2006). Eggs hatch in 2-3 days (Garton and Brandon 1975), and the tadpole stage lasts 5-9 weeks, depending on food availability and temperature (Dodd 2013a).

During the atlas period, Green Treefrogs were found in 119 quads and 375 blocks, representing every county on the Atlantic Coastal Plain, including Caroline County. Atlas records were found near the tidal portions of these counties, but also extended well beyond tidal waters. Several Piedmont localities also were documented during the atlas period, including in Baltimore, Carroll, Frederick, Harford, Howard, and Montgomery counties. These records constitute a range expansion for this species in Maryland since

H. S. Harris's (1975) publication. However, the more distant Piedmont records in Frederick and Carroll counties may represent transplanted individuals rather than breeding populations. This was believed to be the case at least for the Carroll County record (David R. Smith, pers. comm.).

The status of the Green Treefrog in Maryland seems secure. The range appears to be expanding, which bodes well for this coastally distributed species, given the potential habitat loss due to sea-level rise. The conservation of aquatic breeding habitats remains essential. Fortunately, Maryland's tidal and nontidal wetland laws provide protection to many of these wetland habitats. Additionally, the Green Treefrog has adapted to human-made ponds, including those with fish, provided there is emergent vegetation in the pond or along its edges.

GLENN D. THERRES

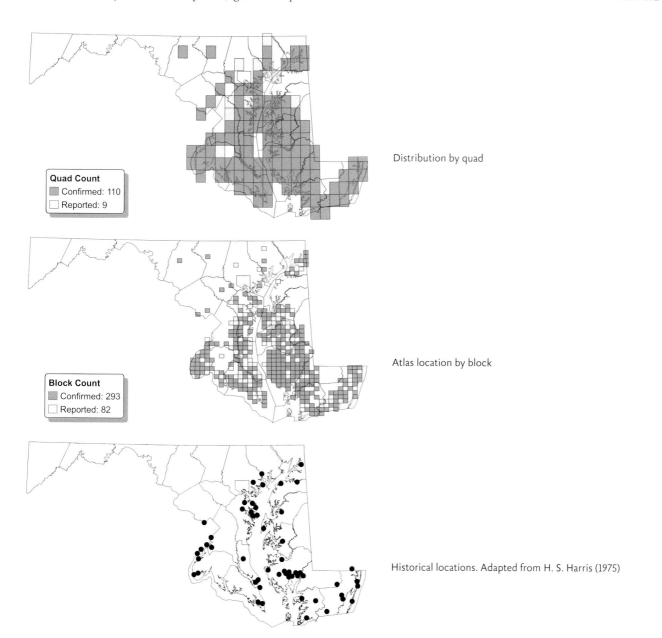

Distribution by quad

Quad Count
Confirmed: 110
Reported: 9

Atlas location by block

Block Count
Confirmed: 293
Reported: 82

Historical locations. Adapted from H. S. Harris (1975)

Photograph by Robert T. Ferguson II

Barking Treefrog
Hyla gratiosa

Maryland's most spectacular frog is the Barking Treefrog. This plump, bright green species is the largest treefrog native to the United States and is primarily a frog of the southeastern part of the country (Dorcas and Gibbons 2008). Body color can vary from light green to brown, depending on body temperature and activity level (White and White 2007). The dorsal surface tends to have a slightly rough appearance and may have dark, circular spots. Under certain conditions, these spots may fade or disappear completely (Dorcas and Gibbons 2008). Small, golden or white dorsal flecks may also be present. A creamy white stripe on the upper lip extends jaggedly down the sides of the body; this stripe may be bordered, in part, by a purplish brown stripe or blotch (White and White 2007). Males have a greenish throat, and females a white throat. The belly is white or cream-colored, occasionally with a tinge of pink, and the groin is yellowish (Dorcas and Gibbons 2008). The Green Treefrog is similar in appearance but is smaller and more slender; it also is smooth skinned and has white side stripes that are straight instead of jagged (White and White 2007). Adult Barking Treefrogs range from 5.1 cm to 6.7 cm in body length.

The Barking Treefrog occurs primarily on the Atlantic Coastal Plain of the eastern United States, from North Carolina throughout much of Florida and west to Louisiana, with isolated populations in Delaware, Kentucky, Maryland, Tennessee, and Virginia. A population found in southern New Jersey in the 1950s was considered likely to have been introduced (Black and Gosner 1958), and until the 1980s, the Barking Treefrog was thought to occur naturally only as

far north as Norfolk, Virginia. This species was not known from Maryland at the time of H. S. Harris's (1975) publication. However, in 1982, an adult female was found in Caroline County by K. Anderson and Dowling (1982); the authors thought the specimen collected may have been a member of a small, relict population or a released captive animal. In 1984, a population of Barking Treefrogs was discovered in southern New Castle County, Delaware (White 1987; Arndt and White 1988). In 1987, choruses of 1–30 individuals were heard calling at the New Castle County vernal pond (Arndt and White 1988). In subsequent years, other apparently isolated populations were located in Maryland in eastern Kent, Queen Anne's, and Caroline counties and in Sussex County, Delaware (White and White 2007). By 1990, Barking Treefrogs were known from four locations in Maryland, and the species was subsequently listed by the MDNR as endangered. The Barking Treefrog populations on the Delmarva Peninsula and in southern New Jersey are now believed to be remnants of a historically larger population and not the result of introductions (White and White 2007).

Barking Treefrogs are found in a variety of Atlantic Coastal Plain woodlands associated with temporary wetlands for breeding (Dodd 2013a). In the southeastern United States, these treefrogs can be abundant in Carolina bays and other temporary wetlands that do not contain fish (Dorcas and Gibbons 2008). In Maryland, they are primarily found in forested areas with open-canopied Delmarva bays. During the breeding season, they move into wetlands from surrounding forest at dusk or after sunset (Dodd 2013a). After the breeding season, the frogs seek shelter in dense forested habitats close to the breeding ponds. Winter months and periods of dry weather are spent underground or in other moist, protected areas (Dorcas and Gibbons 2008).

Barking Treefrogs breed in flooded vernal pools on warm humid nights when nighttime temperatures reach approximately 21 °C, usually during May and June in Maryland (White and White 2007). In Maryland, breeding sites are typically high-quality ephemeral pools (e.g., Delmarva bays) and fishless borrow pits, but during years with high rainfall, these treefrogs will utilize flooded fields. At the breeding pools, males—which greatly outnumber females—attempt to attract mates by calling while floating on the water's surface. Males utter a series of loud, distinctive, doglike barks to attract females (Martof et al. 1980). This loud, hound-like barking advertisement call can be heard almost a mile away (White and White 2007). Male body size is an important determinant in the selection of mates by females (Poole and Murphy 2006). If a female is receptive, amplexus ensues, with the male grasping the female just behind her front legs. The eggs can be laid either singly or in masses, with clutch size averaging about 2,000 eggs (Dodd 2013a). The eggs hatch quickly. The translucent tadpoles have a dark saddle at the base of the tail when young and later develop a dark lateral stripe; metamorphosis occurs after 41-65 days at a length of about 5 cm (Dodd 2013a).

During the atlas project, Barking Treefrogs were found in five blocks within two quads on the Eastern Shore in Caroline and Kent counties. The MARA distribution was similar to the known distribution determined in the 1990s after discovery of this species in Maryland and subsequent surveys.

Given the few populations of Barking Treefrogs in Mary-

land and the species' dependence on large, fishless vernal pools for reproduction, the continued state listing of this species as endangered seems warranted. The protection afforded through the state's Nongame and Endangered Species Conservation Act and the Nontidal Wetlands Protection Act should continue to be employed to help ensure the continued existence of the Barking Treefrog in Maryland.

JAMES F. WHITE, JR., AND GLENN D. THERRES

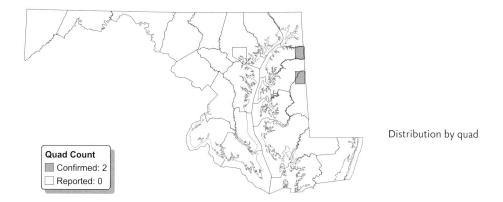

Quad Count
▨ Confirmed: 2
☐ Reported: 0

Distribution by quad

Photograph by Wayne G Hildebrand

Spring Peeper
Pseudacris crucifer

The Spring Peeper is a small frog that is more often heard than seen. Its generic name, *Pseudacris,* is derived from the Greek words *pseudes,* meaning "false," and *akris,* meaning "locust." "False Acris" is a reference to its resemblance to cricket frogs of the genus *Acris.* The specific epithet, *crucifer,* refers to its dorsal markings, and the common name is in reference to the time of year that these frogs typically call. The body coloration is brown, tan, pinkish, gray, or olive with distinctive, dark, diagonal lines that usually cross on the dorsum forming an X pattern. In some individuals, these cross markings may be missing or broken. The belly coloration is yellowish white. The legs have dark banding above and are light beneath and may have light yellow coloration on the underside of the femur. Spring Peepers are treefrogs; however, unlike most other treefrogs, they have toe tips that are only slightly expanded. Male Spring Peepers have a dark vocal sac on the throat, especially prominent during the breeding season. Adult Spring Peepers are 1.9–3.2 cm in body length, and females are slightly larger than males.

The Spring Peeper has a wide distributional range from Canada south to Florida and west to eastern Texas, Iowa, Missouri, and Minnesota. In Maryland, the distribution is statewide (H. S. Harris 1975).

Habitat for Spring Peepers consists of woodlands near wetlands where they breed, including swamps, marshes, vernal pools, flooded fields, ditches, and retention ponds (Dorcas and Gibbons 2008; Dodd 2013a). Though they inhabit a range of forest types, Spring Peepers are most abundant in hardwood and mixed-hardwood forests (Dodd 2013a). They are also able to tolerate areas with moderate urbanization. In developed areas, these frogs may be found in riparian forests (Burbrink et al. 1998). Within woodlands, Spring Peepers are usually found on low vegetation, under logs, or in leaf litter. These small frogs may also use knot holes in trees as refugia (Carr 1940). During late winter through spring, they may be encountered crossing roads near wetlands during warm rains.

Spring Peepers are one of the earliest anurans to breed in Maryland, with Atlantic Coastal Plain populations breeding as early as February. Their breeding season lasts several months, usually ending in May on the Atlantic Coastal Plain. Populations in the western part of the state breed slightly later, from March through July. The breeding call of the Spring Peeper is a single, high-pitched *peep* repeated at approximately 1-second intervals. Up to 25 peeps per minute may be produced (Powell et al. 2016). Large choruses of Spring Peepers sound like sleigh bells from a distance. The males also make an aggressive call that is a trill similar to and often mistaken for that of New Jersey Chorus Frogs and Upland Chorus Frogs. About 200–1,500 eggs are laid individually and are attached to aquatic vegetation or scattered about the bottom of the breeding wetland (Wright and Wright 1949). The eggs take about 5–14 days to hatch. Tadpoles reach a maximum length of 3.5 cm. The tadpole phase of the Spring Peeper lasts 60–120 days. Reproductive age is reached at about 1 year for both males and females (White and White 2007).

After the breeding season, Spring Peepers disperse into forested habitat surrounding the wetlands. Generally, the woodlands that they inhabit are within 100–1,000 m of their breeding site (Herrmann et al. 2005; Eigenbrod et al. 2008). Spring Peepers establish a home range roughly 1.7–5.4 m in diameter around logs, stumps, woody debris, and vegetation (Delzell 1958). Although the home ranges of individuals often overlap, these frogs tend not to interact with each other outside their breeding season. Hibernation occurs in leaf litter or under surface objects. Spring Peepers are capable of surviving freezing temperatures, using glucose as a cryoprotectant (Churchill and Storey 1996). Up to 35% of their body water may be frozen during exposure to freezing temperatures. Spring Peepers have been known to survive up to 2 weeks at below-freezing temperatures (Storey and Storey 1986).

Spring Peepers have color vision (Hailman and Jaeger 1974), suggesting that the majority of feeding occurs during daylight hours. They are generalist predators and tend to forage in the morning and late afternoon (Dodd 2013a). Individuals forage in leaf litter and surface debris (Dodd 2013a) and have also been observed aggregating in bushes along a roadway to feed (McAlister 1963).

During the MARA project, Spring Peepers were documented statewide in 1,135 blocks within 234 quads, more than for any other amphibian or reptile in Maryland. This distribution remained unchanged since H. S. Harris's (1975) publication. Absences reflected in western Maryland may have been due to lack of effort or targeted surveys, whereas gaps in Baltimore City may represent loss of populations.

The Spring Peeper is a common frog over its entire range, except near the boundaries of its distribution (Dodd 2013a).

In Maryland, it is considered common and may be the most abundant frog in the state. Roadside calling surveys suggest that Spring Peepers are stable in some areas of the Northeast, including Maryland, but in decline in others, such as Delaware and Pennsylvania (Weir et al. 2014).

WAYNE G HILDEBRAND

Distribution by quad

Atlas locations by block

Historical locations. Adapted from H. S. Harris (1975)

Photograph by Ed Thompson

Upland Chorus Frog
Pseudacris feriarum

Formerly a member of the Western Chorus Frog (*Pseudacris triseriata*) species complex, the Upland Chorus Frog was elevated to full species status at the recommendation of Hedges (1986), based on the results of starch gel electrophoresis. Platz and Forester (1988) and Platz (1989) supported his recommendation based on properties of the advertisement call and morphometric measurements, and it has more recently been confirmed by extensive DNA analysis (Lemmon et al. 2007; Barrow et al. 2014).

The Upland Chorus Frog is light brown to gray, with three broken to complete longitudinal stripes on the dorsum. A dark lateral stripe runs from the tip of the snout through the eye to the groin. The upper lip is distinctly white, and many individuals have a dark, backward-pointing triangle between the eyes that connects with the middorsal stripe. Breeding males may be distinguished from females by the presence of dark pigmentation on the throat. The venter is white with occasional dark spots on the chest. Adults range from 1.9 cm to 3.5 cm in body length, with females significantly larger than males.

The Upland Chorus Frog is widely distributed from north-central Pennsylvania southward to Georgia, primarily through the Piedmont and Ridge and Valley provinces. Its range dips to the panhandle of Florida along western Georgia in the south, where the species is parapatric with the Southern Chorus Frog (*P. nigrita*). The range extends westward above the Gulf Coastal Plain of Alabama and Mississippi, and northward into Arkansas and Tennessee, where it is parapatric with the Cajun Chorus Frog (*P. fouquettei*) (Barrow et al. 2014). In Maryland, Upland Chorus Frogs are distributed west of the Chesapeake Bay and Susquehanna River to the western edge of the Ridge and Valley Province (H. S. Harris 1975). Historically, most records were from the Atlantic Coastal Plain and eastern Piedmont (H. S. Harris 1975). The species is also widely distributed along the Potomac River, extending north and west into the Ridge and Valley (Forester 1999).

Upland Chorus Frogs inhabit forested areas associated with breeding wetlands such as flooded meadows, forested wetlands, depression wetlands, and marshes (N. B. Green and Pauley 1987; Dodd 2013a). They exhibit a preference for breeding habitats with an open canopy (Albaugh 2008). Following the breeding season, adults and metamorphs disperse up to 200 m into the surrounding forested upland habitat (D. G. Alexander 1965; Kramer 1973). Upland Chorus Frogs are largely terrestrial and virtually never climb trees. Little is known about the home range of this species, but Kramer (1974) tagged adults of the closely related Western Chorus Frog and calculated their mean minimum-area home range to be 2,117 m^2.

In Maryland, Upland Chorus Frogs are among the first anurans to breed each year, in February and early March, and the time of onset of breeding is correlated with elevation: individuals at lower elevations breed earlier than those at higher elevations (Pollio and Kilpatrick 2002). The breeding season usually lasts about one month (Sias 2006) and is over by early June (Hildebrand 2005). Males typically vocalize while sitting in shallow water or while hidden in grassy vegetation (Sias 2006). The male advertisement call is trilled and has been described as a slow *crreeeeek*, similar to the sound created when a finger is dragged over the teeth of a plastic comb (Martof et al. 1980). The length of the call is inversely correlated with temperature, ranging from 1 to 2 seconds in duration (D. Forester, pers. obs.), and includes 14–35 pulses (Dorcas and Gibbons 2008). Calling activity is not restricted to the scotoperiod, and choruses may occur throughout the day (Cope 1889; Dickerson 1906). Amplexed females oviposit 500–1,500 eggs (Wright and Wright 1949), laid in clusters of 40–60 eggs each and attached to submerged vegetation (Martof et al. 1980; N. B. Green and Pauley 1987). Sias (2006) placed two amplexed pairs in a laboratory aquarium and reported that collectively they produced 26 clusters of eggs. Eggs hatch in 3–17 days, and tadpoles grow to 3.4–3.6 cm before metamorphosing, which occurs in 42–91 days, depending on water temperature (Green and Pauley 1987; Pollio and Kilpatrick 2002; Dodd 2013a). Age at maturity is 1 year for both sexes (Dodd 2013a).

Chorus Frogs are size-limited, dietary generalists. Adults eat beetles, grubs, and spiders, while metamorphs primarily consume collembolans and mites (D. G. Alexander 1965; Whitaker 1971). When not foraging, Upland Chorus Frogs rest beneath leaf litter, and in the fall they crawl beneath logs or rocks to overwinter (Green and Pauley 1987). While in torpor, they decrease their respiration, heart rate, and circulation. To avoid freezing, they produce glucose as a cryoprotectant by metabolizing glycogen stores in their liver and concentrating it in their body cells (Storey and Storey 1987).

During the MARA project, Upland Chorus Frogs were recorded in 39 quads and 78 blocks. Although robust populations still persist in southern Maryland and the Ridge and Valley Province, there has been a precipitous decline in this species in the eastern Piedmont since the last comprehensive survey (H. S. Harris 1975).

Regionally, the Upland Chorus Frog has declined in portions of southeastern West Virginia (Sias 2006) and is listed as a species of special concern in Pennsylvania. The decline in Maryland could possibly be attributed to urban sprawl, human-caused degradation, or habitat succession. Additional studies may be warranted to determine the extent and causes of decline before effective conservation measures can be developed.

DON C. FORESTER

Distribution by quad

Quad Count
Confirmed: 18
Reported: 21

Atlas locations by block

Block Count
Confirmed: 24
Reported: 54

Historical locations. Adapted from H. S. Harris (1975)

Photograph by Wayne G Hildebrand

New Jersey Chorus Frog
Pseudacris kalmi

The New Jersey Chorus Frog has the smallest range-wide distribution of any frog found in the northeastern United States. Rarely observed by the general public, this species is much less known than the widespread and congeneric Spring Peeper. Adult dorsal color can be gray, tan, brown, or rust. Three dark, broad, parallel stripes typically run down the back, although some individuals have thin stripes, while others have broken stripes. Individuals with faint or no dorsal stripes also occur. A dark stripe often extends on each side of the head from the nostrils, through the tympanum, and along the flanks to the hind legs. Typically, a distinctly white or light stripe is located on the upper lip that extends to the forelimbs. Individuals may also have a dark, backward-pointing triangle between the eyes that may connect to the middorsal stripe. The New Jersey Chorus Frog has toe pads that aid in climbing but are not as large or well-defined as those of most other treefrogs. This species is virtually identical in appearance to the Upland Chorus Frog, but it has a complementary (nonoverlapping) range. The New Jersey Chorus Frog is diminutive, with adults ranging from 1.9 cm to 3.5 cm in body length.

The global range of the New Jersey Chorus Frog is limited to a very small portion of the Atlantic Coast in the mid-Atlantic United States, extending from Staten Island, New York, south through New Jersey, extreme southeastern Pennsylvania, and to the southern tip of the Delmarva Peninsula. In Maryland, the New Jersey Chorus Frog occurs primarily on the Atlantic Coastal Plain of the Eastern Shore; H. S. Harris (1975) reported one record from the Piedmont in northwestern Cecil County.

The New Jersey Chorus Frog inhabits woodlands and meadows with sandy soils near freshwater nontidal wetlands utilized for breeding, such as seasonal pools, Delmarva bays, ditches, swamps, marshes, swales, and flooded fields. Breeding habitats are typically fishless, and water levels of these habitats draw down significantly during dry months of the year.

The breeding season typically begins in February or March for this species, but males may begin calling in January during mild winters. Males occasionally call into May. The call is best described as the sound made when running a fingernail over the teeth of a comb, but much louder. Females typically enter breeding habitats on warm days or nights from February through March and spend a considerably shorter amount of time in the breeding habitats than do males. The author (Ruhe) observed females entering breeding habitats during daylight hours in Delaware. These females were intercepted by a gauntlet of males, beginning amplexus almost immediately, and several deposited eggs within 20 minutes of entering the breeding habitat. Spent females stayed hidden among floating debris for an additional 60-120 minutes before exiting the breeding habitat and moving into upland leaf litter and debris piles within 10 m of the littoral zone (B. Ruhe, unpubl. data). Egg masses are generally spherical or oval in shape and attached to objects just below the surface of the water to depths of 25 cm. Females deposit several masses containing dozens of eggs. Hatching time and larval development period are unknown but are likely to be similar to those of the closely related Upland Chorus Frog.

New Jersey Chorus Frogs spend the majority of the year in upland habitats. What they do in those habitats is largely unknown as the species is rarely encountered outside the breeding season. Kramer (1973, 1974) studied the movements and home range of the similar Western Chorus Frog (*Pseudacris triseriata*) and found individuals moved up to 200 m from breeding habitats into home ranges of just over 6,000 m^2. Overwintering sites for the New Jersey Chorus Frog are also unknown, and it is not known whether the species is freeze tolerant like several other species of *Pseudacris*.

The MARA project documented the presence of New Jersey Chorus Frogs in 256 blocks within 63 quads. All records were on the Eastern Shore of Maryland, generally similar to those reported by H. S. Harris (1975), except none were documented on the Piedmont. MARA locations revealed that this species is nearly ubiquitous across the region, though gaps in the range exist in southern Cecil, most of Kent, and coastal portions of Dorchester counties, despite concerted effort to document them there (Nathan Nazdrowicz, pers. comm.).

Maryland has no direct conservation efforts specifically targeting the New Jersey Chorus Frog. This species suffers from disturbances to both breeding and nonbreeding habi-

tats and might (like other seasonal pool-breeding amphibians) be quite vulnerable to habitat fragmentation in addition
to outright destruction. Given its very limited global range,
any conservation actions within Maryland are of added sig-
nificance.

BRANDON M. RUHE

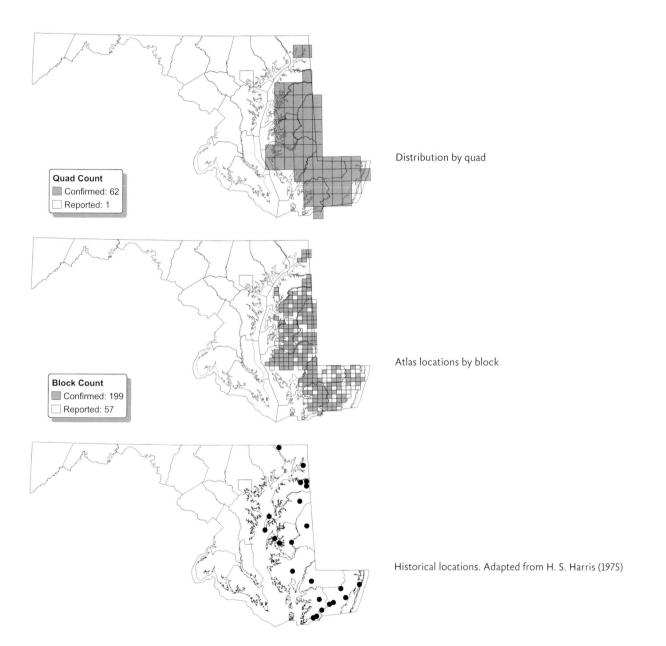

Distribution by quad

Quad Count
Confirmed: 62
Reported: 1

Atlas locations by block

Block Count
Confirmed: 199
Reported: 57

Historical locations. Adapted from H. S. Harris (1975)

Photograph by Andrew Adams

Eastern Narrow-mouthed Toad
Gastrophryne carolinensis

Originally described by John E. Holbrook in 1835, the Eastern Narrow-mouthed Toad was first discovered in Maryland 100 years later in the Cove Point and Solomons Island areas of Calvert County (Noble and Hassler 1936; J. D. Hardy 2009). The well-named Eastern Narrow-mouthed Toad has a distinctively small, pointy-snouted head and disproportionately large, pear-shaped body. Its variable dorsal color ranges from dark or rusty brown to olive, gray, or nearly black, sometimes with broad, lighter-colored stripes flanking the dorsal coloration. The underside is mottled gray. A fold of skin crosses the head behind the eyes, which is thought to shield the eyes from attacking ants and other prey species. Each hind foot has a small spade used for burrowing underground. Males usually have dark throats during the breeding season. Adults reach a body length of 2.2–3.2 cm, and males average slightly smaller than females.

Native to the southeastern United States, the Eastern Narrow-mouthed Toad is distributed throughout the Atlantic Coastal Plain from southern Maryland and northeast Virginia south to Key West, Florida. Its range extends from the Atlantic Coast west to central Texas, eastern Oklahoma, and central Missouri; it is absent from the Appalachian Mountains. Maryland is the northeastern limit of its range, where it is restricted to southern portions of the Atlantic Coastal Plain. Since its discovery in Calvert County, the species was extirpated from the Cove Point area by development and habitat loss in the early 1970s (CREARM 1973). In the 1950s, two new locations were discovered: near

Great Mills, St. Mary's County (J. A. Fowler and Stine 1953) and, the first Eastern Shore record, on Taylor's Island, Dorchester County (Conant 1958b). A second Dorchester County location and a Somerset County record were added in the 1970s (Lee 1973b; H. S. Harris 1975); multiple individuals were found near Nassawango Creek in Worcester County in 1975 (Czarnowsky 1975) and again in 1990 (Grogan 1994). Another St. Mary's County location was added in 1979 (R. W. Miller 1980). Records from Charles County (Hildebrand 2005) and Kent County (J. D. Hardy 2009) have been reported but not verified.

Although Eastern Narrow-mouthed Toads have been reported from a wide variety of habitats throughout their range, in Maryland they are found in forested or open habitats in the vicinity of wetlands, either freshwater or somewhat brackish, or near ant nests. They favor areas with moist, sandy soils with ample ground cover such as dead vegetation, rotting logs, or other debris, often near breeding sites, which include woodland ponds, flooded grassy swales, marshes, temporary vernal pools, and ditches. Wetlands that lack predatory fish are preferred. Like the Eastern Spadefoot, this species is fossorial, digging a burrow or living in burrows of other animals. With only the top of its head near the surface, it ambushes ants, termites, and other prey that wander near (Dodd 2013a).

Breeding usually occurs from late May through July, initiated by heavy precipitation, especially early in the season. When conditions are right, explosive breeding takes place, and an entire localized population breeds in a short period (White and White 2007). Males travel at dusk to shallow (<30 cm) wetlands with cover such as branches, bark, detritus, or vegetation, both for concealment and for stability while they call in a near-vertical position. Their call resembles a nasal, slightly buzzy, bleating sheep sound and lasts for 1–4 seconds (Powell et al. 2016). Females approach stationary calling males. During amplexus, males secrete an adhesive substance from ventral glands to aid the short-limbed species to remain in amplexus for the 1.5–2 hours required for oviposition (Conaway and Metter 1967). A female oviposits 150–1,100 eggs in multiple clusters (P. K. Anderson 1954). Eggs are fertilized as they are laid in floating groups on the water's surface, with about 30 eggs per cluster. Eggs hatch in 1–3 days, and tadpoles metamorphose in 20–70 days (Wright and Wright 1949; CREARM 1973; Martof et al. 1980). Sexual maturity may be reached within 1–2 years, when adults reach 2.1–2.7 cm in body length (Dodd 2013a).

During the MARA project, Eastern Narrow-mouthed Toads were documented by 43 sightings within 16 quads and 35 blocks: 3 blocks in St. Mary's County, 23 blocks in Dorchester County, 1 block in Wicomico County, and 8 blocks in Worcester County. All 35 blocks were verified by photos or recordings. Calls of breeding males on warm, rainy summer nights accounted for most of these detections. This species' limited distribution, restricted to a few counties in the southern Atlantic Coastal Plain, was comparable to its overall historical range (H. S. Harris 1975), although these frogs were documented in many more localities during the atlas project. The historical Calvert County locations were known to be destroyed in the 1970s (CREARM 1973). The locations in Charles and Kent counties reported in the early 2000s were not confirmed during the atlas project, and additional surveys in these areas may be needed.

The Eastern Narrow-mouthed Toad was one of 14 amphibians and reptiles added to Maryland's list of endangered species in 1972, based on the recommendation of the Committee on Rare and Endangered Amphibians and Reptiles of Maryland, an ad hoc committee of the Maryland Herpetological Society (CREARM 1973). These frogs were known from only a handful of locations prior to the atlas study, but are now much more extensively documented in the Pocomoke River drainage of Worcester and Wicomico counties, and especially in the low-lying areas of western Dorchester County. Given that much of southwestern Dorchester County is expected to convert to open water in the next 50–100 years due to subsidence and sea-level rise resulting from climate change, this species may lose the majority of its currently occupied habitat. However, because of the secretive nature of this unusual anuran, additional targeted surveys may still uncover new Maryland populations of the Eastern Narrow-mouthed Toad.

LYNN M. DAVIDSON

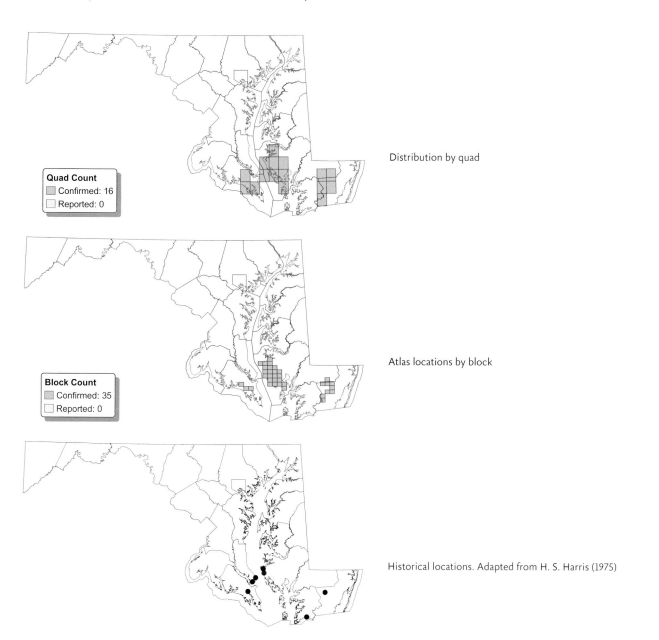

Distribution by quad

Quad Count
Confirmed: 16
Reported: 0

Atlas locations by block

Block Count
Confirmed: 35
Reported: 0

Historical locations. Adapted from H. S. Harris (1975)

Photograph by Robert T. Ferguson II

American Bullfrog
Lithobates catesbeianus

The American Bullfrog's common name refers to the male's breeding call, and the specific epithet, *catesbeianus*, refers to the eighteenth-century naturalist Mark Catesby (Hulse et al. 2001). The coloration of this frog's body is a dull green, brown, or gray. The head is typically greenish, and the belly is whitish or yellowish and mottled with gray. Dorsolateral ridges extending down the sides are absent; instead, a skinfold, which starts at the back of the eye, curves over and around the tympanum. The hind legs may have dark bands, and the hind feet are fully webbed, with the exception of the last section of the fourth toe. Males and females are of nearly equal size and can be distinguished by the male's yellowish throat and a tympanum larger than the eye. In females, the throat is white and the tympanum is smaller than or the same size as the eye. The American Bullfrog is North America's largest frog, with a typical adult body length of 9–15.2 cm.

Due to introductions, the current range of the American Bullfrog extends across much of the continental United States, as well as parts of Asia, Europe, South America, and the Caribbean. Historically, the range-wide distribution was restricted to eastern North America, from southern Canada south to the Gulf Coast and west to the Rocky Mountains. This species remains absent from southern Florida. In Maryland, the range is statewide (H. S. Harris 1975).

American Bullfrogs occupy a wide variety of wetland habitats, although permanent bodies of water such as lakes, marshes, ponds, rivers, and streams are necessary for reproduction (Dodd 2013b). In the eastern United States, American Bullfrogs are frequently associated with forested areas. They favor habitats where forests comprise at least 80% of the area within 1-km radius of their breeding site; outside their breeding season, individuals may remain along the shorelines of breeding wetlands or disperse far distances to other wetland habitats (Dodd 2013b). Adult and juvenile American Bullfrogs tend to favor the same type of habitat, although habitat partitioning by life stages may occur (Dodd 2013b).

American Bullfrogs are generally active in spring through autumn. Individuals emerge from overwintering sites when water temperature reaches 14 °C, typically in April in Maryland (Dodd 2013b). Peak activity occurs during the spring, with terrestrial movements coinciding with night and rainfall (Dodd 2013b). American Bullfrogs do not exhibit freeze tolerance as do some other members of their genus. Winter hibernation generally occurs underwater, where individuals may bury themselves in the mud, construct pits, or occupy existing holes in the substrate (Treanor and Nichola 1972).

Breeding occurs from March through August in Maryland. The breeding call of the male is a deep *jug-o-rum*, with larger males producing deeper calls than smaller males. The calls may carry up to 0.4 km. Early in the breeding season, males establish territories around a selected calling site, which they will aggressively defend from other males (Dodd 2013b). In dense populations, males may occur at 2-m to 6-m intervals around the breeding pond (Dodd 2013b). Female American Bullfrogs lay 6,000–20,000 or more eggs in a foamy, floating sheet that measures up to 0.6 m in diameter on the water's surface (Wright and Wright 1949). The eggs hatch in about 2–5 days. Tadpoles attain a total length of 10–17.1 cm and take 1–2 years to transform. Sexual maturity is reached after several more years for both males and females (Hulse et al. 2001). Longevity in the wild is estimated at 8–10 years (Oliver 1955; Goin and Goin 1962).

American Bullfrogs are notorious for being voracious predators of generally any animal that will fit in their mouth (Schwalbe and Rosen 1988; Dodd 2013b). They prey on invertebrates and vertebrates during the day or night, although insects comprise the majority of their diet. American Bullfrogs have also been known to eat fish, other frogs, lizards, turtles, birds, bats, shrews, and a young alligator (G. R. Brooks 1964; Bury and Whelan 1984; Schwalbe and Rosen 1988; Dodd 2013b). Of course, these frogs are also prey for some of these organisms. Interestingly, American Bullfrogs exhibit resistance to the venom of copperheads (Dodd 2013b).

During the MARA project, American Bullfrogs were identified statewide and documented in 913 blocks within 221 quads—roughly 90% of the total quads with at least 40.47 ha (100 ac) of uplands in Maryland. The atlas distribution was similar to that described by H. S. Harris (1975), including the fewer reports from western Maryland, which may reflect a lack of habitat, particularly on the Appalachian Plateaus.

The American Bullfrog is one of the most common anurans in Maryland. Globally, its status is secure, and these frogs are probably increasing in number (Dodd 2013b). With their preference for human-made ponds, high reproductive

rates, large larval size, and voracious appetites, they often displace other native frogs in areas where they have been introduced. Pollution, road mortality, or harvesting for food may lead to localized population declines, but the strong dispersal abilities of this species most likely ensure the continued existence of American Bullfrogs in the landscape.

WAYNE G HILDEBRAND

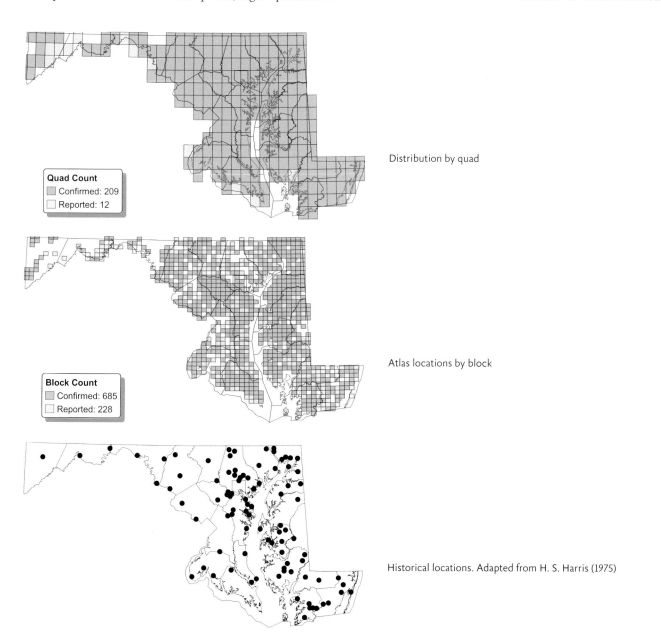

Quad Count
Confirmed: 209
Reported: 12

Distribution by quad

Block Count
Confirmed: 685
Reported: 228

Atlas locations by block

Historical locations. Adapted from H. S. Harris (1975)

Photograph by Scott McDaniel

Green Frog
Lithobates clamitans

The Green Frog is one of the most commonly encountered frogs in Maryland. The common name of this frog refers to its typical body coloration, and the specific epithet, *clamitans*, means "loud calling." The body color is green to brown and is often marked with dark brown or grayish dorsal spots. The upper lip is typically green. The hind legs have indistinct dark barring. The belly is white, with dark mottling under the head and legs. A dorsolateral ridge extends from behind the eye and down the body but does not reach the hind legs. In males, the tympanum is larger than the eye, whereas in females, the tympanum is the same size as or slightly smaller than the eye. During the breeding season, the throat of the male becomes yellow. Adult Green Frogs measure 5.4-7.5 cm in body length, and males and females are about the same size.

The Green Frog ranges through nearly all of eastern North America, extending as far north as southern Canada in the east and as far west as eastern Minnesota in the north and eastern Texas in the south, but it is absent from southern Florida. This species has also been introduced in many areas outside its native range, with established, localized populations in Arizona, Iowa, Montana, Utah, Washington State, British Columbia, the Bahamas, and The Netherlands (Slater 1955; Behle and Erwin 1962; Matsuda et al. 2006; Kraus 2009). The Maryland distribution is statewide (H. S. Harris 1975).

These frogs occupy a variety of aquatic habitats—often exhibiting a ubiquitous presence in regions where they are found (Dodd 2013b). They inhabit shallow freshwater wetlands such as spring-fed fens, streams, rivers, ditches, lakes, and ponds (Dorcas and Gibbons 2008; Dodd 2013b). Typically, Green Frogs favor habitats that are unaffected by urbanization and agriculture, but that does not preclude their presence at such locations (Dodd 2013b). They prefer habitats with long hydroperiods but can be found in smaller ephemeral pools during wet periods in spring and summer. Green Frogs tend to remain near their aquatic habitat throughout the year, living near the banks and margins, but may make longer journeys away from wetlands during wet periods and are often encountered crossing roads on rainy nights.

Green Frogs are active during March through October in Maryland (Smithberger and Swarth 1993). Throughout much of the year, adults remain near the water but will forage in terrestrial areas (Dodd 2013b). Both adults and subadults establish home ranges; however, the microhabitat preferred by each differs (Dodd 2013b). Subadults typically inhabit shallow waters with dense vegetation, while adults prefer deeper waters—provided American Bullfrogs are not present. When displaced from their home range, Green Frogs are able to home to familiar areas by using olfactory cues. However, the return to their home range may take a few days (Dodd 2013b). Male Green Frogs are territorial, and territories are aggressively defended during the breeding season; defending a territory may involve behaviors such as patrolling, splashing displays, vocalizations, chases, attacks, and wrestling (Dodd 2013b). Overwintering typically occurs in aquatic habitats, usually in mud and other bottom debris. Streams appear to be preferred overwintering sites, as the flowing water remains unfrozen throughout winter and is well oxygenated (Dodd 2013b).

Breeding occurs from March through August in Maryland. The male Green Frog makes a call like the plucking of a loose banjo string, *gu gu gu*, which may be repeated briskly 2-3 times (Powell et al. 2016). Females lay about 1,000–5,000 eggs on the water's surface in a clear thin film up to 0.3 m in diameter (Wright and Wright 1949). Hatching occurs in 3-5 days, depending on temperature, and tadpoles grow to a length of 8-10 cm. Metamorphosis may occur at about 70 days but often takes more than a year. Both male and female Green Frogs reach maturity at about 1-3 years of age.

During the MARA project, Green Frogs were identified across Maryland in 1,064 blocks within 231 quads—95% of the total quads with at least 40.47 ha (100 ac) of uplands. This statewide distribution was unchanged from that reported by H. S. Harris (1975). Green Frogs were found in nearly every block on the Atlantic Coastal Plain and Piedmont provinces and in the majority of blocks in the mountainous regions of Maryland.

The Green Frog is secure in Maryland and throughout its entire range, although some localized populations may be threatened by human-induced impacts such as pollution or habitat loss and degradation resulting from development or other land-use changes (Dodd 2013b). However, the Green Frog, like the American Bullfrog, is a habitat generalist and easily adapts to almost any freshwater habitat, including human-made wetlands.

WAYNE G HILDEBRAND

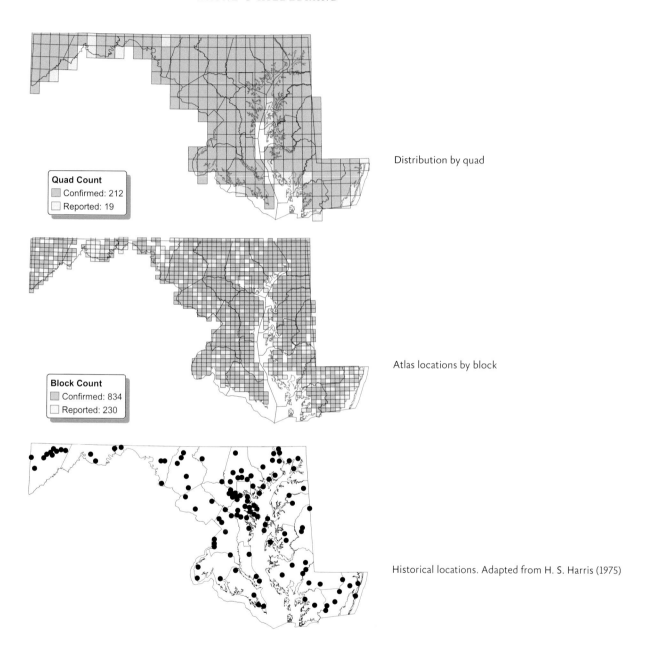

Quad Count
Confirmed: 212
Reported: 19

Distribution by quad

Block Count
Confirmed: 834
Reported: 230

Atlas locations by block

Historical locations. Adapted from H. S. Harris (1975)

Photograph by Kevin M. Stohlgren

Pickerel Frog
Lithobates palustris

The common name for this frog is derived from its use as fishing bait to catch pickerel (*Esox*); the specific epithet, *palustris,* means "of the marsh," a reference to the species' habitat. The dorsal coloration of the Pickerel Frog is olive green to tan with dark squarish spots, often in two parallel rows running down the back. Additional dark squarish spots are present on the sides of the body. A weak dorsolateral ridge is present. A light line highlights the upper lip. The belly is plain whitish, with brilliant yellow to orange under the hind legs; this bright coloration may extend forward along the lower sides to under the forelimbs. The upper surfaces of the hind legs are barred with dark bands, and the forelimbs have dark spotting. Adult Pickerel Frogs reach a body length of 4.4-7.5 cm, with the females being larger than the males. Young Pickerel Frogs appear metallic and lack the bright yellow under-leg coloration.

The distribution of the Pickerel Frog extends from coastal Canada south to Georgia and west to eastern Texas and extreme eastern Minnesota. Extensive gaps in the range exist in northwestern Indiana, much of Illinois, parts of Missouri, and areas in the Southeast, including southern Louisiana, Alabama, and Georgia, and nearly all of Florida. The Maryland distribution is presumably statewide, although H. S. Harris (1975) reported no records for Somerset County.

Pickerel Frogs occupy a wide variety of habitats throughout the year. In early spring, they can be found in and near freshwater wetland habitats, which they utilize for breeding, including woodland pools and ponds, agricultural ponds, floodplain swamps, and marshes. After the breeding season, they may remain near these habitats or disperse to river and stream margins, or they may venture far from water into moist woodlands and meadows. Pickerel Frogs may also be found in springhouses (White and White 2007) or caves that contain water (Schreiber 1952). In fact, the Pickerel Frog may be the most cave-adapted anuran in North America (P. W. Smith 1961; Schaaf and Smith 1970; McDaniel and Gardner 1977; Resetarits 1986). In areas of karst topography, the species can be abundant. Overwintering occurs in springs, on pond bottoms, or under stones in protected areas such as caves and ravines (Dodd 2013b). Pickerel Frogs are often encountered under rocks along woodland streams, in wet grass on floodplains or adjacent to wetlands, or crossing roads during evening rains in early spring.

In Maryland, Pickerel Frogs are most active from late winter through fall, with breeding occurring from March through June. Weather conditions exert a strong influence on the timing of reproduction. In general, movement to breeding sites requires one or more of the following conditions: water temperatures of 7-18 °C, surface soil temperature of 7-12 °C, and air temperature of 10-26 °C (Wright 1914; J. A. Moore 1939; Pope 1944; M. J. Johnson 1984; L. M. Hardy and Raymond 1991). The breeding call of this frog is a low-pitched, slow snore lasting for 1-2 seconds (Powell et al. 2016). Males call from along the water margin and sometimes when submerged underwater. Females lay about 2,000-3,000 brown eggs in a bulbous mass typically attached to submerged aquatic vegetation or twigs (Wright and Wright 1949); the eggs hatch in 7-14 days. Tadpoles metamorphose in approximately 70-90 days. It takes 3 years for the Pickerel Frog to reach maturity (Wright and Wright 1949).

Pickerel Frogs eat mainly insects but also consume snails, worms, spiders, and other small invertebrates (Dorcas and Gibbons 2008; Dodd 2013b). Only a few observations of natural predation on Pickerel Frogs have been reported (Redner 2005); predators include American Bullfrogs, minks, and Bald Eagles (*Haliaeetus leucocephalus*) (Babbitt 1937; Applegate 1990; Beane 1990). The Pickerel Frog produces a skin secretion that is noxious or unpalatable to a number of potential predators (Dickerson 1906; Schaaf and Smith 1970), and several species of snakes that prey upon frogs reject them (Dodd 2013b). The bright yellow-orange coloration on the concealed surfaces of the frog's hind legs may serve as a warning to predators (Wright and Wright 1949).

During the MARA project, Pickerel Frogs were identified in 661 blocks within 187 quads, which remained similar to the distribution reported by H. S. Harris (1975). The frogs were reported from every county in Maryland, though fewer reports were from western Maryland and the lower Eastern Shore.

Pickerel Frogs are considered common frogs in Maryland. They are abundant in adjoining states and throughout their entire distribution (N. B. Green and Pauley 1987; Hulse et al. 2001; Beane et al. 2010). Their overall population is probably stable; however, due to an apparently scattered distribution in some areas, including Maryland's Eastern Shore, some populations may be vulnerable (Dorcas and Gibbons 2008; Dodd 2013b). In fact, roadside calling surveys suggest that the Pickerel Frog may be in decline in much of the northeastern United States (Weir et al. 2014).

WAYNE G HILDEBRAND

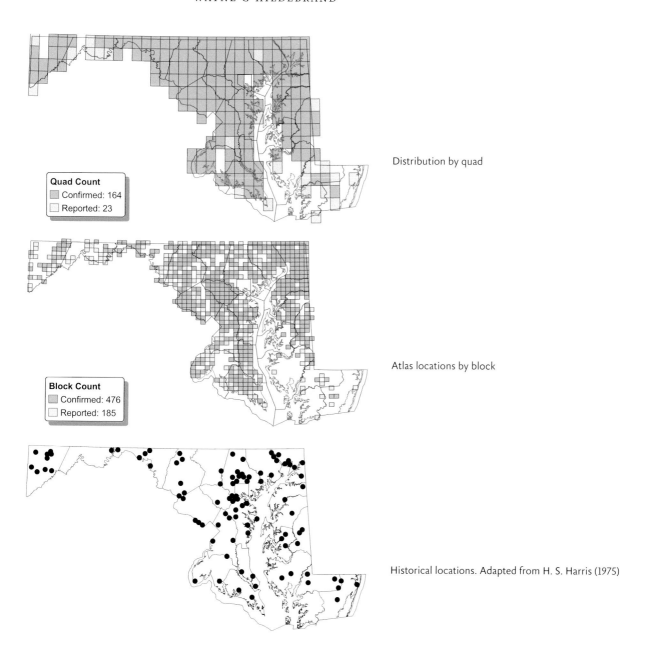

Distribution by quad

Quad Count
Confirmed: 164
Reported: 23

Atlas locations by block

Block Count
Confirmed: 476
Reported: 185

Historical locations. Adapted from H. S. Harris (1975)

Photograph by Robert T. Ferguson II

Southern Leopard Frog
Lithobates sphenocephalus

The Southern Leopard Frog is one of three leopard frog species native to the Northeast. The frog's common name refers to its spotted body, and the specific epithet, *sphenocephalus*, means "wedge headed," referring to its pointed snout. The body coloration ranges from brown to green with dark, roundish spots irregularly to randomly distributed on the dorsum. Conspicuous, light-colored dorsolateral ridges are present. The belly is white, and dark banding is present on the hind legs. Distinguishing characteristics of the Southern Leopard Frog are a central light spot on the tympanum and light patterning on the inner dorsal surface of the thighs. Male Southern Leopard Frogs are smaller than females. Adults measure about 5.1-9 cm in body length.

The Southern Leopard Frog can be found from New Jersey south through Florida. The range extends west to central Texas and north to central Illinois in the Mississippi drainage. Historically, Southern Leopard Frogs were considered a species restricted to the Atlantic Coastal Plain in Maryland (H. S. Harris 1975). More recently, however, they have been identified in Frederick County on the Piedmont (Hildebrand 2003, 2005).

Southern Leopard Frogs are habitat generalists and occupy a wide variety of mesic habitats ranging from forests to vegetated fields, as well as most wetland habitats (Dodd 2013b). Breeding sites include natural and human-made, permanent and temporary, and freshwater and brackish-water wetlands, but these frogs prefer fishless habitats (Dodd 2013b). After the breeding season, Southern Leopard Frogs may venture great distances and occupy more terrestrial sites. They are primarily active at night and are often encountered on roads in large numbers on rainy nights during the spring (Grogan and Bystrak 1973b). During hot weather, Southern Leopard Frogs seek refuge in moist habitats (Butterfield et al. 2005). They hibernate underwater in the mud (Hulse et al. 2001). Typically, these overwintering sites are located in large permanent bodies of water that are well-oxygenated (Butterfield et al. 2005).

In Maryland, Southern Leopard Frogs may be active from February through October (Smithberger and Swarth 1993) and breed from February through June (Grogan and Bystrak 1973b). The breeding call, which is a series of clucks repeated four or five times in quick succession, is produced by the male while floating in the water (Powell et al. 2016). The call resembles the sound made by an old car trying to start. A raspy growl, likened to rubbing two balloons together, may be interspersed. Oval egg masses containing several hundred to 2,000 eggs are attached to submerged plants or twigs (Wright and Wright 1949; Dodd 2013b). The eggs hatch in about 7-14 days, and tadpoles attain a length of 6.5-8.5 cm. Transformation into juvenile frogs occurs at 60-90 days. Southern Leopard Frogs reach maturity in 2-3 years.

Southern Leopard Frogs are primarily insectivorous but have also been known to feed on aquatic invertebrates, including crayfish (Force 1925). They are preyed upon by a variety of predators, including several species of snakes, Snapping Turtles, wading birds, raccoons, and foxes (Dodd 2013b). In general, Southern Leopard Frogs are alert and agile, making them difficult to capture (Wright 1932). When near water, they usually escape terrestrially. If fleeing into water, however, they dive in and arc back underwater, resurfacing near the shoreline, sometimes continuing their escape terrestrially (Carr 1940).

During the MARA project, Southern Leopard Frogs were documented in 566 blocks within 141 quads. They were found mostly on the Atlantic Coastal Plain up to the Fall Line, as previously reported by H. S. Harris (1975). They also were documented in three blocks from Frederick County. A new species, the Mid-Atlantic Coast Leopard Frog (*Lithobates kauffeldi*), was described during the atlas period (Feinberg et al. 2014), but due to its cryptic appearance and uncertain occurrence, as well as the timing of the publication, it was not recognized as a separate species for the MARA project. The range of this new species in Maryland warrants investigation, and ascertaining its presence may alter the distribution of the Southern Leopard Frog where the species are not syntopic.

The Southern Leopard Frog is considered common in Maryland. It is abundant in adjoining states and throughout its entire distribution (N. B. Green and Pauley 1987; Hulse et al. 2001; Beane et al. 2010; Dodd 2013b). However, this species may experience significant mortality due to vehicular strikes when wetland habitats are adjacent to roadways

(Dodd 2013b), and roadside calling surveys suggest that the Southern Leopard Frog may be in decline in much of the Northeast (Weir et al. 2014). Its ubiquitous occurrence on the Atlantic Coastal Plain in Maryland and its ability to rapidly colonize new wetlands (Merovich and Howard 2000) suggest that the Southern Leopard Frog can persist despite these anthropogenic impacts.

WAYNE G HILDEBRAND

Distribution by quad

Quad Count
Confirmed: 131
Reported: 10

Atlas locations by block

Block Count
Confirmed: 434
Reported: 132

Historical locations. Adapted from H. S. Harris (1975)

Photograph by Scott McDaniel

Wood Frog
Lithobates sylvaticus

The Wood Frog is one of the first species of frogs encountered in Maryland each year as adult frogs migrate to breeding ponds (White and White 2007). Its common name refers to its preferred habitat of forested areas. Likewise, the specific epithet, *sylvaticus*, refers to this frog's sylvan (forest) preference. The body coloration is a uniform brown, ranging from pinkish to reddish to tan to dark brown, with males frequently appearing darker than females. A dark mask on the sides of the head from the eye through the tympanum is always present. Wood Frogs have dorsolateral ridges that extend from the eye to the groin along the sides. The belly is light colored, and the legs have dark-colored bands. Adult Wood Frogs reach a body length of 3.5-7 cm. In Maryland, lowland Wood Frogs have a smaller body size than Wood Frogs from the mountains (Berven 1982).

The Wood Frog has the largest native geographic range of any North American frog and a range that extends farther north than that of any other North American amphibian or reptile, terminating at the Arctic Ocean (Dodd 2013b). It can be found throughout most of Canada to Alaska, and its range extends south to the Appalachian Mountains of northern Georgia. To the west, isolated populations are found in Alabama, Tennessee, Arkansas, Missouri, and Illinois. The Maryland distribution is presumably statewide, although H. S. Harris (1975) reported no records from Dorchester and St. Mary's counties.

Wood Frogs inhabit moist deciduous woodlands and are found away from water except during their short breeding period. These frogs are sensitive to a reduced canopy and edge effects associated with particular silviculture practices

(deMaynadier and Hunter 1998; Gibbs 1998a, 1998b). In the late winter to early spring, Wood Frogs are usually encountered as they migrate to their breeding ponds, which are typically vernal pools or fishless ponds (Dorcas and Gibbons 2008; Dodd 2013b). Breeding habitats are typically within or adjacent to adult habitat. Hibernation occurs on land, often buried in leaf litter. Wood Frogs tolerate freezing temperatures by increasing their levels of blood glucose, acting as a cryoprotectant, to prevent ice crystal formation that would cause tissue damage (Dodd 2013b).

The breeding season begins with the first warm rains of winter in February and extends through April. The breeding call is a hoarse, duck-like quack repeated several times by the male while floating on the water's surface. A chorus of Wood Frogs has an auditory resemblance to a flock of ducks in the distance. Wood Frogs are explosive breeders, and all adults in a given area breed over a few days. Large egg masses are attached to submerged vegetation or branches close to the surface, often communally in large conglomerations. Individual egg masses may contain 3,000 eggs (Wright and Wright 1949). Hatching occurs within 10-21 days, depending on the temperature, and the tadpole stage lasts 44-113 days. Wood Frog tadpoles achieve a total length of 5-6.6 cm. Male Wood frogs are able to breed at 2 years of age, whereas females breed at 3 years (White and White 2007). A longevity of 3 years for males and 4 years for females is typical (Berven 1982).

Wood Frogs are active during the day and at night (Dodd 2013b). While these frogs prefer forested environments, they will forage in open areas at night or in overcast weather conditions (Heatwole 1961). As adults, Wood Frogs prey on a variety of invertebrates, including ants, beetles, gastropods, and spiders (Dodd 2013b). The coloration of Wood Frogs offers protection against predators by camouflage; the dark eye mask acts as a disruptive coloration to break up contours of the frog (Dodd 2013b). Death feigning in response to threats has been observed in some individuals (Dodd 2013b).

During the MARA project, Wood Frogs were identified in 629 blocks within 183 quads. H. S. Harris (1975) did not report any Wood Frogs from St. Mary's County, but the MARA project documented a countywide distribution. In general, during the atlas project, more localities of Wood Frog were documented than were reported by Harris (1975). However, Wood Frogs were not documented in Somerset County during the atlas period, compared with two localities reported by Harris. None were reported from Dorchester County, as in Harris's report.

Wood Frogs remain common frogs in Maryland. They are abundant in adjoining states and throughout their entire distribution (N. B. Green and Pauley 1987; Hulse et al. 2001; Beane et al. 2010). Their overall population is probably stable, but some populations are in decline (Dodd 2013b). Individuals migrating to breeding ponds often cross roads and are vulnerable to vehicular strikes (Dorcas and Gib-bons 2008). Additionally, an emerging disease, ranavirus, has resulted in larval die-offs in breeding ponds (Brunner et al. 2011). Roadside calling surveys suggest that Wood Frog populations are decreasing in some areas of the northeast-ern United States such as Maine and increasing in others, in-cluding New Hampshire and Pennsylvania (Weir et al. 2014).

WAYNE G HILDEBRAND

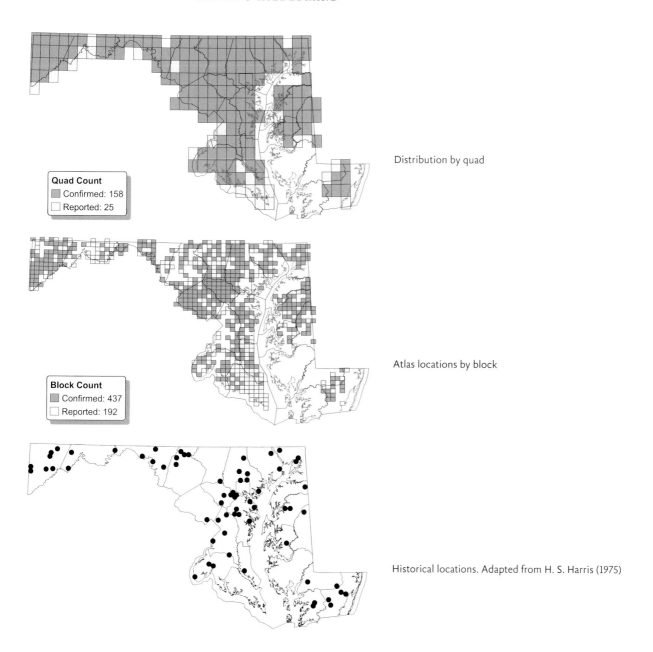

Distribution by quad

Quad Count
Confirmed: 158
Reported: 25

Atlas locations by block

Block Count
Confirmed: 437
Reported: 192

Historical locations. Adapted from H. S. Harris (1975)

Photograph by Don C. Forester

Carpenter Frog
Lithobates virgatipes

The Carpenter Frog is one of the less understood and more unusual frogs in Maryland. The specific epithet, *virgatipes*, meaning "striped foot," refers to the markings on the hind feet (Hulse et al. 2001), though this species is best identified by the two yellowish, reddish brown, or light brown longitudinal stripes running parallel along each side of the body—one beginning from behind the eye and the other from the snout, continuing to the groin. These stripes are lighter in color than the background bronze, brown, or muddy olive dorsum. No dorsolateral ridge is present. The groin, underside of legs, and throat are often vividly patterned or mottled, contrasting with the otherwise cream underside. Males and females are the same size, ranging from 4.1 cm to 6.7 cm in body length, but males have a larger tympanum.

The Carpenter Frog is found exclusively on the Atlantic Coastal Plain from southern New Jersey to northeastern Florida. Within this area, it has a discontinuous distribution because it is a habitat specialist. The first Maryland record was from 1947 in Dorchester County (Conant 1947a). These frogs were subsequently discovered in Worcester County (Meanley 1951), Caroline County (R. W. Miller 1980), and Queen Anne's County (Given 1990). This paucity of known populations led to the Carpenter Frog being state-listed as in need of conservation. However, surveys from 1993-2005 by Hildebrand (2005), Otto (2006), Otto et al. (2007a), William Grogan (unpubl. data), and the MDNR (unpubl. data) found the species to be much more widespread than previously believed, confirming 50 extant breeding ponds in five counties (Caroline, Dorchester, Queen Anne's, Wicomico, Worcester). The species was subsequently removed from state listing in 2005. Reed (1957d) and Hildebrand (2005) re-

ported Carpenter Frogs in Charles and St. Mary's counties, respectively, but neither observation has been verified.

This highly aquatic species is associated with acidic freshwater wetlands, particularly with mats of sphagnum or emergent grass-like vegetation. These habitats include Delmarva bays, pond and river edges, sphagnum bogs, and swamps, usually surrounded by sandy pine flatwoods (Wright and Wright 1949; Hulse et al. 2001; White and White 2007; Elliott et al. 2009; Dodd 2013b). While some researchers have noted that the Carpenter Frog's breeding ponds are typically permanent wetlands (Given 1988; Zampella and Bunnell 2000; Dodd 2013b), in Maryland, the majority of known breeding sites are seasonal wetlands with intermediate hydroperiods (Otto et al. 2007a). Otto et al. (2007a) found that the frequency of occurrence of these frogs showed a positive relationship with forest cover and wetland size and a negative relationship with pH and hydroperiod. Carpenter Frogs have been found in microhabitats with pH as low as 3.5 (Gosner and Black 1957) but are not present when pH is above 4.5-5.0 (Given 1999; Zampella and Bunnell 2000; Otto et al. 2007a). This extreme acidity excludes some potential anuran competitors and is lethal to many other anurans' eggs and larvae (Zampella and Bunnell 2000; Dodd 2013b). Competition with acid-tolerant sympatric species is reduced through habitat partitioning (Given 1990). For example, when in the presence of the more aggressive Green Frog, male Carpenter Frogs will call from out in the wetland (Standaert 1967; Given 1990), while Green Frogs call from the vegetated shoreline (Given 1990), although female Carpenter Frogs are more likely to be close to shore (Standaert 1967). Given (1988) found that the number of submerged stems was positively correlated with the number of male Carpenter Frog calling sites, so vegetative structure is an important part of breeding habitat (Mitchell 2005).

Carpenter Frogs emerge from aquatic hibernation in March. They begin calling on warm, humid evenings from early to mid-April through mid-July and may also call infrequently during the daytime (Otto el al. 2007a; Simpson 2009; Dodd 2013b). The loud, resonating call, from which this frog gets its common name, sounds like two carpenters hammering nails slightly out of sync: *pu-tuck pu-tuck pu-tuck* (Powell et al. 2016). Egg masses are oblong and are laid in shallow water near the surface, attached to vegetation or debris (Dodd 2013b). The number of egg masses laid per female is unknown, but Wright and Wright (1949) estimated 200-600 eggs per mass. Incubation period is also unknown. The distinctively striped tadpoles hatch from May to September. The larval period may last up to 1 year, with tadpoles overwintering in breeding ponds (Mitchell 2005), but larval period is unknown in the seasonal breeding ponds primarily used in Maryland. It is possible that successful reproduction in Maryland occurs only in wet years, when hydroperiods are extended. Sexual maturity occurs the following spring for individuals that metamorphose in summer, but not until late summer for those that overwinter as tadpoles (Dodd 2013b).

Carpenter Frogs are cryptic, sitting still among the sphagnum, submerging quickly, and/or hiding in submerged vegetation to avoid detection by potential predators (Dodd 2013b). Larger males exclude smaller males from their 0.5-m to 6.5-m diameter territories through aggressive interactions and vocalizations (Given 1987, 1988). Longevity in the wild rarely exceeds 3 years (Standaert 1967; Given 1988; Dodd 2013b), but a captive individual lived 6 years (Snider and Bowler 1992; Mitchell 2005).

During the MARA project, Carpenter Frogs were recorded in 10 quads and 21 blocks in five counties, including Somerset, representing a new county record. The MARA results covered a greater area than depicted by H. S. Harris (1975), who noted that Carpenter Frogs were restricted to the southern part of the Eastern Shore, in only three coun-

ties. However, the MARA project documented fewer reports than in prior surveys from 1993 to 2005, which represented 22 quads and 41 blocks, suggesting that the MARA coverage was incomplete. Although not depicted on the distributional maps, an unverified report of a calling Carpenter Frog from the Point Lookout CW block in St. Mary's County during the atlas period, along with the two unverified Western Shore historical reports (Reed 1957d; Hildebrand 2005), suggests that additional surveys are needed in this area.

Although MARA data underreported the known Carpenter Frog distribution in Maryland, several new confirmed locality records were submitted. These data highlight the importance of projects such as MARA in documenting a secretive frog that is often overlooked. Also, given that more than half of the locations documented from 1993 to 2005 were not recorded during the atlas project, additional surveys may be needed to confirm that these populations are still present. Although the Carpenter Frog is no longer a state-listed species, it may still warrant some conservation attention due to its specialized habitat needs and scattered populations. The future of this species depends on a full understanding of its distribution, as well as conservation of a network of acidic wetlands within extensive forest cover.

SCOTT A. SMITH

Distribution by quad

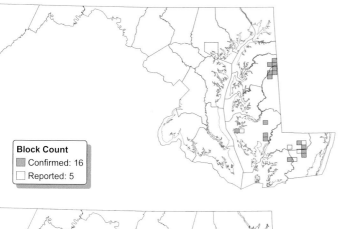

Atlas locations by block

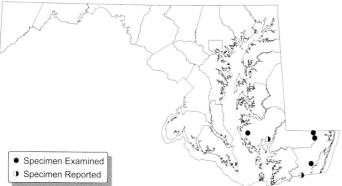

Historical locations. Adapted from H. S. Harris (1975)

Photograph by Andrew Adams

Eastern Spadefoot
Scaphiopus holbrookii

The Eastern Spadefoot is a common yet seldom encountered frog. With its fossorial habits, the species spends a majority of its active period underground, much like the Eastern Narrow-mouthed Toad. The Eastern Spadefoot has a dorsal, light yellow, hourglass pattern on a dark, purplish brown body and often has a smattering of small pink or orange tubercles (Lang et al. 2009). The soles of the hind feet have a black, sickle-shaped, horny process that gives the Spadefoot its name. Parotoid glands are present but are less conspicuous than in toads (i.e., Bufonidae) (Dickerson 1906). Spadefoots have large eyes and, unlike other frogs in Maryland, have vertical pupils. Males generally have dark throats, whereas females have light or white throats (Tyning 1990). This species is moderately sized, ranging from 4.4 cm to 5.7 cm in adult body length.

The Eastern Spadefoot is found from eastern Massachusetts south to Florida and west to the Mississippi River, primarily on the Atlantic Coastal Plain and the Gulf Coastal Plain. Isolated and disjunct populations occur on the Piedmont, Ridge and Valley, and Appalachian Plateaus. In Maryland, the species has been found mostly on the Atlantic Coastal Plain (Mansueti 1947; H. S. Harris 1975) but was also known from two localities on the Piedmont in Frederick County (Stine et al. 1956). Subsequent to H. S. Harris's (1975) publication, Eastern Spadefoots were documented on the Atlantic Coastal Plain in two additional counties: Baltimore (H. S. Harris and Crocetti 2008) and Harford (R. W. Miller 1977).

This species occupies habitats with loose sandy soils in which it can easily burrow, including hardwood and pine forests, agricultural fields, pastures, and floodplains. Breeding sites are ephemeral, and any low-lying area that fills with water after heavy rain can be utilized. Observing Spadefoots can be difficult even though they may be locally abundant. Being fossorial, they stay buried for extended periods, depending on local conditions. They exit burrows only at night after rains or when nighttime humidity is high (Dodd 2013b). Eastern Spadefoots may burrow 7-30 cm below the surface (Pearson 1955). Their burrows are not open tunnels; rather, loose soil collapses around them as they burrow deeper (Dodd 2013b). They tend to return to the same burrow after nightly forays (Dodd 2013b), but multiple burrows may also be used (Pearson 1955; K. A. Johnson 2003). In Maryland on the Atlantic Coastal Plain, these frogs are most commonly encountered at night on roadways after rains, when foraging or moving to breeding pools.

Breeding is explosive and triggered by heavy rain. In Maryland, Eastern Spadefoots typically breed from late March through October when air temperatures exceed 10 °C (Dorcas and Gibbons 2008). Males call while floating on the water's surface. The call is an explosive grunt, repeated several times (Powell et al. 2016). Amplexus in these frogs, unlike other frogs in Maryland, is inguinal, with the male clasping the female just above the hind legs. Females lay 2,000-5,000 eggs (Dorcas and Gibbons 2008) in small strands of about 200 eggs each on submerged vegetation, sticks, or leaves (Dodd 2013b). Eggs hatch in as little as 1-15 days, depending on water temperature (Dodd 2013b). The tadpoles are small, seldom exceeding 3 cm (Hulse et al. 2001). Tadpoles often aggregate and swim in large schools. The larval period can last 9-63 days, depending on water temperature and hydroperiod, but typically is 15-30 days (Dodd 2013b). Eastern Spadefoots can live from 5 to 10 years (Tyning 1990; Dorcas and Gibbons 2008).

Adult Eastern Spadefoots mostly eat insects and other invertebrates such as caterpillars, moths, beetles, spiders, earthworms, and snails (Whitaker et al. 1977; Dorcas and Gibbons 2008). Spadefoots have only a few known predators, which include American Bullfrogs, Common Watersnakes, North American Racers, and owls (Dodd 2013b). When threatened, the Spadefoot arches its back, pulls its head down, and tucks its feet under its body. Toxins, which are noxious or distasteful to predators, exude from glands on the dorsal surface and have an odor reminiscent of peanut butter.

During the atlas project, Eastern Spadefoots were recorded in 94 quads and 314 blocks. Nearly all records were from the Atlantic Coastal Plain, with one block on the Piedmont in Montgomery County and three blocks in the Great Valley region of the Ridge and Valley Province in Washington County. Spadefoots were not detected in central and western Kent County or western Talbot and Dorchester counties, despite targeted efforts to locate them there (Nathan Nazdrowicz, pers. comm.). Also, none were reported from Baltimore or Frederick counties, where they were previously documented. The most recent observation from Baltimore County was from 2008 (H. S. Harris and Crocetti 2008). Aside from these differences, the results from the atlas project were similar to the distribution presented by H. S. Harris (1975).

No conservation actions appear necessary for the Eastern Spadefoot at this time. Habitat destruction has affected some breeding populations, but the results of the atlas project demonstrate these frogs are still widespread in many areas. However, because this species' specialized life history relies on appropriate weather conditions and wetland hydroperiods for successful breeding, the potential impacts of a changing climate may warrant additional monitoring in the future.

GEORGE M. JETT

Distribution by quad

Quad Count
Confirmed: 89
Reported: 5

Atlas locations by block

Block Count
Confirmed: 279
Reported: 35

Historical locations. Adapted from H. S. Harris (1975)

Reptiles

Photograph by Robert T. Ferguson II

Spiny Softshell
Apalone spinifera

The Spiny Softshell is unmistakable and easily distinguished from every other turtle in Maryland: it is the only freshwater turtle in Maryland with a leathery carapace and plastron. Its name refers to the numerous spine-like projections along the anterior edge of the carapace (Mitchell 1994). The carapace is flat and round with a rough, sandpaper-like texture, and the posterior margin is extremely flexible (Hulse et al. 2001). Coloration of the carapace ranges from olive to tan to olive brown with scattered black, irregular to ring-shaped spots and a black line along the margin (Hulse et al. 2001). This patterning is lost in older females, which become mottled with dark blotches. The plastron is a solid white to yellowish color (Mitchell 1994). The head has a pair of dark-bordered, yellowish stripes running from the snout down the neck. The snout is elongated and snorkel-like, and the neck is long. Females are much larger than males, with an adult carapace length of 18-54 cm, compared with 12.5-31 cm in males. Juveniles have reduced spiny projections and more distinctive dark markings of the carapace, and the marginal black line is paired with a yellow line (Mitchell 1994).

The Spiny Softshell has an extensive range throughout the United States, primarily in the Mississippi River watershed, but also extending across the southern Atlantic Coastal Plain from southern North Carolina to north Florida, and along the Gulf States to Texas and northern Mexico. The western extent of its range is restricted to river valleys. In the Northeast, the range is disjunct in southern Quebec and western Vermont (Ernst and Lovich 2009). The species has been introduced in parts of the mid-Atlantic region, including southeastern Pennsylvania, southern New Jersey, and

Norfolk, Virginia (Hulse et al. 2001; Powell et al. 2016). In Maryland, Spiny Softshells are found only in the Youghiogheny River on the Appalachian Plateaus in Garrett County. Although they were long suspected to occur in the Ohio River drainage in Maryland, the first specimens were not collected until 1970 (Wemple 1971). Reports of Spiny Softshells from other areas in Maryland—Anne Arundel County (Mansueti and Wallace 1960) and St. Mary's County (Rambo 1992)—are believed to be the result of releases, not natural occurrences. Eighteen Spiny Softshells were released into the Potomac River near Cumberland in 1883 (Dukehart 1884) in an attempt to establish a population, but this was unsuccessful.

This species is highly aquatic and inhabits slow-moving rivers and streams, lakes, ponds, and impoundments with well-oxygenated water. It prefers habitats with soft bottoms and submerged aquatic vegetation and fallen woody debris (Ernst and Lovich 2009). The Spiny Softshell can stay submerged for extended periods of time and often selects resting sites in water just deep enough so its nostrils can reach the water's surface when the neck is fully extended (Ernst and Lovich 2009). Despite its aquatic lifestyle, this species basks frequently on banks, shorelines, mudflats, sand bars, rocks, and logs, or by floating at the water's surface (Ernst and Lovich 2009).

Spiny Softshells are active from April through October (Ernst and Lovich 2009), and most activity takes place during the day. At night, they take shelter buried in the substrate or within submerged woody debris (Hulse et al. 2001; Ernst and Lovich 2009). The breeding season is in April and May; nesting occurs in late May to August, and eggs are laid near the water, up to 100 m away, in mud banks and gravel or sand bars (Ernst and Lovich 2009). When suitable nesting habitat is limited, nests may be concentrated in small areas. Up to 2 clutches of white, brittle, round eggs can be produced each year, with each clutch ranging from 4 to 39 eggs but on average containing 12-18 eggs (Mitchell 1994; Ernst and Lovich 2009). Incubation time varies with temperature but usually lasts 75-85 days (Ernst and Lovich 2009). Hatching takes place from late August to October, and in Maryland, hatchlings have been recorded in late September (Wemple 1971). Unlike most turtle species, the Spiny Softshell does not have temperature-dependent sex determination; rather, sex is determined genetically by chromosome composition, resulting in a nearly equal sex ratio over a broad range of incubation temperatures (Ernst and Lovich 2009). Males reach maturity at a body weight of about 130 g, but females do not mature until about 1,500 g (Robinson and Murphy 1978).

These turtles are primarily carnivorous and feed on a wide range of prey, including mollusks, crayfish, isopods, aquatic insect larvae, and fish (Hulse et al. 2001; Ernst and Lovich 2009). Although few natural predators of adults have been reported, their nests may be raided by skunks and raccoons, and the young are preyed on by fish, turtles, snakes, birds, and mammals (Ernst and Lovich 2009).

Spiny Softshells were confirmed in two quads and two blocks during the MARA project. One block was in Garrett County near the historical locality plotted by H. S. Harris (1975). The other block was in Frederick County, undoubtedly a released individual.

The Spiny Softshell is listed as in need of conservation in Maryland (MDNR 2016b). This species is sensitive to pollu-tion and stream degradation, and populations have declined in some parts of its range (Ernst and Lovich 2009). Conservation efforts in Maryland should focus on protecting the Youghiogheny River and maintaining suitable nesting areas. Due to its restricted range in Maryland, this species may warrant and benefit from being listed as threatened or endangered.

SETH METHENY

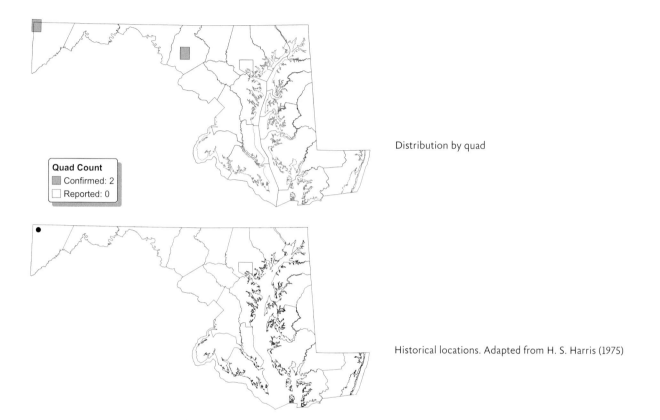

Distribution by quad

Historical locations. Adapted from H. S. Harris (1975)

Photograph by NOAA

Loggerhead Sea Turtle
Caretta caretta

The Loggerhead Sea Turtle is Maryland's second-largest sea turtle and is the largest of all living hard-shelled turtles (Ernst and Lovich 2009). Its distinctively massive, block-like head is responsible for the name "Loggerhead"—given to the species by English sailors. The large, shield-shaped carapace, head, and flippers are reddish brown and are sometimes encrusted with barnacles and algae. The vertebral scutes of the carapace are slightly keeled but become progressively smoother with age. The first of the five pairs of pleural scutes contact the cervical scute, a distinguishing feature shared with Kemp's Ridley Sea Turtle (Leviton 1971). The plastron is yellow, and the bridge usually has three inframarginal scutes. Two pairs of prefrontal scales are located between the eyes. Males have longer tails than females, extending beyond the posterior edge of the carapace (Gibbs et al. 2007). The typical adult Loggerhead has a carapace length of 79-114 cm and a weight of 80-200 kg; those in Maryland are usually in the 70- to 135-kg range (F. J. Schwartz 1967).

In the Atlantic Ocean, Loggerheads have been recorded from Newfoundland and the British Isles south to Argentina (Beane et al. 2010). They also occur in the Pacific and Indian oceans. Of the five sea turtle species found in Maryland, the Loggerhead is by far the most common. From 1979 to 1986, 91% of all identifiable sea turtles in the Chesapeake Bay were Loggerheads (Ernst et al. 1994). In the following decade, Loggerheads made up more than 90% of reported strandings and live individuals in Maryland waters (J. Evans et al. 1997; Norden et al. 1998). Graham (1973) reported a nesting attempt on Maryland's coastal beach near Ocean City in 1972. In the late 1960s, a multiyear effort to establish a rookery at Assateague

Island by relocating nests was unsuccessful (Graham 1973). The first successful nesting event with emergence of young turtles was confirmed in 2017 at Assateague Island National Sea Shore (NPS, unpubl. data).

Although sea turtles are primarily inhabitants of tropical and subtropical oceans and seas, several species, including the Loggerhead, are known as regular summer visitors in Maryland's more temperate region (Groves 1984). They inhabit warm waters on continental shelves and areas among islands, but prefer estuaries, coastal streams, and salt marshes for feeding (Gibbs et al. 2007). Loggerheads are powerful swimmers and can attain speeds of 20 km per hour and cover 28-70 km per day (Ernst and Lovich 2009). Tagged individuals have been taken well over 1,600 km from the point of marking. They may be found as far out as 240 km offshore but generally stay in the waters of the western Gulf Stream current. Incidents of cold stunning occur after sudden drops in environmental temperature. Water temperatures below 6.5 °C may be lethal (Ernst and Lovich 2009).

Loggerhead Sea Turtles nest on sandy beaches from North Carolina to Florida. They nest sparingly as far north as Chincoteague Island, Virginia, but more commonly through the Carolinas (Beane et al. 2010). Nests in Maryland and farther north are rare and are usually not successful. Ninety percent of nests within the United States occur in Florida (Gibbs et al. 2007). Female Loggerheads nest every 2-3 years, from April through August (CREARM 1973), and females return to the same nesting beaches year after year (Ernst and Lovich 2009). They lay as many as 6 clutches throughout a season, averaging more than 100 eggs in each (Alderton 1988). The incubation period lasts 49-76 days (Dodd 1988), and hatching success is about 60% to 85% (Ernst and Lovich 2009). Loggerheads, like other sea turtles, exhibit temperature-dependent sex determination, with warmer nest temperatures favoring females and cooler temperatures favoring males (Ernst and Lovich 2009). Hatchlings drift in ocean currents, sometimes in floating mats of *Sargassum* (Alderton 1988). The estimated maximum lifespan of a wild female Loggerhead is 47-62 years (Dodd 1988).

Loggerheads are omnivorous but largely carnivorous, feeding on mollusks, crustaceans, fish, and other marine animals. While they are known to eat fish and eelgrass (*Zostera marina*), invertebrates such as sponges, jellyfish, mollusks, and crabs are their most important food sources (Burke et al. 1993). Eggs and hatchlings suffer high predation rates from raccoons, foxes, and gulls, whereas adults are killed by sharks, Orcas (*Orcinus orca*), and humans.

Atlas efforts yielded a total of 51 sightings in 40 blocks within 22 quads. Seven of the sightings were of live individuals and 44 were of remains. Of the remains, distribution was nearly equal between the Chesapeake Bay and Atlantic Coast. Two sightings of live individuals were of nesting events in 2012 and 2013 on Assateague Island National Seashore. These sea turtles had a distribution generally similar to that shown by H. S. Harris (1975), albeit much more common, particularly along the Western Shore of the Chesapeake Bay and the Atlantic Coast. During the MARA project, an independent offshore survey documented Loggerhead Sea Turtle as the most commonly occurring sea turtle at least three miles offshore of Maryland in the Atlantic Ocean (Pallin et al. 2015).

The Loggerhead Sea Turtle is listed as threatened under the ESA and by the state of Maryland (MDNR 2016b). Although the species has been listed under CITES, Appendix II, there is ample evidence of continued commercial harvest and international trade, especially in Latin American countries. Loggerheads, like other sea turtles, are regularly caught in commercial fishing gear. The invention of turtle excluder devices (TEDs) has substantially reduced mortality caused by the shrimp trawling industry in waters of the south Atlantic and Gulf of Mexico, although channel dredging and offshore blasting continue to take a toll. In the Chesapeake Bay and other Maryland coastal waters, Loggerheads are most often caught in pound nets and purse seines—relying on the diligence of watermen for a safe release.

Conservation efforts to improve nesting success occur when nesting attempts are discovered in Maryland. Marylanders can also act responsibly to reduce Loggerhead mortality in local waters. With boat strikes being a substantial source of adult mortality, educational material distributed and posted at marinas can alert boaters to the seasonal presence of sea turtles in local waters. Cleaning up and properly disposing of plastic trash, which sea turtles ingest, as well as educating citizens to prevent plastics and other litter from reaching Maryland waterways, are other important ways to help reduce sea turtle mortality.

KYLE RAMBO

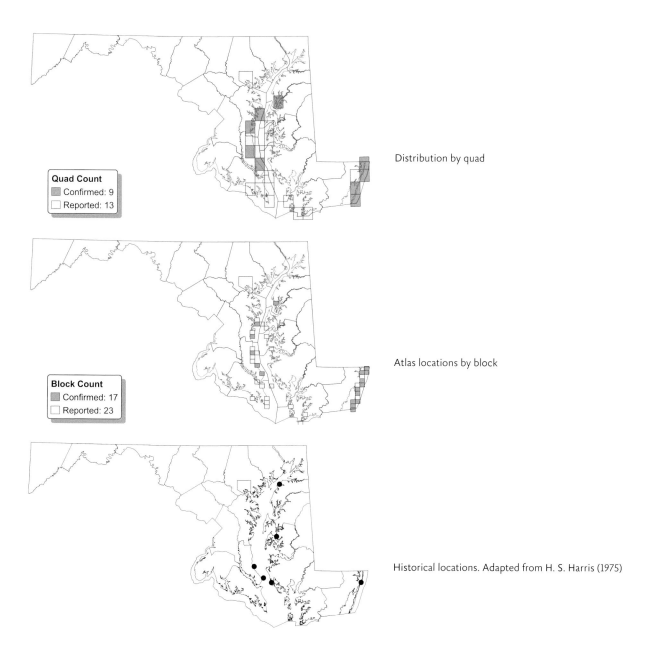

Quad Count
Confirmed: 9
Reported: 13

Distribution by quad

Block Count
Confirmed: 17
Reported: 23

Atlas locations by block

Historical locations. Adapted from H. S. Harris (1975)

Photograph by Mark Sullivan / NOAA

Green Sea Turtle
Chelonia mydas

The Green Sea Turtle is perhaps the most economically important reptile in the world as a food source, famous as the base for turtle soup. The common name of this sea turtle refers to the color of its fatty flesh rather than its external coloration (H. C. Robertson 1947; Parsons 1962). Like other sea turtles, it has limbs that are modified into flippers, and the streamlined carapace tapers posteriorly to a slightly serrated edge. Brown, olive, or black above, the scutes may be patterned with mottled or radiating lighter colors. Green Sea Turtles lack a vertebral keel, and the cervical scute does not contact the pleural scutes. Four inframarginal scutes are present on the bridge of the plastron. The head has one pair of elongated prefrontal scales and four postocular scales. Males are slightly larger than females and have a longer, prehensile tail for staying with the female during copulation. Females have a more rounded carapace, with a tail that is shorter than the carapace rim. Green Sea Turtle adults reach a carapace length of 90-122 cm and a weight of 113-204 kg.

Green Sea Turtles are found throughout much of the world's oceans and seas, especially in warmer regions. In the western North Atlantic, mostly juveniles and subadults have been found as far north as Nova Scotia, while adults usually occur farther south, especially around Florida and in the Caribbean and Gulf of Mexico (Ernst and Lovich 2009). In Maryland, Green Sea Turtles occur in both the Atlantic Ocean and Chesapeake Bay, although they are considered uncommon. H. C. Robertson (1947) reported a Green Sea Turtle found in 1934 near Cove Point, Calvert County, that weighed nearly 80 kg. H. S. Harris (1975) mapped an additional Calvert County record and one from Worcester

County. From 1991 to 2009, the Maryland Marine Mammal and Sea Turtle Stranding Network recorded carcasses from Sinepuxent Bay near the Ocean City Inlet, Worcester County, and from the Chester River, Queen Anne's County (MDNR, unpubl. data). During this time period, a cold-stunned, live turtle from Assateague Island, Worcester County, was also documented (National Aquarium, unpubl. data).

Although Green Sea Turtles can travel thousands of kilometers across open ocean between breeding and foraging areas, they live primarily in shallow-water lagoons and coastal shoals near sandy beaches, coral and hard-bottom reefs, and beds of aquatic vegetation in tropical and subtropical waters (F. J. Schwartz 1967; Ernst and Lovich 2009). Hatchlings can drift a year or more in the Gulf of Mexico and Caribbean Sea among floating mats of *Sargassum* (Carr and Meylan 1980; Carr 1987). Activity cycles and movements vary by nesting colony, with water temperature being an important factor in seasonal abundances. Some Green Sea Turtles appear to be residents near breeding beaches, while others migrate from breeding sites to distant feeding areas. The species is susceptible to cold stunning and reduced mobility when water temperatures drop to 10 °C or lower, and these individuals can die in 9-12 hours at temperatures below 6.5 °C (F. J. Schwartz 1978). Adults are primarily herbivorous, feeding on sea grasses and algae. Juveniles will also feed on mollusks, crustaceans, and other invertebrates.

For females breeding in the southeastern United States, nesting occurs in March to October, usually every 3 years, when females return to natal beaches located primarily in Florida but, rarely, as far north as Virginia Beach, Virginia (Kleopfer et al. 2006). Crawling with synchronous limb movements, females take about 2 hours to crawl onto the beach at night, dig a roughly 80-cm deep hole in the sand, lay about 115-130 eggs, cover the nest, and return to the water (Ernst and Lovich 2009). Females lay an average of 3-4 clutches per season. Incubation lasts 50-60 days, and hatching success averages 65% to 70% (Ernst and Lovich 2009). Hatchlings synchronously emerge at night, 4-7 days after hatching (M. H. Godfrey and Mrosovsky 1997). Females reach sexual maturity in 27-30 years (Spotila 2004).

Of the five sea turtle species known to occur within Maryland's bays or in the adjacent Atlantic Ocean, the Green Sea Turtle was the rarest species encountered during the MARA project, given that there were no reports of Hawksbill Sea Turtles (*Eretmochelys imbricata*) in Maryland waters. During the atlas period, two sightings of Green Sea Turtles were documented from two atlas blocks within the same quad. In July 2010, tracks in the sand with characteristics resembling those of a Green Sea Turtle were found on the ocean beach at Assateague Island. In November 2013, a cold-stunned turtle was rescued near the Ocean City US Coast Guard Station, rehabilitated at the National Aquarium, and released in Florida the following spring. Sightings along the Atlantic Ocean were similar to the historical distribution shown by H. S. Harris (1975). Harris's historical records also included localities for this species from the Chesapeake Bay, but none were recorded from the Bay during the atlas period.

Like other sea turtle species, Green Sea Turtles have undergone substantial population declines. This species faces a variety of natural threats from predation, diseases, and natural deterioration of nesting beaches. Compounding its plight are numerous human-created problems, including entanglement in fishing gear, degradation and loss of nesting beaches, beachfront lights that disorient hatchlings, boat strikes, and marine pollution such as oil spills and plastic de-

bris that resembles prey. The primary cause of decline, however, has been human exploitation and overharvesting for eggs, meat, and oil. The first protective legislation for the Green Sea Turtle was passed in Bermuda in 1620 to prohibit the killing of young turtles (Zwinenberg 1975). Presently, conservation efforts to address this species' many threats are underway, such as the protection and monitoring of nesting beaches, creation of lighting ordinances, and predator control at nesting beaches (NMFS and USFWS 1991; Ernst and Lovich 2009). The Green Sea Turtle is listed as threatened under the ESA and by the state of Maryland (MDNR 2016b). All sea turtles are protected jointly by the USFWS and the NMFS.

LYNN M. DAVIDSON

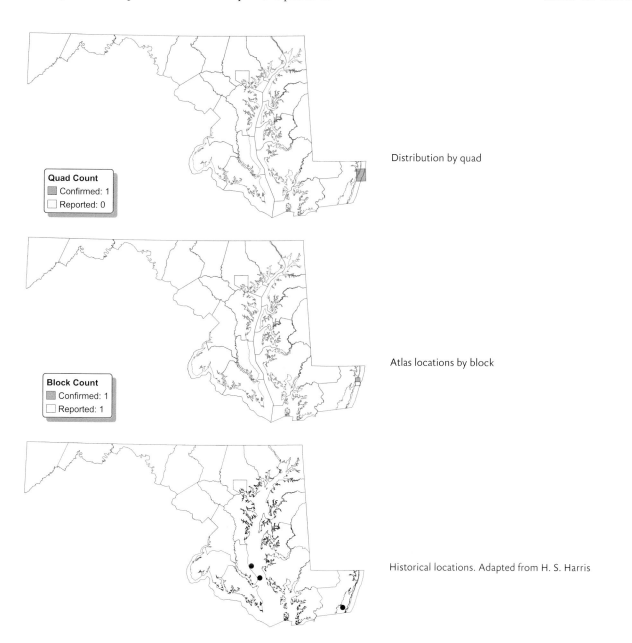

Distribution by quad

Quad Count
Confirmed: 1
Reported: 0

Atlas locations by block

Block Count
Confirmed: 1
Reported: 1

Historical locations. Adapted from H. S. Harris

Photograph by Jereme Phillips / USFWS Southeast Region

Kemp's Ridley Sea Turtle
Lepidochelys kempii

The smallest sea turtle in the world, Kemp's Ridley Sea Turtle was first described in 1880 from a Florida specimen provided by Richard Kemp. The origin of "ridley" is unknown but may be from "riddle," used by locals who were uncertain of the species' identity (Dundee 2001). Like all sea turtles, Kemp's Ridley has forelimbs modified into flippers and hind limbs shaped as paddles. The bony carapace is heart shaped, grayish green, and often more rounded than that of other sea turtles. It usually has five pairs of pleural scutes, with the first pair contacting the cervical scute, and four inframarginal scutes on the bridge that have associated pores. Two pairs of prefrontal scales are located between the eyes. Compared with females, males are slightly smaller, have a longer tail, and have curved claws on the forelimbs. Adults have a carapace length of 58-70 cm and usually weigh 36-45 kg.

Adult Kemp's Ridley Sea Turtles are mostly limited to the Gulf of Mexico, primarily along the southwestern Florida coast. Juveniles are more widespread, traveling to northern South America, across the Atlantic Ocean, and as far north as Newfoundland and Ireland. They regularly migrate along the eastern United States from southern waters to estuaries such as the Chesapeake Bay, Long Island Sound, and coastal waters of New England (Ernst and Lovich 2009). Early Chesapeake Bay records include one Kemp's Ridley collected in 1932 near the mouth of the Potomac River, two carcasses found in Calvert County, and one from Baltimore Harbor (Reed 1956d; J. D. Hardy 1962). Norden and Norden (1998) reported an additional Calvert County record from the early 1950s and another just north of Sandy Point State

Park in Anne Arundel County in 1979. From 1991 to 2009, 17 Kemp's Ridley carcasses were reported to the Maryland Marine Mammal and Sea Turtle Stranding Network from Baltimore, Calvert, Dorchester, Queen Anne's, St. Mary's, and Worcester counties (J. Evans et al. 1997; MDNR, unpubl. data), and six live individuals from Calvert, Dorchester, Somerset, and Worcester counties were reported to the National Aquarium (Norden et al. 1998; National Aquarium, unpubl. data).

Kemp's Ridleys usually inhabit shallow waters less than 50 m deep but can dive to depths exceeding 400 m. They can remain underwater up to 4 hours, but dive durations of 12-18 minutes are more typical for the species (Ernst and Lovich 2009). Hatchlings drift in Atlantic currents in the open ocean for the first 2 years of life, often on floating mats of *Sargassum*, before moving to shallow coastal waters (White and White 2007). The Chesapeake Bay is an important summertime habitat for Kemp's Ridley juveniles. Here, they occupy shallow waters with beds of eelgrass (*Zostera marina*), where they feed primarily on crabs and other marine invertebrates (Lutcavage and Musick 1985).

This marine turtle nests in large groups primarily in the Gulf of Mexico, nearly all at Rancho Nuevo, Tamaulipas, Mexico. Adults congregate to mate in ocean waters off nesting beaches in March, just prior to the nesting season, which lasts from April through July. The females crawl onto beaches en masse during daylight in an event termed an *arribada* and dig 45-cm deep nests in fine sand, roughly 5-60 m inland. One to four clutches of 90-110 white, round eggs are laid per season. Incubation lasts 40-60 days, and hatching success averages 65% to 75% (Ernst and Lovich 2009). Hatchlings synchronously emerge at night, and emergence often coincides with strong northerly winds (Pritchard and Marquez M. 1973). For the few that survive to adulthood, sexual maturity is reached in about 10-16 years; average life span is unknown but may be at least 35 years (Ernst and Lovich 2009).

Of the five sea turtle species known to occur within Maryland's bays or in the Atlantic Ocean off Maryland, the Kemp's Ridley was the second most documented during the atlas period, after the Loggerhead Sea Turtle. Sixteen records were documented from 13 atlas blocks within 10 quads. Seven records were from the beaches of Worcester County, and nine were from the Western Shore of the Chesapeake Bay. The two northernmost records were from Baltimore. Seven records were of remains, and nine were of live individuals. Of the live individuals, at least two were released from fishing gear, one had boat strike injuries, and two unhealthy individuals were discovered in the fall and later died (National Aquarium, unpubl. data). Data collected from the shorelines of the Chesapeake Bay were similar in extent to the historical distribution of the species in the state (H. S. Harris 1975). Although Harris (1975) plotted no records along the Atlantic Coast, Kemp's Ridleys were documented there during the atlas period and previously by the Maryland Marine Mammal and Sea Turtle Stranding Network and the National Aquarium.

The Kemp's Ridley Sea Turtle is critically endangered (MTSG 1996). On a single day in 1947, 42,000 Kemp's Ridley Sea Turtles nested at the beach in Rancho Nuevo, Mexico; by 1989, numbers of nesting females had dwindled to fewer than 550 (Ross et al. 1989; Pritchard 1990). Numerous threats to this species' survival have been identified, from animal predation and natural nesting beach deterioration to various human-created problems, including exploitation for eggs and meat, entanglement in fishing gear, degradation and loss of nesting beaches, beachfront lights causing disorientation in hatchlings, boat strikes, and marine pollution such as oil spills and plastic debris that resembles prey. Presently, Kemp's Ridley Sea Turtle is listed as endangered under the ESA and by the state of Maryland (MDNR 2016b). Joint protection for marine turtles is provided by the USFWS and the NMFS. Conservation efforts include protection of nesting beaches, creation of lighting ordinances, and translocation of hatchlings to additional protected beaches at Padre Island, Texas, and elsewhere. From 1978 to 1992, more than 22,000 juveniles were released at Florida and Texas beaches, and the population began to recover—the count of adult females in the Gulf of Mexico was estimated at 5,000 early in the twenty-first century (Spotila 2004). However, the number of nests has declined since 2010, raising concerns that recovery may have stalled or even reversed (NMFS 2016).

LYNN M. DAVIDSON

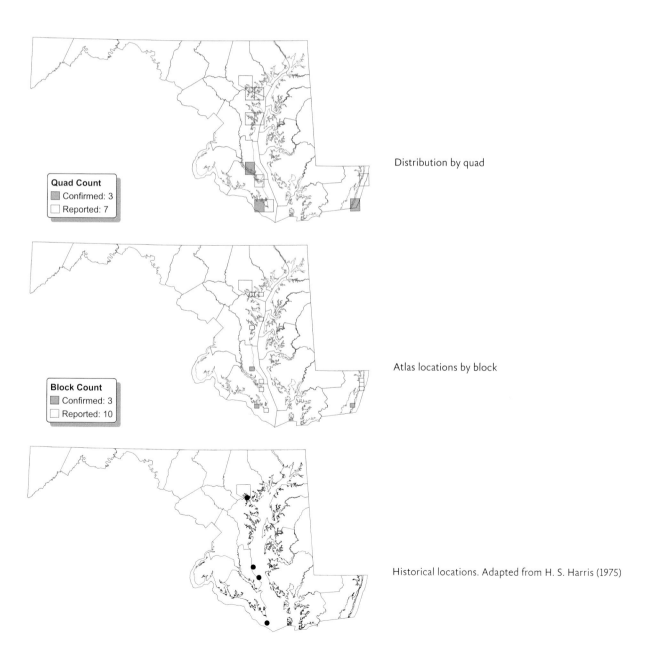

Distribution by quad

Quad Count
Confirmed: 3
Reported: 7

Atlas locations by block

Block Count
Confirmed: 3
Reported: 10

Historical locations. Adapted from H. S. Harris (1975)

Photograph by George M. Jett

Leatherback Sea Turtle
Dermochelys coriacea

As the only living member of the Family Dermochelyidae, which dates back more than 100 million years, the Leatherback Sea Turtle is truly a unique species. It is one of the few turtles lacking a hard, bony shell. Its torso is covered by oil-saturated connective tissue over interlocking dermal bones, giving it a leathery, smooth, and flexible shell, with seven dorsal and five ventral longitudinal ridges. These ridges, along with a tapering carapace at the rear, provide a hydrodynamic structure. The shell is blackish with pale spots above and pinkish white below. The Leatherback Sea Turtle's paddle-like front flippers are larger in proportion to body size than those of any other sea turtle. Another unique characteristic of this species is the lack of claws on all limbs. As the largest turtle species in the world, most Leatherback Sea Turtles reach an adult carapace length of 135-178 cm and a weight of 295-544 kg. Individuals from Maryland have been estimated to reach 180 cm in length (Scarpulla 1989) and weigh up to 500 kg (*Maryland Conservationist* 1932).

With the widest distribution of any reptile, the Leatherback Sea Turtle is found in every ocean of the world except the Arctic and Southern oceans. In the Atlantic Ocean, it occurs as far north as Newfoundland (Ernst and Barbour 1972; Rebel 1974). The first documented specimen from US waters was taken in 1811 from the Chesapeake Bay (Carr 1952). Numerous sighting and stranding records along the mid-Atlantic US coast indicate that the species is seasonally common in this area (Ernst and Gilroy 1979). In Maryland, early documentation of Leatherback Sea Turtles includes four Chesapeake Bay records in Calvert and Dorchester counties from 1932 to 1967 (J. D. Hardy 1969; H. S. Harris

1975). Fifty stranding records from 1991 to 2009 include 6 from the Chesapeake Bay at Kent Island, Smith Island, and Crisfield, and 44 carcasses on beaches or in near-shore coastal waters in Worcester County (J. Williams 1995; J. Evans et al. 1997; Norden et al. 1998; MDNR, unpubl. data; National Aquarium, unpubl. data). Live individuals reported before 2010 include two in the Atlantic Ocean off the coast of Maryland near the Isle of Wight Shoal in 1982 (Scarpulla 1989) and two disentangled from fishing gear: one in Assawoman Bay in 2001 and the other 32 km off Ocean City in 2002 (National Aquarium, unpubl. data). Potential nesting activity was observed on Assateague Island in Maryland in 1996, but no eggs were found (Rabon et al. 2003).

The Leatherback Sea Turtle is a pelagic species and primarily occupies open oceans. It can inhabit colder waters than other sea turtles due to its large body size and thick layer of fat, which allows it to maintain a higher internal body temperature than its surroundings. Leatherbacks dive deeper than any other sea turtle, up to 1,230 m. These sea turtles may remain underwater for more than 70 minutes, but dive durations typically average 5-20 minutes (Ernst and Lovich 2009). Leatherbacks nest on sandy beaches circum-globally, mostly in the tropics. In the eastern United States, natal areas include Texas, eastern Florida, and, infrequently, Georgia, South Carolina, and North Carolina. Leatherback Sea Turtles travel an average of 6,000 km between feeding and breeding areas, farther than any other sea turtle, but they may remain near shore during migrations (Ernst and Gilroy 1979; Ernst and Lovich 2009). They feed primarily on soft-bodied animals, especially jellyfish, but also consume floating seaweed and algae.

Nesting occurs primarily from April through July, usually every 2-4 years. Females crawl onto the beach and dig a nest within 15 m of the high-tide zone. In a nesting season, Leatherback females lay 5-7 clutches of about 80-90 white, round eggs, with an average of 10 days between each clutch. Eggs incubate for 50-78 days, after which hatchlings synchronously emerge at night and dash for the surf (Ernst and Lovich 2009). For the few that survive to adulthood, sexual maturity is reached in about 6-15 years. The average lifespan is unknown but may be at least 30-45 years (Ernst and Lovich 2009).

The Leatherback was the third most documented sea turtle during the atlas period. Eight records were obtained from seven blocks within five quads. The northernmost of these was off Kent Island near the Chesapeake Bay Bridge. Two records were of swimming individuals, two others were of live turtles entangled in fishing gear, and three were of remains, which showed evidence of damage from boat propellers or fishing gear lacerations. As also reported by H. S. Harris (1975), Leatherbacks were observed in the Chesapeake Bay during the atlas period, but they were also recorded along the Atlantic Coast during the project. An independent offshore survey conducted during the period of the MARA project documented Leatherbacks as the second most commonly occurring sea turtle at least three miles offshore of Maryland in the Atlantic Ocean (Pallin et al. 2015).

The Leatherback Sea Turtle is listed as endangered under the ESA and by the state of Maryland (MDNR 2016b). Marine turtles are protected jointly by the USFWS and the NMFS. Throughout their migratory pathways, Leatherback Sea Turtles face a number of threats. Natural challenges include predation and habitat deterioration at natal beaches. Human-generated problems for Leatherbacks include exploitation for eggs and meat, entanglement in fishing gear, degradation and loss of nesting beaches, beachfront lights causing disorientation of hatchlings, and boat strikes. Marine pollution such as oil spills and floating plastic debris, especially plastic bags that resemble jellyfish prey, also pose very series threats to the species. Conservation efforts to address some of these threats are underway, including the protection of nesting beaches and creation of lighting ordinances. However, most conservation actions will require international cooperation and significant societal changes.

LYNN M. DAVIDSON

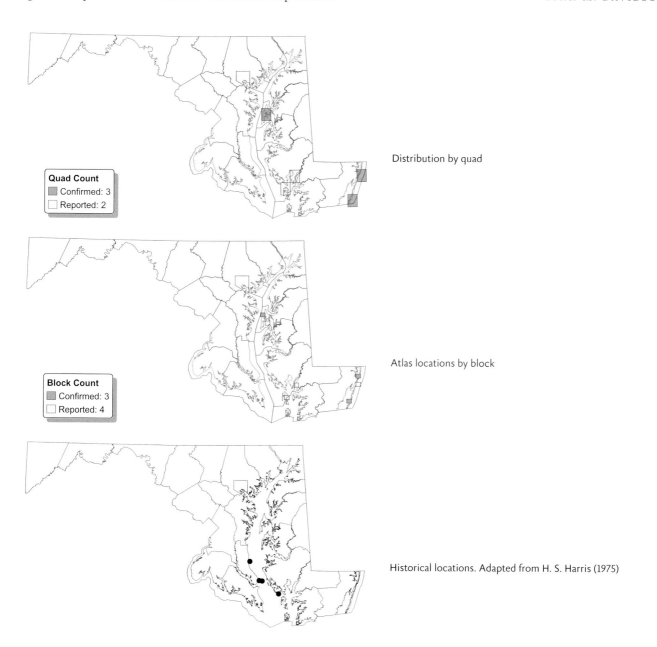

Distribution by quad

Atlas locations by block

Historical locations. Adapted from H. S. Harris (1975)

Photograph by Andrew Adams

Snapping Turtle
Chelydra serpentina

The Snapping Turtle is the largest freshwater turtle in Maryland. This species has a robust stature, with a large head, powerful jaws, strong limbs, and heavy claws. The carapace and dorsal surfaces of the body are black, gray, tan, or brown. The scutes of the carapace are sculpted, each with a low keel that forms three longitudinal ridges, and the posterior margin of the carapace is strongly serrated. In older individuals, the ridges are less pronounced, and the shell becomes smooth and may be covered by algae. The plastron is yellowish to tan and covers only about half the ventral surface. The eyes are set high on the head, and the jaws have a sharp beak. The chin has a small pair of fleshy protuberances, called barbels. The neck is long, about one-half to three-quarters the carapace length, and can "snap" in any direction at a potential predator (Mitchell 1994). The tail is thick and nearly as long as the carapace and bears a dorsal row of enlarged scales. Males are generally larger than females (Ernst and Lovich 2009). Adults typically reach a carapace length of 20.3–36 cm but may reach a maximum length of 49 cm and a weight of 34 kg.

The geographic range of the Snapping Turtle is east of the Rocky Mountains, from southern Nova Scotia to southern Florida and west to southeastern Alberta and southern Texas. In Maryland, the distribution is statewide (McCauley 1945; H. S. Harris 1975), including several barrier islands (Mitchell and Anderson 1994; White and White 2007).

Snapping Turtles inhabit relatively shallow, slow-moving freshwater and brackish-water habitats with soft mud or sand bottoms and extensive aquatic vegetation. They will lie buried in the substrate with only their eyes and nos-

trils exposed and then extend their necks to the surface to breathe (Ernst and Lovich 2009). Submerged woody debris, human-made objects, aquatic plants, beaver lodges, and beaver and muskrat bank burrows serve as shelter areas. Snapping Turtles lack lachrymal salt glands to excrete extra salt, and individuals inhabiting brackish marshes move between areas of higher and lower salinities to limit their exposure to harmful salinity levels (Dunson 1986). On Long Island, New York, Kinneary (1993) found Snapping Turtles in habitats with salinities up to 24.5 ppt. In saline environments, there is an inverse relationship between turtle body mass and net water loss, giving larger adults an advantage over smaller individuals (Dunson 1986). In general, Snapping Turtles rarely bask out of water (Boyer 1965); individuals in northern regions bask more than their southern counterparts (G. P. Brown and Brooks 1993). More commonly, these turtles thermoregulate by moving into preferred water temperatures of about 24–28 °C (Schuett and Gatten 1980; Spotila and Bell 2008), or they float at the water's surface to bask (Ernst and Lovich 2009). Hibernation occurs in ponds, small streams, and muskrat and beaver lodges (Meeks and Ultsch 1990). Although Snapping Turtles are primarily diurnal, nesting females may be active at night (Ernst and Lovich 2009). During drought conditions, turtles may aestivate in mud or move to a wetter habitat (Ernst and Lovich 2009).

Mating occurs in the water from April to November (F. J. Schwartz 1967). Females can store sperm up to several years, and DNA fingerprinting confirms that clutches of eggs can have multiple paternity (Galbraith et al. 1993). The eggs are white and spherical with a leathery shell, are generally laid from mid-May to mid-June, and typically hatch in 75–95 days; clutch sizes range from 4 to 109 eggs but average about 35 eggs (Ernst and Lovich 2009). Snapping Turtles have temperature-dependent sex determination, and eggs positioned higher in a nest, which is warmer, produce more females (Wilhoft et al. 1983). Longevity in the wild can exceed 40 years (Galbraith and Brooks 1989).

Snapping Turtles are omnivorous. Individuals eat algae, aquatic plants, insects and other invertebrates, fish, amphibians, turtles, snakes, birds, and mammals, including carrion (Ernst and Lovich 2009). Adults frequently ambush their prey, while young individuals actively forage. Raccoons, foxes, and skunks are among the many predators of Snapping Turtle eggs. A high percentage of nests are destroyed or raided each season, and hatchlings and juveniles are preyed upon by large fish, bullfrogs, adult Snapping Turtles, watersnakes, large wading birds and raptors, and mammalian predators (Ernst and Lovich 2009). Adult mortality results from commercial harvest, drowning in hoop (fyke) nets set for fish and on trotlines, and being hit by vehicles (Ernst and Lovich 2009).

During the MARA project, Snapping Turtles were found in 757 blocks within 221 quads. The MARA distribution matched the historical statewide range in Maryland (H. S. Harris 1975).

In Maryland, legislation was adopted in 2007 to authorize the MDNR to establish regulations to manage Snapping Turtles, including commercial harvest. The Snapping Turtle Workgroup developed recommendations for regulations that took full effect in 2009. These regulations (Code of Maryland Regulations 08.02.06.01) include a prohibition on possession of Snapping Turtles under 28 cm (11 in) in carapace length, which was designed to protect more than

60% of females. In addition, Snapping Turtles may be harvested only from tidal waters, commercial harvest reports are required, and harvest gear restrictions are now in place. Regulations pertaining to personal use and pets allow possession of one Snapping Turtle. Conservation efforts should concentrate on reducing mortality, primarily by enforcing regulations to prevent overharvesting and bycatch. The Snapping Turtle was added to CITES, Appendix III, effective 21 November 2016.

DAVID E. WALBECK

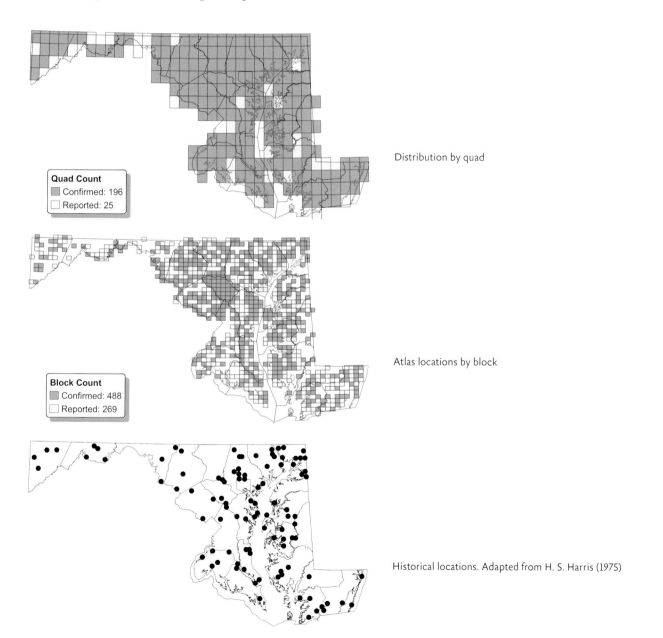

Distribution by quad

Quad Count
Confirmed: 196
Reported: 25

Atlas locations by block

Block Count
Confirmed: 488
Reported: 269

Historical locations. Adapted from H. S. Harris (1975)

Photograph by Andrew Adams

Painted Turtle
Chrysemys picta

The Painted Turtle is the most widely distributed turtle in North America, occurring from the Atlantic to the Pacific oceans. The generic name, *Chrysemys*, means "golden turtle" (Hulse et al. 2001), possibly referring to the yellow plastron; *picta* means "to paint" and highlights this species' colorful skin and carapace markings (Mitchell 1994).

The dark, olive black carapace of the Painted Turtle is smooth, flattened, and keel-less. Red bars or crescents mark the edge of the marginal scutes, and the anterior edges of the vertebral and pleural scutes are bordered with light olive to yellow. The thin scutes overlying the shell are shed annually (Oliver 1955). Unique among all Maryland turtles, in eastern populations, the vertebral and pleural scute seams align, and their markings form light-colored bands that cross the shell. In western populations, however, these scutes do not align and instead are staggered and lack the light-colored borders. The plastron is solid yellow in eastern populations, whereas western populations have a dark purplish blotch in the center. Between these extremes in Maryland, Painted Turtles exhibit a broad range of intermediate shell characteristics of carapace scute alignment and central blotching on the plastron. The head, neck, legs, and tail are black with yellow striping that changes to red closer to the body. An elongated, yellow spot is present behind each eye. Male Painted Turtles have long, thick tails and elongated foreclaws that are used in courtship displays (Ernst and Lovich 2009). Adult Painted Turtles attain a carapace length of 11.5–20 cm, with adult females being noticeably larger than males. Juveniles have a round, slightly keeled carapace but are otherwise miniature replicas of adults.

The Painted Turtle ranges from southern Canada through Maine and south to Georgia. In the north, the range extends west to Oregon and Washington State. Scattered disjunct populations occur in the Southwest. Painted Turtles are distributed throughout Maryland (H. S. Harris 1975), although populations occurring on Assateague Island may have been introduced (Mitchell and Anderson 1994).

These turtles are found in almost any freshwater habitat in Maryland, including streams, rivers, ponds, lakes, swamps, and ditches, and they will also enter brackish water (Mitchell and Anderson 1994; Ernst and Lovich 2009). Typical aquatic habitat has emergent and submerged vegetation for cover, a soft muddy bottom for hibernation, and basking sites. Often, many turtles can be found basking on the same log, and the slightest disturbance will send them diving into the water. Painted Turtles may be active during any month of the year and can be seen basking during warm periods in the winter or even swimming beneath ice (Oliver 1955; White and White 2007). They are omnivores, eating almost any plant or animal matter in their habitat. They actively forage for living prey but will also readily feed on carrion.

Courtship typically occurs in March through June and begins with the male swimming after a prospective mate; once he overtakes the female, he swims around to face her and waves the backside of his elongated foreclaws alongside and lightly touching the sides of her face (Ernst and Lovich 2009). If the female is receptive, she will respond with a reciprocating nail-waving gesture, touching his forelegs. After copulation, the female may store sperm in her oviduct for several years; therefore, a given clutch of eggs may represent multiple fathers. Nesting begins in May, typically in late afternoon or early evening (Ernst and Lovich 2009). Females lay eggs in loose soil in sunny locations, sometimes several hundred meters from water. Clutches contain 2–14 elliptical eggs, and multiple clutches may be laid each year. The sex of offspring is temperature dependent: warmer temperatures (29–32 °C) result in females, and cooler temperatures (21.5–27 °C) produce males; equal sex ratios are produced at temperatures of 20 °C and 28 °C (Ernst and Lovich 2009). Hatching occurs in about 65–80 days, although hatchlings may overwinter in the nest, emerging the follow spring. Hatchlings measure 2.7 cm in carapace length (Beane et al. 2010). Females mature in 6–10 years and males in 2–4 years (Ernst and Lovich 2009).

During the MARA project, Painted Turtles were documented in 771 blocks within 217 quads. They were found on all provinces in Maryland, similar to the distribution documented by H. S. Harris (1975). The fewer records, both historical and current, from the Appalachian Plateaus may be due to lack of suitable habitat.

Painted Turtles are common in Maryland and throughout their entire distribution. Often, they are the most abundant turtle species observed in suitable habitats (Ernst et al. 1994). Their ability to adapt to human-made habitats, however, makes them vulnerable to road mortality (Buhlmann et al. 2008).

WAYNE G HILDEBRAND

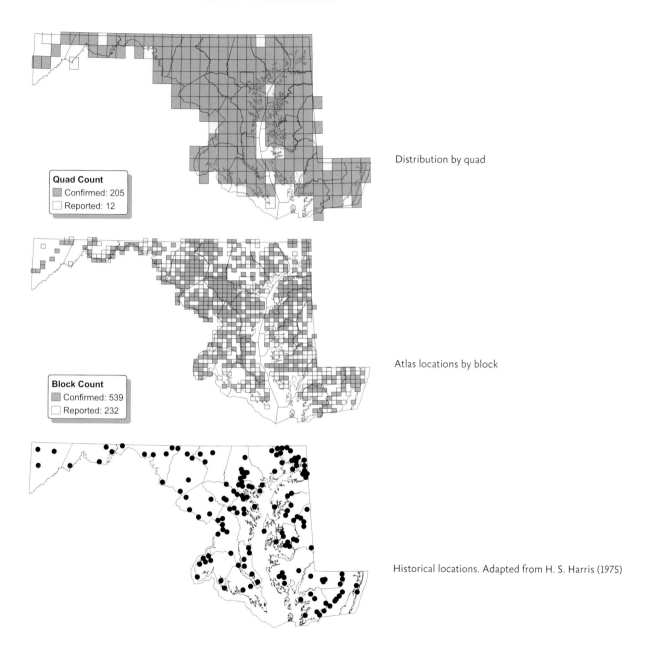

Distribution by quad

Quad Count
Confirmed: 205
Reported: 12

Atlas locations by block

Block Count
Confirmed: 539
Reported: 232

Historical locations. Adapted from H. S. Harris (1975)

Photograph by Andrew Adams

Spotted Turtle
Clemmys guttata

Like stars against the night sky, yellow to orange spots scattered across a jet black carapace make the Spotted Turtle one of the most recognizable and sought-after turtle species in Maryland. The carapace is oval, smooth, and evenly arched and has one or more yellow or light orange spots per scute. The plastron is yellowish with a large, black blotch near the margin of each scute; these blotches often coalesce in older individuals. The upper surfaces of the head, limbs, and tail are black; the undersides of the limbs and tail are orange. Yellow to orange spots are prominent on the head and may also be present on the upper surface of the limbs. In older individuals, spots may become faint or fade entirely. Female Spotted Turtles have bright orange eyes, a conspicuous yellow stripe on the lower jaw, and orange mottling or streaks on the throat. Males have dark brown eyes, lack the lower jaw stripe and throat coloration, and have a longer and thicker tail and a slightly concave plastron (McCauley 1945; F. J. Schwartz 1967). The Spotted Turtle is a small turtle, with adults attaining a carapace length of 9-11.5 cm. Hatchlings are more circular and flattened than adults, with a slight vertebral keel and a single yellow spot on each scute (McCauley 1945).

The Spotted Turtle is primarily distributed along the Atlantic Coast states from southern Maine south to northern Florida and in the Great Lakes region. In Maryland, the species has been documented in every county (H. S. Harris 1975). Most historical records were reported from the Atlantic Coastal Plain and Piedmont provinces, where these turtles were considered very common (McCauley 1945). Few records exist from western Maryland (McCauley 1945;

H. S. Harris 1975), and Harris (1975) noted that the species was probably absent from southwestern Garrett County.

Spotted Turtles inhabit quiet, sluggish, muddy-bottomed waters. Ideal habitats are small, shallow streams, edges of ponds, vernal pools, ditches, swamps, and marshes. The species may also enter brackish conditions (McCauley 1945). Spotted Turtles may be seen walking along the bottom below the water's surface or sunning themselves on logs or tree roots extending over the water. Individuals are often wary and drop off logs or stumps into the water at the slightest disturbance. Once in the water, Spotted Turtles are slow moving and easily captured. A captured individual often withdraws its head with a sharp hiss but rarely attempts to bite (McCauley 1945).

These turtles are generally most active in the spring, but early-emerging individuals can be seen basking in late winter. As temperatures rise in midsummer, Spotted Turtles become scarce, hiding in thick vegetation and mud (Mitchell and Anderson 1994). They may aestivate in late summer if their vernal habitats dry out. Mating occurs in the water in the spring. Females nest in the morning or late afternoon from May through July (Ernst and Lovich 2009). They typically lay 2-8 elliptical eggs in a flask-shaped nest constructed in well-drained soils or deposit them in grass tussocks, sphagnum moss, or decaying wood in full sun (Ernst and Lovich 2009). Eggs hatch in August and September after 70-83 days of incubation, but hatchlings may overwinter in the nest and emerge the following spring (Mitchell and Anderson 1994). Sexual maturity is attained for both sexes in 7-10 years (Ernst and Lovich 2009). In captivity, a Spotted Turtle lived for 42 years (F. J. Schwartz 1967).

As omnivores, Spotted Turtles eat an assortment of insects and other invertebrates, frogs and tadpoles, and salamanders, as well as aquatic algae and grasses (Mitchell and Anderson 1994). Predators include ratsnakes and watersnakes, large wading birds, and mammals (Ernst and Lovich 2009). Predation rates may vary with population; in Pennsylvania, 13.5% of turtles showed signs of predator attacks or injury from mowing, such as missing limbs and scarring of the shell (Ernst and Lovich 2009).

Spotted Turtles were documented in 104 quads and 166 blocks during the atlas period. As reported earlier by H. S. Harris (1975), they were found throughout the Atlantic Coastal Plain and Piedmont, suggesting that this species is still common in these regions. Only a single sighting was recorded from Washington County, and none were recorded from Allegany and Garrett counties, reflecting the rarity of this species in the western portion of the state.

Although seemingly still common, Spotted Turtle populations may be in decline overall due to habitat loss and overcollection for the pet trade (White and White 2007; Ernst and Lovich 2009). Populations have experienced substantial declines with the degradation of freshwater habitats

(Mitchell and Anderson 1994), and because of its pleasant demeanor, nonaggressive temperament, and aesthetic appeal, the Spotted Turtle has been valued as a popular pet (F. J. Schwartz 1967). The MDNR has adopted regulations prohibiting the taking of any Spotted Turtles from the wild in Maryland. Regulations require that a state-issued permit be obtained to possess a Spotted Turtle that was captively produced or obtained legally from another state, with proper documentation.

BRIAN C. GOODMAN

Distribution by quad

Quad Count
Confirmed: 88
Reported: 16

Atlas locations by block

Block Count
Confirmed: 122
Reported: 44

Historical locations. Adapted from H. S. Harris (1975)

Photograph by Michael Kirby

Wood Turtle
Glyptemys insculpta

The Wood Turtle, locally known as "red-legs" or "red-legged snapper," is a medium-sized semiaquatic turtle. Wood Turtles have a rough, oval-shaped carapace that is brown, olive brown, or grayish brown in color. Each scute has a series of concentric growth rings and ridges that give the appearance of a piece of sculpted wood—hence its name. The plastron is cream to yellow in color with a large dark blotch along the outer margin of each scute. The skin on the neck and interior portions of the legs is a bright orange or orange-red. Males have a wider head and longer, more robust tail than females, and a concave plastron. In a given population, the largest individuals are males. Adults have a carapace length ranging from 14 cm to 20 cm, with the largest just over 25 cm. Hatchlings look considerably different than adults. From above, their shape looks almost circular, and their tail is extremely long, often as long as the carapace. The bright orange coloration is also lacking for the first 3-4 years of life (E. L. Thompson, pers. obs.).

The Wood Turtle has a decidedly northern distribution and ranges from Nova Scotia south to northern Virginia and west into the Great Lakes region to Minnesota and Iowa. In Maryland, this species primarily occurs above the Fall Line through the Piedmont west to the Appalachian Plateaus (H. S. Harris 1975). McCauley (1945) reported that the Wood Turtle was entirely absent from the Atlantic Coastal Plain of Maryland; however, a record has since been reported from Cecil County (H. S. Harris 1975), as well as several from Prince George's and Anne Arundel counties (Norden and Zyla 1989). A record reported by J. Norman (1939) and an-

other by H. S. Harris and Crocetti (2007) from Talbot County are suspected releases.

These turtles require a flowing stream or river as the main feature of their habitat. Wood Turtles are typically found in clear streams with moderate currents and sand or gravel substrates. They may occupy swamps, bogs, and wet meadows during the active season. Adjacent upland habitats such as forest, fields, and pastures are used, particularly during summer months, but they usually do not stray far from water (Ernst and Lovich 2009). However, some individuals will make long upland forays, even traveling over mountain ridges to other streams (Michael Jones and Thomas Akre, pers. comm.). During research in Maryland, the author (Thompson) has found a few marked individuals several kilometers from where they were originally captured. Wood Turtles hibernate in flowing water, choosing deep pools with an undercut bank, or among the roots of a large tree growing on the edge of the stream, where they lie on the substrate or bury in mud. Wood Turtles often congregate at favored hibernacula sites (Bloomer 1978; Niederberger 1993).

Seasonal activity begins in March or April, depending on latitude. The first warm, sunny days in March stimulate activity in Maryland. Early in the season, individuals may be encountered lying motionless on the bottom of a stream, often half buried in mud. Mating commences as soon as they become active and occurs underwater. Mating pairs can be observed throughout early spring. Mating also occurs in the fall, when the turtles start congregating for winter hibernation (Kaufmann 1992). In Maryland, females lay eggs from late May through mid-June. Clutch size can range from 3 to 20 eggs but averages about 9-11 eggs. Hatching occurs in mid-August to October, with an average incubation period of 67 days. In Maryland, the author has observed recently hatched young on 30 August.

Wood Turtles are relatively long lived. Ernst (2001) reported that Wood Turtles in the wild can live for more than 40 years, and a captive individual lived 58 years (Ernst and Lovich 2009). Wood Turtles can be aged with some degree of accuracy by counting growth rings on the scutes of the carapace or plastron. However, this method becomes difficult or inaccurate for turtles older than 15-20 years (Ernst and Lovich 2009).

Wood Turtles are omnivores, feeding on various plants, berries, slugs, worms, and other invertebrates (Farrell and Graham 1991). The author has witnessed turtles feeding on slugs and fresh green vegetation along a stream. Their nests and hatchlings are often heavily preyed upon by skunks, raccoons, and foxes (Harding and Bloomer 1979).

During the atlas period, Wood Turtles were documented in 65 blocks within 38 quads and nine counties. Overall, the distribution of Wood Turtles remained unchanged since H. S. Harris's (1975) report, except for an absence of records from Cecil County. Several records were reported from along the Fall Line, and two on the Atlantic Coastal Plain in Prince George's and Anne Arundel counties. For the Atlantic Coastal Plain records, it is unknown whether the observations represent populations or released individuals, and these warrant further investigation.

Wood Turtle populations are known to be declining throughout their range (Ernst and Lovich 2009), and Maryland is no exception. The viability of populations on the Piedmont, Blue Ridge, and Appalachian Plateaus in Maryland is questionable at this time and needs additional study. Currently, the Ridge and Valley region of Maryland, where several apparently viable populations remain, is the stronghold for the species.

Wood Turtles are threatened by urban sprawl and other development, stream degradation, vehicular strikes, and agricultural machinery. Additionally, an unfortunate underground market exists for Wood Turtles, and populations are known to have been affected by poaching. Given that these turtles are long lived and have low recruitment, removal of adults can have a serious impact on the long-term viability of a population. Well-documented population declines have occurred after stream habitats have been opened to recreation (Garber and Burger 1995). Due to these various sources of pressure and habitat degradation, the long-term future of Wood Turtles in Maryland is uncertain.

ED THOMPSON

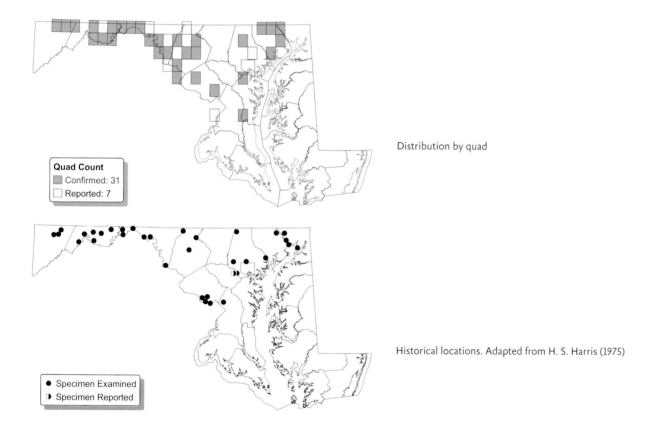

Distribution by quad

Quad Count
Confirmed: 31
Reported: 7

Historical locations. Adapted from H. S. Harris (1975)

Specimen Examined
Specimen Reported

Photograph by Jim White

Bog Turtle
Glyptemys muhlenbergii

The Bog Turtle is a tiny, secretive turtle best known for the large, reddish orange to yellow blotch on each side of its head (USFWS 2001; Ernst and Lovich 2009). The dark brown to ebony carapace is longer than it is wide and is moderately domed with a low keel. Younger individuals have growth annuli on the scutes, giving them a rough appearance, whereas older turtles have smooth shells. Some individuals have a "starburst" pattern on each scute: a faint yellow ivory or reddish central spot with radiating lines of matching color. The plastron is dark brown to black and sometimes bears irregular yellow markings. The upper jaw is medially notched. The skin is mottled brown, black, and reddish orange. Males have long, thick tails, with the cloaca beyond the carapace margin and a concave plastron. Females have short tails, with the cloaca within the carapace margin and a flat plastron (Ernst and Lovich 2009). This is one of the smallest turtles in the world; adult carapace length is only 7.5-9 cm, with males slightly larger than females.

The Bog Turtle has a discontinuous range through 12 states in the eastern United States, divided into northern and southern populations separated by a 402-km hiatus (Bury 1979). The northern population extends through seven states from western Massachusetts to northeastern Maryland, occurring in 44 counties, predominantly in New York, New Jersey, and Pennsylvania (Lee and Norden 1996; USFWS 2001). The southern population extends through five states from southwestern Virginia to north Georgia, occurring in 31 counties, with the majority in North Carolina (Stratmann 2015). In Maryland, Bog Turtles occur primarily on the Piedmont of Baltimore, Carroll, Cecil, and Harford

counties (McCauley and Mansueti 1944; J. E. Cooper 1949; Reed 1956a; Campbell 1960; H. S. Harris 1975). Historically, Bog Turtles were reported from the Atlantic Coastal Plain in Cecil County (H. S. Harris 1975; R. W. Miller 1984a) and adjacent New Castle County, Delaware (Arndt 1977), but currently only one extant Atlantic Coastal Plain population is known.

These turtles are semiaquatic, as indicated by their weakly webbed feet (USFWS 2001). They occur in small, open-canopy, spring-fed wetlands, including sphagnaceous bogs, black spruce swamps, wetlands that support an abundance of grassy and mossy cover, and open marshy meadows (Bury 1979; Lee and Norden 1996; USFWS 2001; Ernst and Lovich 2009). In Maryland, larger populations occur in circular drainage basins with spring-fed pockets of shallow water, a substrate of soft mud, abundant low grasses and sedges, and interspersed dry hummocks (Chase et al. 1989). G. J. Taylor et al. (1984) described Maryland Bog Turtle wetlands as permanently wet, small (≤0.81 ha) sedge meadows with moderate to dense herbaceous vegetation and saturated soils. Bog Turtles may use streamside and stream habitats as travel corridors, and stream banks as overwintering habitat (Somers et al. 2007). In Maryland, average home ranges vary from 0.04 ha to 0.77 ha (Chase et al. 1989; Dinkelacker 2000; Morrow et al. 2001; Byer 2015; S. A. Smith, unpubl. data).

Bog Turtles are primarily diurnal and active at air temperatures of 16-31 °C. Spring activity begins in March and April, when air temperatures remain above 21 °C, with peak activity occurring in May (Lovich et al. 1992; Ernst and Lovich 2009). Activity declines through the summer with relative inactivity in August, followed by a modest increase in September as turtles move to hibernacula. Winter dormancy begins in late September to November (Lovich et al. 1992) and lasts up to 6 months (Ernst et al. 1989). Bog Turtles overwinter singly or in groups of up to 140 individuals in muskrat burrows, in subterranean rivulets, under sphagnum mats, or beneath the roots of maples and alders (Holub and Bloomer 1977; Ernst et al. 1989; Ernst and Lovich 2009).

This species is polygamous; mating occurs in April through June (Ernst and Lovich 2009). Nesting occurs in June through early July, with most nesting activity in the late afternoon and evening. Bog Turtles typically lay a single clutch of 1-6 eggs per year in elevated sedge tussocks and sphagnum mats, but they may use other elevated areas such as rotting tree roots and stumps, the edges of adjacent pastures, and railroad embankments (Ernst and Lovich 2009). The incubation period is 42-80 days, depending on temperature. Hatchlings emerge in late August and September but may overwinter in the nest. Sexual maturity is attained when plastron length is 7.0 cm or more, at about 6 years of age (Ernst 1977), though this may vary geographically (Lovich et al. 1998). Bog Turtles can live 40 or more years (USFWS 2001).

Bog Turtles are omnivorous, feeding on a diversity of animal and plant matter; slugs are a major part of the diet (Ernst and Lovich 2009). These turtles are prey for various species of reptiles, birds, and mammals, with raccoons as the most effective predator (Ernst and Lovich 2009). Bog Turtles burrow into the mud when threatened. Nest predation rates range from 12% to 100%, with small to medium-sized mammals and ants the primary predators (Ernst and Lovich 2009; Byer 2015).

During the MARA project, Bog Turtles were confirmed in 13 quads and 28 blocks, all in the four counties of historical occurrence on the Piedmont. The MARA results were roughly equivalent to the distribution shown by H. S. Harris (1975), except for a lack of reports from Elk Neck in Cecil County. No targeted efforts for Bog Turtles were conducted on the Atlantic Coastal Plain during the MARA project.

Extensive population surveys conducted from 1992 to 1994 found that Maryland Bog Turtle populations had declined 54% since the late 1970s (S. A. Smith 2004). The species was state listed as threatened in Maryland in 1994, and the northern population was federally listed as threatened in 1997 (USFWS 2001). Bog Turtles are threatened by wetland habitat loss, collection for the pet trade, emerging infectious diseases, and increased predation from human-subsidized predators (Lee and Norden 1996; Buhlmann et al. 1997; USWFS 2001). A metapopulation approach to Bog Turtle conservation (Buhlmann et al. 1997) has been implemented in Maryland but needs to be accelerated to give this small, rare turtle a fighting chance.

SCOTT A. SMITH

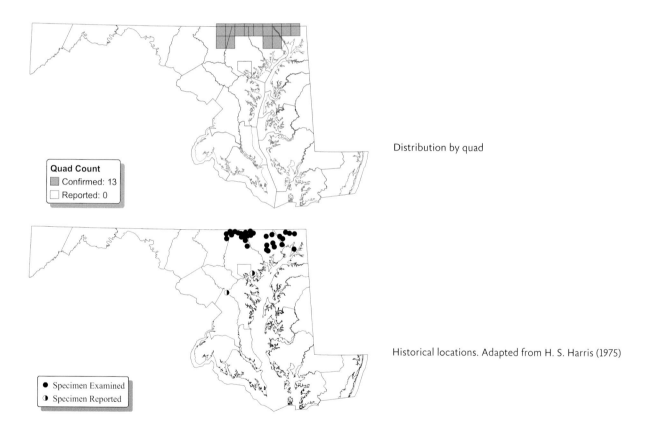

Quad Count
Confirmed: 13
Reported: 0

Distribution by quad

Specimen Examined
Specimen Reported

Historical locations. Adapted from H. S. Harris (1975)

Photograph by Scott McDaniel

Northern Map Turtle
Graptemys geographica

The Northern Map Turtle is a shy inhabitant of large rivers and lakes. These turtles are gregarious baskers (Lindeman 2013) but extremely wary: they dive to safety at the slightest disturbance. They are distinguished from other *Graptemys* species by a triangular, yellow postorbital spot, a distinct vertical line between this spot and the eye, upward-curving lower lateral neck stripes, and a low carapacal keel with blunted points (Lindeman 2013). The olive green carapace has a network of yellow concentric and connecting ovals that resemble the topographic lines of a map—from which the species gets its common and scientific names (Ernst and Lovich 2009). The plastron is pale yellow in adults; juveniles have black pigment along the scute seams (Lindeman 2013). The skin is olive to brown black, and the head and legs have yellow to greenish yellow stripes (Ernst and Lovich 2009). Map Turtles have large jaw surfaces, which appear as light-colored "lips" (White and White 2007). Compared with females, males have narrower heads, a more oval-shaped carapace, and longer, thicker tails (Ernst and Lovich 2009). Adult females dwarf males, ranging from 18 cm to 27.3 cm in carapace length, compared with a length of 9-16 cm for males. Hatchlings have a boldly patterned carapace and a conspicuous midline keel (Hulse et al. 2001).

The Northern Map Turtle has the widest geographic distribution of its genus (Lindeman 2013). Its range extends from southern Quebec west to central Minnesota and south to northwestern Georgia through northern Louisiana. There are isolated populations in the Delaware River in Pennsylvania and in the Hudson River in New York (Ernst and Lovich 2009). Maryland's population is restricted to the

Susquehanna River and the lower reaches of its larger tributaries in Cecil and Harford counties (Ernst and Lovich 2009). Occasionally, individuals are observed in the upper Chesapeake Bay and Northeast River (H. S. Harris 1975; White and White 2007). The first Northern Map Turtle from Maryland was collected in 1882 near Havre de Grace, Harford County (McCauley 1945). The species was subsequently reported in Cecil County along the shore of Elk Neck (Conant 1945), in the Northeast River at Rodney Scout Camp on Elk Neck (McCauley 1945), and in the Susquehanna River near Darlington (McCauley 1945). Later efforts documented Northern Map Turtles in the entire lower Susquehanna River, both above and below the Conowingo Dam to Havre de Grace, and the lower reaches of Broad Creek and Deer Creek (Richards-Dimitrie 2011; Richards-Dimitrie et al. 2013; K. P. Anderson 2014; K. P. Anderson et al. 2015). Richards-Dimitrie (2011) reported Map Turtle concentrations in an island complex opposite Susquehanna State Park and at the mouths of Deer Creek and Octoraro Creek.

This aquatic species spends much of its time underwater (Carriere et al. 2009; Richards-Dimitrie 2011). The turtles inhabit large bodies of water such as rivers and lakes that have suitable foraging areas, abundant basking sites, nocturnal resting areas, nearby nesting sites, and hibernacula (Moll and Moll 2004; Ernst and Lovich 2009). Optimal nesting sites are sandy beaches; however, human-altered habitats, including lawns of shoreline residences, roadsides, railroad rights-of-way, and developed parkland, may also be used (Ernst and Lovich 2009; Richards-Dimitrie 2011). Northern Map Turtles congregate at hibernacula in deep riverine pools with well-oxygenated water, where they lie motionless on the bottom substrate, fully exposed (Richards-Dimitrie 2011). Peak activity occurs in April through early November (Ernst and Lovich 2009). While males are fairly sedentary, females will move up to 15.4 km to nest (Ernst and Lovich 2009). In Maryland, Richards-Dimitrie (2011) recorded maximum movements of 4.9 km.

Mating typically occurs near hibernacula in both the fall and spring (Ernst and Lovich 2009). Females typically lay 2 clutches per year from late May to early July, with a mean clutch size of 10-13 eggs (Ernst and Lovich 2009; Lindeman 2013). Eggs hatch in August and September, but most hatchlings overwinter in the nest and emerge the following spring (Lindeman 2013). Map Turtles have temperature-dependent sex determination, with more males produced at incubation temperatures of 25 °C and more females at or above 30 °C (Ernst and Lovich 2009). Females reach sexual maturity at 8-13 years, whereas males mature in 4 years (Lindeman 2013). Individuals usually live more than 20 years (Ernst and Lovich 2009).

Northern Map Turtles forage in the morning and late afternoon and typically bask at midday (Ernst and Lovich 2009). They feed on a variety of algae, vascular plants, invertebrates, and dead fish; gastropods are their dietary staple (Ernst and Lovich 2009). Richards-Dimitrie et al. (2013) found that adult males fed primarily on small trichopterans, gastropods, and invasive Asiatic Clams (*Corbicula fluminea*), while adult females fed primarily on pleurocerid snails.

During the MARA project, Northern Map Turtles were confirmed in five quads and seven blocks in Cecil and Harford counties. These results were comparable to those of H. S. Harris (1975), although the atlas data included more records spanning the entire Susquehanna River and an additional record from Broad Creek in Harford County. The MARA results were more comprehensive than Harris's due to the activity of Towson University researchers who were concurrently studying Northern Map Turtles during the atlas project and determined that the Maryland population is larger than previously believed.

The Northern Map Turtle is state listed as endangered in Maryland (MDNR 2016b) because of its restricted range and solitary population. Threats to this species include human-subsidized predators, water pollution, siltation, waterfront development, automobile traffic, boat propellers, habitat destruction and alteration, dams, channelization, removal of snags and logjams, disturbance resulting from recreational activities, collection for the pet trade or as food, and genetic pollution due to release of pet-trade turtles (M. J. C. Moore and Seigel 2006; Ernst and Lovich 2009; Lee 2012). In Maryland, these turtles contend with extremely variable flow rates and water levels daily caused by the Conowingo Dam (Richards-Dimitrie 2011; Richards-Dimitrie et al. 2013; K. P. Anderson 2014). High recreational use of the area by fishermen and boaters also reduces basking (K. P. Anderson 2014), and unique activities such as artifact collecting have disturbed and destroyed communal nesting sites. Furthermore, nest predation rates in Maryland exceed 95% (Richard Seigel, pers. comm.). The Northern Map Turtle's future in Maryland is looking brighter, however, with recent public-private partnerships protecting, restoring, and creating nesting and basking habitat and a greater awareness of conservation by the public and by local governments—most notably Towson University and the Town of Port Deposit.

SCOTT A. SMITH

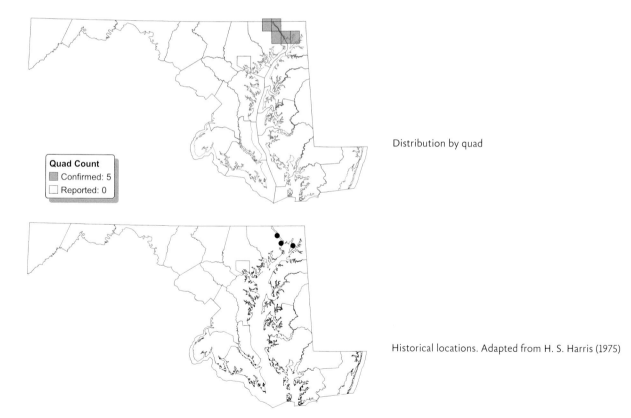

Quad Count
Confirmed: 5
Reported: 0

Distribution by quad

Historical locations. Adapted from H. S. Harris (1975)

Photograph by Lance H. Benedict

Diamond-backed Terrapin
Malaclemys terrapin

The Diamond-backed Terrapin occupies a special status in the Free State because it is the designated Maryland state reptile and the official mascot of the University of Maryland. "Terrapin" is derived from the Algonquin Indian word for turtle, and the common name is derived from the prominent grooves and ridges that form concentric rings on the cara-pace scutes. The oblong carapace of the Diamondback, as it is often called, is gray, brown, or black, with a vertebral keel. Posterior marginal scutes are somewhat serrated and curved slightly upward on their outer edges. Typically, the plastron is green to yellow in color. The gray legs and neck are intri-cately patterned with dark spots or flecks but may be com-pletely dark on some individuals. Many Diamondbacks have dark, mustache-like markings on the face, and the eyes are black or gray. Females have considerably shorter tails and proportionately larger heads. Adult male Diamondbacks measure 10-14 cm in carapace length, whereas females average about a third larger at 15.2-23.8 cm.

The Diamond-backed Terrapin occurs in coastal waters from New England south to Florida and along the Gulf Coast to southeastern Texas. In Maryland, this species most likely occurs in all coastal areas throughout the state (H. S. Harris 1975). It was once very abundant along the Atlantic Coast and inland bays and in Chesapeake Bay marshes, ranging perhaps to the limits of tidal water. However, there is some question as to the validity of records in the Potomac River near Washington, DC (McCauley 1945).

These turtles occupy salt and brackish waters in all coastal areas of the Atlantic Coastal Plain. The Diamond-back is the state's only true estuarine and salt marsh turtle,

inhabiting larger, open waters as well as tidal creeks, guts, and marshes, especially those lined with cordgrass (*Spar-tina* sp.) (Beane et al. 2010). This species possesses lachry-mal glands for the excretion of excess salts in body fluids (Mitchell 1994), and with a behavioral trait of drinking from thin films of fresh rainwater at the water's surface (Ernst and Lovich 2009), Diamondbacks are quite comfortable in large, open bodies of saltwater. Here, they spend a considerable amount of time basking; however, they will penetrate far into the shallows. These turtles are frequently encountered by boaters and beachgoers, when they see Diamondback heads and pale-colored jaws protruding from the water or encounter nesting females along sandy beaches.

Diamondbacks are generally active from April through October and overwinter in hibernacula in the mud bottoms of channels and tidal flats (Ernst and Barbour 1972). Numer-ous mark-recapture studies have demonstrated a relatively sedentary lifestyle, with small-scale movements and remark-able site fidelity. Courtship and mating take place in April and May, and nesting occurs from early June to late July (Jeyasuria et al. 1994), usually during the day and above the high-tide line. Nests are constructed in shrubland, dune, and mixed grassland habitats, with exposed sand along trails or beaches most heavily used (Feinberg and Burke 2003). Two clutches are often laid (Gibbs et al. 2007), each containing 5-29 pinkish white eggs. Incubation time, like sex deter-mination, is temperature dependent and varies from 50 to 120 days in Maryland (Jeyasuria et al. 1994). Hatching peaks in September, but hatchlings may overwinter in the nest and not emerge until early the following spring. Juveniles spend the first few years of life within and beneath tidal debris near the shoreline (Gibbs et al. 2007). In Maryland, male Diamond-backed Terrapins mature at 4-7 years of age, while females mature at about 8-13 years (Roosenburg 1992); age of sexual maturity is related to egg size (Roosenburg and Kelley 1996). Longevity in this species perhaps exceeds 40 years (Schwartz 1967).

Diamond-backed Terrapins are omnivorous, with a pref-erence for mollusks and crustaceans (Carr 1952), and they have a special affinity for Mud Snails (*Nassarius obsoletus*) (Mitchell 1994) and Marsh Periwinkle Snails (*Littoraria irro-rata*) (Beane et al. 2010). Cochran (1978) reported an apparent attraction to both fresh and cured fish as well.

During the MARA project, Diamond-backed Terrapins were found in 170 blocks within 72 quads—essentially most areas with estuarine waterways, except for the upper reaches of the Chesapeake Bay. This distribution was com-parable to that reported by H. S. Harris (1975), but with the addition of atlas records along the Potomac River shores of St. Mary's County. Inland records such as those from Balti-more and Montgomery counties are presumed releases.

Consumer demand for turtle soup spurred commercial harvesting of Diamond-backed Terrapins with traps and seines and winter dredging of hibernacula. These practices drove Diamondback populations to perilously low levels in the early part of the twentieth century. Commercial harvesting is now prohibited in Maryland (as of 1 July 2007), but Diamondbacks still face significant mortality from accidental drowning in commercial and recreational crabbing and fishing devices, nest and hatchling predation, and vehicular strikes of nesting adults. Diamondback eggs are also destroyed by beach grass (*Ammophila breviligulata*) roots, which can grow into eggs and kill the embryos (Lazell and Auger 1981). Alteration of shorelines by development and by efforts to prevent erosion has adversely affected and restricted access to nesting habitat (Roosenburg 1991). In Maryland, conservation efforts should focus on reducing drowning mortality in commercial and recreational fishing devices such as bank traps, crab and eel pots, and nets. Bycatch-reduction

devices installed in crab and eel pots are a welcomed improvement to trap design (R. C. Wood 1997; Radzio and Roosenburg 2005). These devices are now required on all recreational crabbing pots in Maryland, but Radzio et al. (2013) found less than 35% compliance. Abandonment or loss of pots is also a serious problem (Grosse et al. 2009). These "ghost pots" continue to capture and drown Diamondbacks throughout the life of the trap, unless they are found and removed. Lastly, improved management of road-crossing areas and habitat restoration of nesting sites could benefit many local populations. Following nesting habitat restoration efforts on Poplar Island in the Chesapeake Bay, Roosenburg et al. (2014) determined that nest survivorship was significantly higher than in mainland nests due to the lack of mammalian predators on the island.

KYLE RAMBO

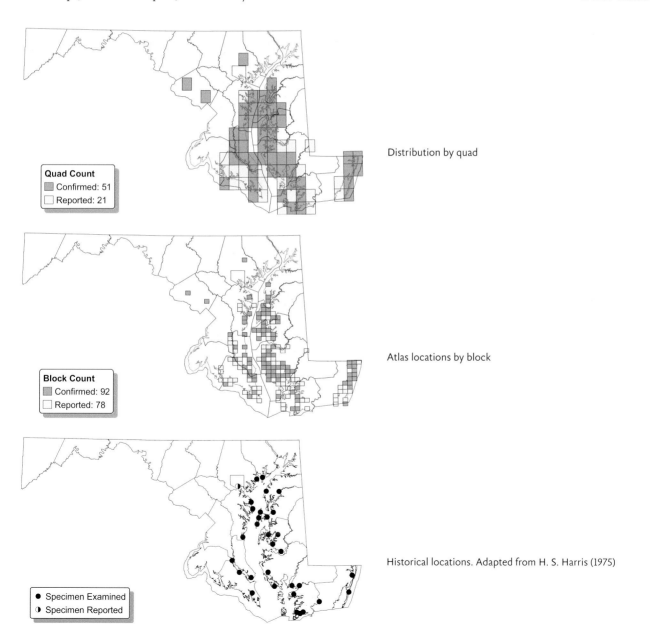

Distribution by quad

Quad Count
■ Confirmed: 51
□ Reported: 21

Atlas locations by block

Block Count
■ Confirmed: 92
□ Reported: 78

Historical locations. Adapted from H. S. Harris (1975)

● Specimen Examined
◐ Specimen Reported

Photograph by Lance H. Benedict

Northern Red-bellied Cooter
Pseudemys rubriventris

The Northern Red-bellied Cooter is Maryland's largest basking freshwater turtle. The robust, elongate, relatively flattened carapace is brown to black with vertical reddish markings on the marginal and pleural scutes. The plastron varies from pinkish to reddish orange and sometimes retains faint dark smudges (Ernst and Lovich 2009). The carapace may become covered with a veneer of green algal growth and silt late in summer. The head and limbs are black with thin yellow stripes. On top of the head at the snout, a yellow prefrontal "arrow" is formed by a central sagittal stripe that passes anteriorly between the eyes and meets the supratemporal stripes extending from the eyes. Coloration is often difficult to distinguish when viewed from a distance, giving this turtle the appearance of a dark head and carapace. The upper jaw has a prominent notch at the tip with a tooth-like cusp on each side, which helps distinguish this species from other pond turtles of similar size. Males are distinguished by long, straight front claws that can exceed 2.4 cm in length, whereas foreclaw length in females averages 1.4 cm (Swarth 2004). Northern Red-bellied Cooters are large turtles measuring 25.4–32 cm in carapace length, and females are the larger sex. Hatchlings have prominent yellow stripes on the head and neck and a brilliant green, keeled carapace with forked yellowish markings on the pleural scutes and a vertical yellow line through each marginal scute. The plastron is bright red with black to purplish swirls and spots that disappear with age.

The Northern Red-bellied Cooter is found in the mid-Atlantic region primarily on the Atlantic Coastal Plain from Long Island, New York, south to the Jacksonville area of North Carolina. A small, disjunct population occurs in Plymouth County, Massachusetts. In Maryland, these turtles are found on the Atlantic Coastal Plain and Piedmont and enter the Ridge and Valley along the Potomac River, where they have been documented as far west as Cumberland in Allegany County (McCauley 1945). H. S. Harris (1975) considered their distribution as primarily occurring on the Atlantic Coastal Plain. Historical records from the southern end of Assateague Island are considered to be of introduced individuals (Mitchell and Anderson 1994).

These turtles inhabit slow-moving, relatively deep bodies of water with soft, silty bottoms and numerous basking locations, such as broad and slow-moving rivers, freshwater tidal wetlands, and beaver and mill ponds. Around the Chesapeake Bay, Northern Red-bellied Cooters occur in tidal waters where the salinity may reach 2–3 ppt (Arndt 1975). An abundance of aquatic plants is required for this herbivorous turtle. Nesting habitats include meadows, pastures, road edges, and other open, sunny habitats within several hundred meters of its aquatic habitats.

In Maryland, Northern Red-bellied Cooters are active from mid-April to mid-November. They spend the winter underwater, buried in soft sediments beneath permanent bodies of water. Emergence from overwintering sites takes place in April. Basking commences as early as 9:00 a.m. and reaches a peak between 11:00 a.m. and 1:00 p.m., when water temperatures are about 23 °C (Swarth 2004). When basking, Northern Red-bellied Cooters are wary and difficult to approach. On warm afternoons in favorable locations, a dozen or more may be seen sunning on logs or pilings, sometimes sharing basking sites with Painted Turtles and Wood Ducks (*Aix sponsa*). In favored locations, 75 turtles or more may be observed basking at one time (C. Swarth, pers. obs.).

The nesting season begins in late May, peak nesting is in mid-June, and nesting typically concludes by mid-July. Of 127 nest-searching females observed over a five-year period in grassy open areas at Jug Bay Wetlands Sanctuary on the Patuxent River in Anne Arundel County, 43 were observed between 5 and 19 June (Swarth 2004). Females often return to the same nesting area year after year, and some females lay 2 clutches in the same season (Swarth 2004). Nest excavation begins as early as 7:00 a.m. and appears to be most common in the morning, but some females excavate nests and lay eggs at night. The turtles may take several hours to search for and discover a suitable nesting site, but excavation, egg laying, and covering of the nest altogether takes only about 40 minutes. The clutch size of 18 nests at Jug Bay averaged 12 eggs (Swarth 2004). The incubation period lasts approximately 75 days, and hatchlings emerge in the fall or overwinter in the nest and emerge in the spring. On emergence from the nest, hatchlings move immediately in the direction of water and will travel as fast as 39 m per hour on this journey (Dickey et al. 2009).

Unlike other freshwater turtles in Maryland, Northern Red-bellied Cooters are almost entirely herbivorous as adults, their diet consisting of a variety of aquatic plants (Ernst and Lovich 2009). The waterweed *Hydrilla verticillata* and the water lily *Nuphar lutea* are important in the adult diet on the Patuxent River (Fogel et al. 2002).

During the atlas project, Northern Red-bellied Cooters were recorded in 139 quads and 315 blocks. Documented in all counties except Garrett, this species appears to have greatly expanded its range since H. S. Harris's (1975) publica-

tion, particularly throughout the Piedmont, and is no longer confined to lowland river valleys as Harris suggested.

This turtle has no special protected status in Maryland; however, populations in Pennsylvania are state threatened, and the Massachusetts population is federally endangered. Ernst et al. (1999) reported shell disease in Northern Red-bellied Cooters from the Rappahannock River, Virginia, but this has not been observed in Maryland turtles. Nests located in agricultural fields may be destroyed if fields are plowed in June and July. In areas with motorboat traffic, these turtles are often struck, and some adults display spectacular carapace scars from boat propellers.

CHRISTOPHER SWARTH

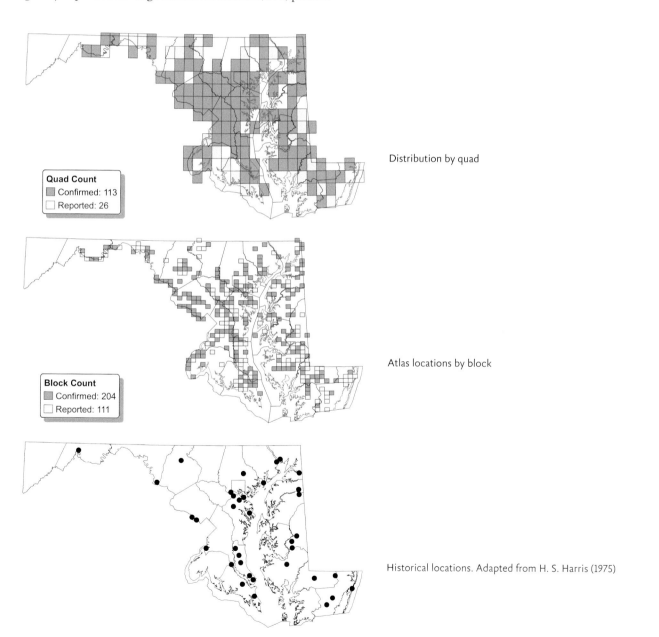

Distribution by quad

Quad Count
◼ Confirmed: 113
☐ Reported: 26

Atlas locations by block

Block Count
◼ Confirmed: 204
☐ Reported: 111

Historical locations. Adapted from H. S. Harris (1975)

Photograph by Andrew Adams

Eastern Box Turtle

Terrapene carolina

Almost everyone has encountered the small, mild-mannered Eastern Box Turtle. This species is easily recognized by its domed, patterned shell and a habit of "boxing up" tightly when disturbed. The hinged plastron allows the turtle to close up completely by withdrawing its head, legs, and tail within the shell. The carapace is brownish, with variable yellow and orange lines, spots, bars, or splotches (Ernst and Lovich 2009). The colors and patterns on the carapace are unique and permanent; mature individuals can be confidently identified by these patterns. The age of a turtle can be approximated up to about 18 years by counting the growth rings on the scutes (Dodd 2001). One ring represents one season of growth. In males, the posterior lobe of the plastron is concave, whereas in females it is flat. The iris in males is red, but in females is usually brown. Males have a thicker, longer tail, and the hind claws are stout and curved. Females have a short, thin tail, and the hind claws are slender, straighter, and more sharply pointed. Adult Eastern Box Turtles measure 11.5-15.2 cm in carapace length. Older males achieve a somewhat greater size than females of a similar age and may exceed 15 cm in carapace length (Ernst and Lovich 2009). In a study at the Jug Bay Wetlands Sanctuary in Anne Arundel County, the average carapace length was 13.0 cm for adult males and 12.4 cm for adult females (C. Swarth and Mike Quinlan, unpubl. data). Hatchlings have a flattened carapace and noticeable keel, with a yellowish spot on each scute. The plastral hinge does not develop until juveniles reach approximately 5.0 cm in carapace length (Ernst and Lovich 2009).

The Eastern Box Turtle is found primarily in the eastern United States, from southern Maine through southern Illinois and south to Georgia and northern Alabama. In Maryland, these turtles have been recorded in every county (H. S. Harris 1975), but they may be uncommon on the Appalachian Plateaus (McCauley 1945).

This species inhabits mixed or deciduous forests, agricultural fields, meadows, suburban backyards, stream valleys, and the margins of freshwater tidal and nontidal wetlands. Preferred habitats contain dense vegetation such as shrubs, vines, or abundant downed trees and branches. During their active season, Box Turtles travel between several habitats. They are most often seen during warm rains in the spring in deciduous forests. They spend a lot of time beneath leaf litter or under shrubs and logs, and they often seek shady places or wetlands to keep cool. Box Turtles are not well suited for swimming but will enjoy a good soak in a stream on a hot day. At dusk, they excavate a shallow "form" in soft soil, leaf litter, or under a log, where they spend the night. For nesting, gravid females seek open, sunny areas with loose soil and low vegetation, such as farm fields, meadows, lawns, or suburban gardens. Juvenile Box Turtles may spend their first 7-10 years in less than a hectare of forest, hidden beneath the leaves.

Eastern Box Turtles emerge from hibernation in April and remain active until fall. The first severe fall frost triggers individuals to excavate a 15-cm to 30-cm depression in a wooded area with a thick layer of fallen leaves. By November or December, Box Turtles are safely underground in hibernacula (Savva et al. 2010). Mating occurs opportunistically throughout the active season but peaks in spring. When a female is encountered, the male walks circles around her while biting at her shell. If receptive, the female opens the posterior end of her plastron, and the male mounts her and hooks his feet into the plastral opening. The female then closes her plastron on the male's feet, allowing him to lean back and copulate. Nesting usually takes place in June, clutch size is typically five eggs, and hatchlings emerge from the nest after approximately 75 days of incubation. Eastern Box Turtles exhibit delayed sexual maturity, high adult survivorship, and low fecundity. Sexual maturity is reached in 8-10 years (Ernst and Lovich 2009) at a carapace length of approximately 10.7 cm (Stickel 1950). Eastern Box Turtles can live 50 years or more (Dodd 2001).

Eastern Box Turtles tolerate neighboring conspecifics and many may share the same area. For example, population density in prime habitats in Anne Arundel County averaged 5.25 turtles/ha (Quinlan et al. 2003). Although they are not territorial, adult turtles inhabit a fixed home range, which in Maryland varies from an average of 1.5 ha for males to 8.3 ha for females (Swarth et al. 2011). Box Turtles are omnivores, feeding on a variety of plant and animal matter, including mushrooms, berries, invertebrates, and carrion. At Jug Bay Wetlands Sanctuary, one was observed feeding on a dead Virginia Opossum (*Didelphis virginiana*).

During the MARA project, Eastern Box Turtles were documented in 218 quads and 794 blocks. This species was found in all 23 counties as well as Baltimore City, which was similar to the historical distribution described by H. S. Harris (1975).

Although Eastern Box Turtles are often considered "common," populations are declining across much of their range (Dodd 2001; Kiester and Willey 2015). For example, a population at the Patuxent Wildlife Research Center in Maryland declined by 75% from 1955 to 1995. Possible causes for this decline were flooding, changing vegetation, and increased vehicular traffic (Stickel 1978; Hall et al. 1999). Even low levels of mortality can cause populations to undergo slow, steady declines (Seigel 2005).

Habitat loss and degradation are the most serious threats to populations. In some regions of Maryland, much of the suitable Eastern Box Turtle habitat has been converted to suburbs and roads. Observations of adult Box Turtles in urban or suburban areas do not mean a population is vi-

able. One population in Rock Creek Park in Washington, DC, has exhibited little or no recruitment of juveniles for many years (Ferebee and Henry 2008). Road mortality, disease, and pet collecting also pose serious threats. Many female turtles are crushed on roads while seeking nesting areas (Gibbs and Steen 2005). An emerging disease, ranavirus, has caused recent deaths of wild Box Turtles in Maryland (Farnsworth and Seigel 2013; Currylow et al. 2014; Adamovicz et al. 2015). The MDNR allows possession of only one Eastern Box Turtle without a state-issued permit. Low reproductive rates, low population density, and the potential for high rates of road mortality combine to place populations at serious risk. Conservation of Eastern Box Turtles can best be achieved through ongoing habitat protection and public education.

CHRISTOPHER SWARTH

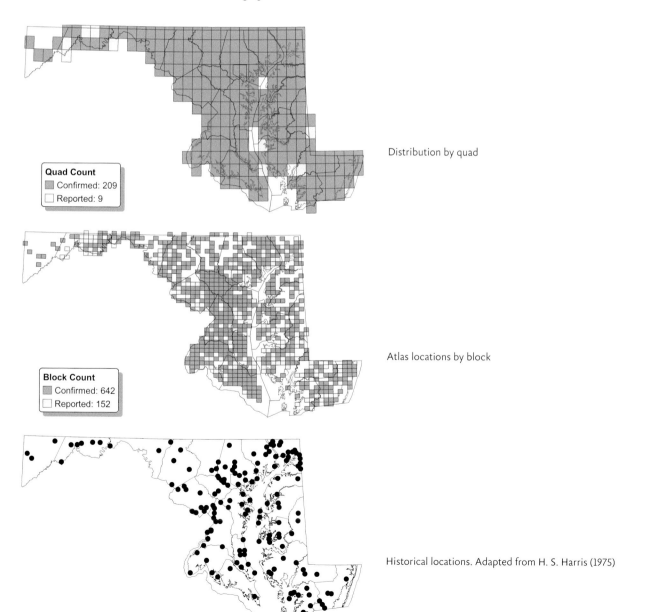

Quad Count
Confirmed: 209
Reported: 9

Distribution by quad

Block Count
Confirmed: 642
Reported: 152

Atlas locations by block

Historical locations. Adapted from H. S. Harris (1975)

Photograph by Lance H. Benedict

Red-eared Slider
Trachemys scripta elegans

The "traditional pet turtle," Red-eared Sliders have been known to generations of youngsters because of their great popularity as pets—sold as hatchlings in pet stores and at county fairs across the country. Unfortunately, when the novelty fades, these turtles have been released by the tens of thousands into aquatic habitats far from their native range, where they have flourished (Somma et al. 2018).

A subspecies of the Pond Slider, the Red-eared Slider has a prominent red to orange mark on the side of the head behind each eye. This "red ear" may be yellow or absent in some individuals (Powell et al. 2016). Body coloration is green to olive brown, with additional yellow stripes along the head as well as on the limbs. The oval, slightly domed carapace has a weak middorsal keel and slightly serrated posterior marginal scutes. The coloration of the carapace is green to olive brown with a transverse yellow bar flanked by elongate black blotches on each pleural scute. The plastron is typically yellow with dark thumbprint smudges in each scute (Ernst and Lovich 2009). The bridge is yellow with dark blotches, and there is a black spot on the ventral side of each marginal scute. Older individuals may become darker or melanistic (Ernst and Lovich 2009). Males have longer, thicker tails, with the cloaca positioned beyond the carapace edge, a flatter carapace, and much longer foreclaws than females. Females are larger than males and have elongated hind claws (Ernst and Lovich 2009). Adults measure 12.5–20.3 cm in carapace length. Hatchlings are marked similarly to adults but are greener overall (White and White 2007).

The native range of the Red-eared Slider extends through the central and lower Mississippi River Valley region to the Gulf of Mexico and west to Kansas, Oklahoma, and New Mexico. This species has been farmed in large quantities in the southern United States for global trade as pets, for food markets, and for religious observances (van Dijk et al. 2011). Due to escapes and releases, there are countless feral populations throughout the United States and in other parts of the world. In Maryland, the species has been introduced on the Ridge and Valley, Blue Ridge, Piedmont, and Atlantic Coastal Plain (H. S. Harris 1975) but is most widespread on the central Piedmont (Powell et al. 2016). H. S. Harris (1975) documented Red-eared Sliders in Anne Arundel, Baltimore, Frederick, Harford, Montgomery, Prince George's, and Washington counties, as well as Baltimore City. A population was later documented in Wicomico County (Grogan 1994).

Red-eared Sliders inhabit permanent freshwater habitats with a minimum depth of 1-2 m, soft and muddy bottoms, and an abundance of aquatic vegetation (Ernst and Lovich 2009), and they occasionally inhabit brackish water (J. D. Hardy 1972). On sunny days, these turtles are frequently seen basking, either floating in the water or on exposed woody debris and shorelines. Red-eared Sliders may bask together in large numbers and with other species, such as Painted Turtles and Northern Red-bellied Cooters (Grogan 1994). Red-eared Sliders occasionally exhibit aggressive behavior by gaping, nipping, or shoving other turtles (Ernst and Lovich 2009). Adults are opportunistic omnivores, whereas juveniles are predominantly carnivorous (Mitchell 1994). In general, Red-eared Sliders are diurnal during warm months and hibernate during winter. The turtles can survive for 6 months submerged underwater at near freezing temperatures (Mitchell 1994).

Courtship and mating occur in the spring and fall. Nesting occurs from April through July, with females producing 1-5 clutches annually. Number of eggs per clutch is correlated with female body size and averages about 11 eggs (Ernst and Lovich 2009). Temperature of the nest determines the sex of the hatchlings and the duration of the incubation period. Hatchlings disperse from nesting sites in late summer and early fall after an incubation period of approximately 77 days but may overwinter in the nest (Mitchell 1994; Ernst and Lovich 2009). Although eggs and hatchlings have the greatest predation risk, there is a roughly constant mortality rate regardless of age. Red-eared Sliders may live upward of 30 years (Ernst and Lovich 2009).

Although a non-native species to Maryland, the Red-eared Slider was the sixth most commonly encountered turtle during the atlas period, with 350 sightings recorded. The species was documented in 237 blocks within 118 quads, in 18 counties and Baltimore City. These data suggest a range expansion in all directions from known historical locations (H. S. Harris 1975; Grogan 1994), although it is unknown which records represent established, reproducing populations. Red-eared Sliders were found throughout the Piedmont and onto the Blue Ridge and Ridge and Valley, with new county records for Allegany, Carroll, Cecil, and

Howard counties. One record in western Allegany County was the first record for the Appalachian Plateaus in Maryland. Expansion also occurred southward onto the Atlantic Coastal Plain into Calvert, Charles, and St. Mary's counties. There were also new records on the Eastern Shore in Caroline, Dorchester, Kent, and Queen Anne's counties. Red-eared Sliders were most frequently observed in Anne Arundel, Baltimore, Frederick, Montgomery, and Prince George's counties and were much more widespread in these counties than they were historically (H. S. Harris 1975). Considering that feral populations of Red-eared Sliders are the result of released or escaped individuals, it is not surprising that their prevalence in these counties mirrors the 2010 Maryland census data, which list Anne Arundel, Baltimore, Howard, Montgomery, and Prince George's as the most populous counties in the state (MSA 2015). Accordingly, Red-eared Sliders were not observed in Garrett, Somerset, Talbot, or Worcester, which were among the counties with the fewest residents reported for Maryland in 2010 (MSA 2015).

Given that Red-eared Sliders are non-native, conservation efforts should focus on limiting or reducing further spread of this species. Although not identified as a vertebrate invasive species of concern for the state (MISC 2005), Red-eared Sliders are a potential threat to native turtle species in Maryland. This species may have a competitive advantage over Maryland's native turtles due to differences in nesting habitat and a greater reproductive output (Ernst and Lovich 2009). There is also a risk for disease, parasites, and antibiotic-resistant pathogens to be transferred from captive turtles to wild individuals (Ernst and Lovich 2009; ISSG 2010). Educational and enforcement campaigns directed at the sale, ownership, and release of Red-eared Sliders (and other pet turtles) could help prevent or limit the introduction of individuals to the wild.

RACHEL GAUZA GRONERT

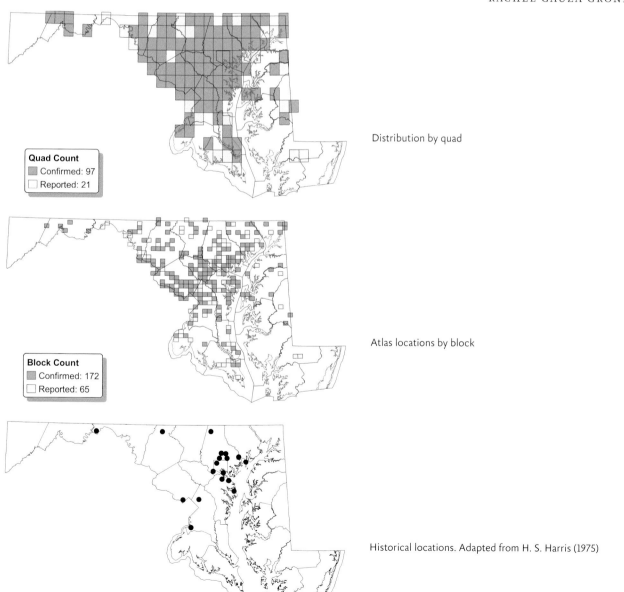

Quad Count
Confirmed: 97
Reported: 21

Distribution by quad

Block Count
Confirmed: 172
Reported: 65

Atlas locations by block

Historical locations. Adapted from H. S. Harris (1975)

Photograph by Scott McDaniel

Eastern Mud Turtle
Kinosternon subrubrum

The small, inconspicuous Eastern Mud Turtle is the most terrestrial of Maryland's aquatic turtles. The smooth, unpatterned, oval carapace is brown black to reddish brown and may be slightly depressed along the midline. The plastron is double hinged, which may be an adaptation to allow turtles to close tightly if aquatic habitats dry up (Ernst and Lovich 2009). Light yellow mottling, sometimes forming two stripes, is often apparent on the sides of the head and neck behind the eyes. Males are distinguished from females by their larger tail with a blunt nail on the tip. Males also have a shorter plastron with a deep posterior notch, and tilted, roughened patches on the upper, inner side of the hind legs. Both sexes have musk glands at the base of their legs from which they emit a foul-smelling, yellowish liquid when disturbed. Eastern Mud Turtles can be confused with the similar-sized Eastern Musk Turtle, but are distinguished by a fully developed plastron with triangular pectoral scutes. Adults have a carapace length that ranges from 7 cm to 12.1 cm. At the Jug Bay Wetlands Sanctuary in Anne Arundel County, adult turtles averaged 9.4 cm, with a maximum length of 11.7 cm (C. Swarth, unpubl. data). Hatchlings have a keeled carapace and reddish orange to reddish yellow markings on the plastron with a dark blotch centrally and extending along the scute seams.

The range of the Eastern Mud Turtle extends from Long Island, New York, south to Florida, west across the Gulf States to eastern Texas, and north to southern Illinois and southern Indiana. In Maryland, these turtles occur on the Atlantic Coastal Plain and extend into the Piedmont along lowland river valleys, but they have not been documented in Allegany, Carroll, or Garrett counties (H. S. Harris 1975).

The Eastern Mud Turtle is semiaquatic and occupies both aquatic and terrestrial habitats. The turtles prefer shallow, slow-moving waterways with soft bottoms and abundant aquatic vegetation (Ernst and Lovich 2009). They forage along the bottom of wetlands and ponds or among aquatic plants on the edges of slow-moving streams. Eastern Mud Turtles are particularly common in brackish and freshwater tidal regions of the Chesapeake Bay. In these areas, they can be seen swimming over bare mud or moving slowly within emergent or submerged aquatic vegetation. Eastern Mud Turtles are well adapted to the ebb and flow of tidal waters. During low tide, these turtles will bury in the mud for several hours until flooding waters again cover the flats (Cordero and Swarth 2010). Occasionally, individuals are seen basking at high tide (in tidal habitats) on aquatic plants or fallen logs. Eastern Mud Turtles also frequently move overland, particularly during or after rains (White and White 2007) or during nesting forays by females. Both sexes generally depart their aquatic environment in the fall and travel to terrestrial hibernacula (Frazer et al. 1991; Ernst and Lovich 2009).

In Maryland, Eastern Mud Turtles are active as early as mid-March (C. Swarth and Mike Quinlan, unpubl. data), although more commonly they emerge from overwintering sites in April. They are active during the day and at night. At the Jug Bay Wetlands Sanctuary, the mean aquatic home range for 10 male and female Eastern Mud Turtles based on radiotelemetry measurements was about 17 ha (Cordero et al. 2012b). Mating usually takes place underwater in the spring, and eggs are laid in May and June (Ernst and Lovich 2009). The mean clutch size is 3 eggs, and hatching takes place after an incubation period of about 95 days (Ernst and Lovich 2009). Hatchlings emerge in the fall or may overwinter in the nest. By mid-September, most return to upland sites to overwinter. The Eastern Mud Turtle is one of few aquatic turtles that overwinter on land (D. H. Bennett et al. 1970; Ultsch 2006). For winter refuge, these turtles will excavate a shallow burrow up to 15 cm deep in forest soils (D. H. Bennett 1972; Cordero et al. 2012a).

Eastern Mud Turtle adults have few predators. Common Watersnakes, large wading birds, and Bald Eagles (*Haliaeetus leucocephalus*) have been documented taking adults and juveniles (Ernst and Lovich 2009). On the Patuxent River, an empty Eastern Mud Turtle shell with a functioning radio transmitter was relocated beneath an Osprey (*Pandion haliaetus*) nest, suggesting that this fish-eating raptor may occasionally catch and eat a turtle (C. Swarth and Susan Blackstone, pers. obs.). Eastern Mud Turtles have an omnivorous diet consisting of mollusks, crustaceans, insects, amphibians, small fish, carrion, and aquatic vegetation (Ernst and Lovich 2009).

During the atlas project, Eastern Mud Turtles were documented in 106 quads and 242 blocks. The overall distribution was similar to that described by H. S. Harris (1975), except for no findings of Eastern Mud Turtles in Frederick or Howard counties.

As with many turtle species in North America, habitat loss caused by human development is considered the main cause for population declines. Populations of Eastern Mud Turtles appear secure in Maryland, provided their aquatic habitats and adjacent upland habitats are preserved. This semiaquatic species requires terrestrial habitats both for nesting and for overwintering, making adequate upland buffers critically important (Burke et al. 1994). Nutrient runoff and excessive eutrophication caused by wastewater treatment plants can degrade the tidal wetlands where sizable numbers of turtles are found (Swarth and Kiviat 2009). Increasing salinity resulting from sea-level rise is another factor that may cause ecological changes to the turtles' habitats.

CHRISTOPHER SWARTH

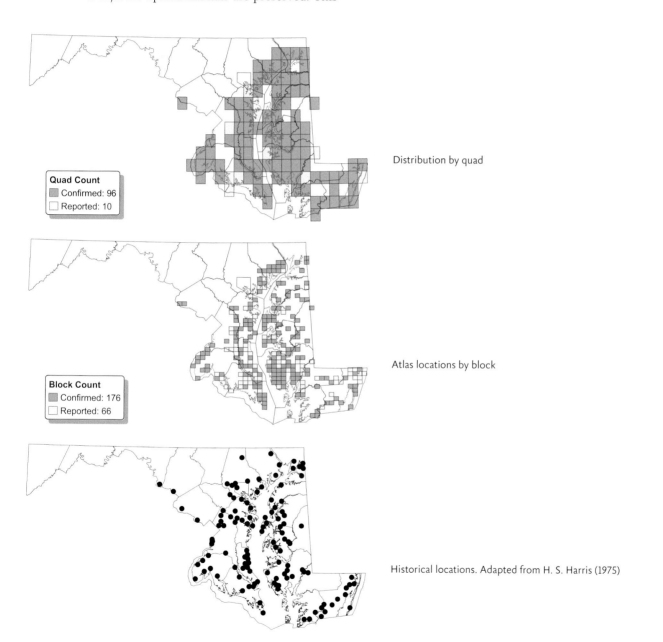

Quad Count
Confirmed: 96
Reported: 10

Distribution by quad

Block Count
Confirmed: 176
Reported: 66

Atlas locations by block

Historical locations. Adapted from H. S. Harris (1975)

Photograph by Lance H. Benedict

Eastern Musk Turtle
Sternotherus odoratus

Malodorous and pugnacious, the Eastern Musk Turtle, or Stinkpot, is named for the foul-smelling secretions produced by this small turtle. The scientific name also calls attention to this characteristic. The Eastern Musk Turtle has a smooth, arched carapace that varies in color from gray brown to almost black. The plastron has a single, nonfunctional hinge and is quite reduced, exposing broad areas of skin between the plastral scutes and leaving the gray, fleshy legs very exposed. The head has two light yellow stripes that extend backward from the snout, passing above and below the eyes. Paired barbels are present on the chin and throat. Males have a longer, thicker tail, with a blunt nail on the tip, and more exposed skin between the plastral scutes than females. Eastern Musk Turtles and Eastern Mud Turtles are similar in size, shape, and habits and thus can be difficult to distinguish without close inspection. The Eastern Musk Turtle has a much smaller plastron, and the pectoral scutes are rectangular rather than triangular as in the Eastern Mud Turtle, which also usually lacks the light facial stripes. Adult Eastern Musk Turtles usually range from 5.1 cm to 11.5 cm in carapace length. Turtles measured at the Jug Bay Wetlands Sanctuary in Anne Arundel County had an average adult carapace length of 9.8 cm; the largest individual was 11.9 cm (C. Swarth, unpubl. data). Hatchlings have a high-keeled carapace and a whitish marking on the edge of each marginal scute.

The Eastern Musk Turtle is found from Maine south to Florida and west to Wisconsin and Texas but is rare throughout the Appalachian Mountains. In Maryland, this species has been recorded in every county (H. S. Harris 1975), but most occurrences are on the Atlantic Coastal Plain. Records from the Piedmont, Blue Ridge, and Ridge and Valley provinces were associated with lowland river valleys such as the Potomac (H. S. Harris 1975). R. W. Miller (1980) reported two specimens from near Rocky Ridge, Frederick County, in the Monocacy River drainage. A single record exists for Garrett County on the Appalachian Plateaus (McCauley 1945).

These turtles prefer aquatic habitats with slow-moving water and soft bottoms. They can be very abundant in freshwater streams, lakes, ponds, and wetlands, but often are not observed because they spend most of their time patrolling the bottom. They are rarely found far from water. The small size of the plastron allows greater mobility of the legs, which may contribute to their reputation as agile climbers. Only very rarely is one seen basking. Canoers paddling along quiet waterways under low overhanging branches have reported Eastern Musk Turtles dropping into their canoe from an overhead perch. Eastern Musk Turtles can be quite abundant in some areas. Fifty-one were captured in a 1-ha beaver pond in Jug Bay Wetlands Sanctuary in the summer of 2010 (Giannini 2010).

Eastern Musk Turtles hibernate underwater or in objects near water such as rotten logs, in muskrat lodges, or under streamside banks. They usually emerge by April and are active until November. Mating can occur throughout the active period but peaks in spring and fall (Ernst and Lovich 2009). Eggs are laid from May to July. Females typically deposit 2–4 eggs in a shallow nest in or under rotting vegetation, fallen logs, or the walls of muskrat or beaver lodges within or near the margins of the wetland (Ernst and Lovich 2009). Females may nest communally or in close proximity to one another: one researcher found 16 nests under a small log (Cagle 1937). The incubation period lasts about 75 days, with hatchlings emerging from the nest in late August to November; these tiny turtles measure an average of only 2.1 cm in carapace length upon hatching and resemble blackened nuts (Ernst and Lovich 2009). The turtles may live to 30 years in the wild, but one on display at the Philadelphia Zoo lived for 54 years and 9 months (Ernst and Lovich 2009).

Eastern Musk Turtles are omnivores, feeding on aquatic vegetation, earthworms, clams, snails, crayfish, insects, and carrion (Ernst and Lovich 2009). They secrete a foul-smelling yellowish fluid from glands located near the bridge of the carapace when disturbed. The odor serves as a defense against predators and may play a role in courtship by allowing individuals to identify the sex of potential mates (Ernst and Lovich 2009).

During the atlas project, Eastern Musk Turtles were found in 80 quads and 124 blocks. They were reported from every county except Garrett, and their overall distribution remained similar to that reported by H. S. Harris (1975).

Pollution and habitat destruction are the major threats to populations. As with most other turtles, if their habitat is protected from development and wetlands are preserved, this goes a long way toward maintaining populations.

CHRISTOPHER SWARTH

Distribution by quad

Atlas locations by block

Historical locations. Adapted from H. S. Harris (1975)

Photograph by Bonnie Ott

Eastern Fence Lizard
Sceloporus undulatus

The Eastern Fence Lizard is among the most common and conspicuous of Maryland's lizard species and is often seen perched on trees, logs, rocks, and, of course, fences. Its pointed, keeled dorsal scales give rise to the common name for members of this genus, the Spiny Lizards. The body is somewhat dorsoventrally flattened and is marked with wavy, dorsal crossbars. This species exhibits a moderate degree of sexual dimorphism in coloration and pattern. Females have a grayish dorsal background that contrasts with the dark crossbars, whereas males are a darker brown to black and have less distinct crossbars. Males have a pair of laterally elongate, glossy, greenish blue to bright blue ventral patches that are bordered by black toward the midline of the belly and a dark blue to black throat patch, which brighten during the breeding season. In contrast, females have a cream-colored belly with only a tinge of light blue to greenish blue on the throat. The total length of adults typically ranges from 10 cm to 18.4 cm. While females are slightly larger than males, their tails are proportionately shorter (H. M. Smith 1946). Juveniles are patterned similarly to females, but the belly and throat are a uniform white to cream color.

The Eastern Fence Lizard is restricted to the eastern United States from southern New Jersey west to the Mississippi River and south to the Gulf Coast and central Florida. In Maryland, McCauley (1945) and H. S. Harris (1975) reported the distribution as statewide except for Dorchester and Garrett counties; however, R. W. Miller (1979) pointed out that Harris had overlooked Mansueti's (1958) report for Garrett County. Given these findings, together with Gro-

gan's (1981) observation in Dorchester County, this lizard has been recorded in every county in the state. Notably, however, Eastern Fence Lizards are most abundant east of the Fall Line, and while present in the western part of the state, there is only one record from the Appalachian Plateaus—which is surprising, considering the number of records for adjacent areas in Pennsylvania (Hulse et al. 2001).

For this mostly arboreal species, a major requirement is access to abundant sunlight. Although most commonly occurring in xeric, open, mixed pine-dominated forests, often near water, Eastern Fence Lizards can also be found in open, nonforested areas such as in and among woodpiles, logs, stone piles, and old abandoned buildings. Individuals can be found in open oak-dominated areas, especially those associated with rocky outcrops or forest edges, and, of course, wherever rail fences remain (McCauley 1945; Mitchell 1994; White and White 2007). These lizards are generally sedentary, moving little except to thermoregulate, capture prey, or escape predators. They are usually observed basking in open or spotted sunlit areas, where they blend in very well with the background substrate. Often, a good way to detect these lizards is to listen for the scratching of their claws on tree trunks as they scurry out of sight as you walk past.

These lizards are generally active from late spring through fall, with seasonal activity influenced by local climate; in colder areas, the period of activity is reduced. At night, individuals retreat to available crevices or holes, then emerge in the morning as the day warms up. Eastern Fence Lizards mate in the spring shortly after emergence (Mitchell 1994). Males establish territories where they defend against incursion by other males and court females. Males display to females and to other males with a series of push-ups and lateral flattening that reveals their blue ventral and throat coloration. Courtship displays also include shallow head nods called "jiggling" (Rothblum and Jenssen 1978). After mating, females deposit 4-17 semispherical, cream-colored eggs in sunlit areas in moist soil or under rotten logs, stumps, or debris (Mitchell 1994; Mitchell and Anderson 1994; White and White 2007). Eggs hatch in late summer to early fall, following an incubation period of about 70 days that varies with local temperatures (Tinkle and Ballinger 1972). Martof et al. (1980) and Shaffer (1991) reported the possibility of multiple clutches per season. Sexual maturity is reached in 2-3 years.

Eastern Fence Lizards are sit-and-wait predators that feed preferentially on insects, especially beetles; they also consume spiders, millipedes, and snails (McCauley 1945). These lizards are, in turn, eaten by racers, kingsnakes, ratsnakes, copperheads, predatory birds, foxes, opossums, and raccoons, as well as domestic dogs and cats (McClellan et al. 1943; Mitchell and Beck 1992). They are frequently parasitized by orange mites, which attach under the scales, particularly along the neck adjacent to the tympanum (McCauley 1945).

During the MARA project, Eastern Fence Lizards were recorded in 144 blocks within 72 quads and 19 counties. Because it may have been overlooked at low densities, the species may occur elsewhere. Overall, the MARA distribution remained similar to the historical distribution in Maryland (McCauley 1945; H. S. Harris 1975), although the species was less frequently observed on the Piedmont and not at all in Kent, Garrett, Harford, and Talbot counties.

Eastern Fence Lizard populations are likely to be influenced by human development and habitation. The species' preference for open or edge-type habitats may enable it to persist in areas with human influences, but the presence of dogs and cats is expected to reduce populations (Loss et al. 2013). The paucity of reports above the Fall Line and onto the Appalachian Plateaus suggests continued relative rarity in the western region, with the exception of Ridge and Valley Province. Given the development and urban sprawl that has occurred in Maryland over the past few decades, it is encouraging that the general distribution of this lizard has not appreciably changed. Eastern Fence Lizards are likely to persist if appropriate habitat is available and anthropogenic impacts are limited.

GEORGE MIDDENDORF

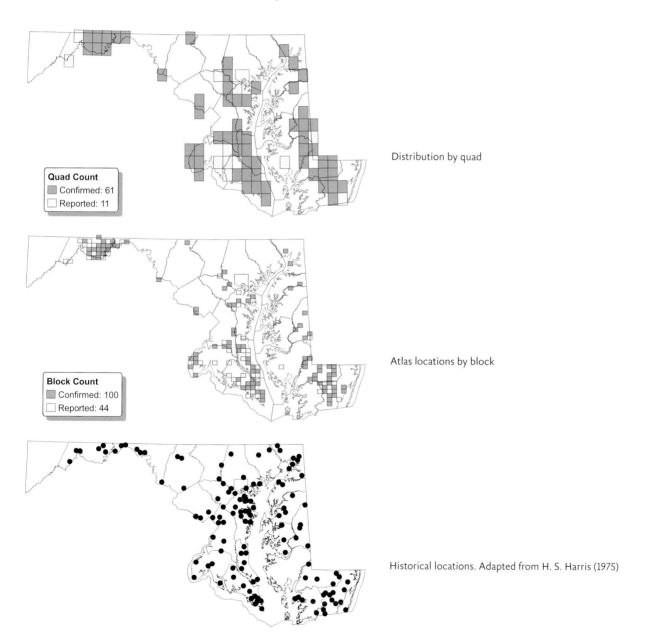

Quad Count
Confirmed: 61
Reported: 11

Distribution by quad

Block Count
Confirmed: 100
Reported: 44

Atlas locations by block

Historical locations. Adapted from H. S. Harris (1975)

Photograph by Ed Thompson

Coal Skink

Plestiodon anthracinus

The Coal Skink was first documented in Maryland in 1949 (Lemay and Marsiglia 1952) and, with very few observations since, is considered the rarest lizard in the state (CREARM 1973). This moderately sized skink has an elongate, cylindrical body and is covered in smooth, cycloid scales. The dorsum is bronze brown from head to tail and separated from wide, dark brown sides by light-colored stripes, each paired with an additional light stripe below. All four stripes extend onto the tail but not the head. From chin to tail tip, the ventral color is gray to bluish gray. Males are slightly smaller than females (Mitchell 1994; Hotchkin et al. 2001) and develop orange to red coloration on the sides of the head during the breeding season. Total length ranges from 12.5 cm to 17.8 cm for adults, with a tail length about twice that of the body. Juveniles have a bright blue tail that gradually fades with age.

The Coal Skink's geographic distribution is extensive but discontinuous. Isolated populations occur from western New York south to western Georgia and west to western Tennessee and northern Mississippi. Its distribution is more contiguous in the Midwest from Missouri south to eastern Texas and Louisiana and along the Gulf Coast in Mississippi, Alabama, and the panhandle of Florida. Historically, in Maryland, Coal Skinks were known from only a handful of observations in Allegany and Garrett counties (CREARM 1973), restricting the known range to the Appalachian Plateaus and Ridge and Valley provinces of western Maryland.

These lizards are terrestrial and secretive and are generally associated with humid wooded areas near springs and creeks (Conant and Collins 1998), but they will occupy a wide range of habitats, including rocky outcrops, particularly limestone ledges, flat sandstone slabs, and large rocks in open areas. McCauley (1945), Lemay and Marsiglia (1952), and Thompson (1984a) reported individuals in Maryland inhabiting open, boggy terrain, where they utilize spaces under large rocks and woody debris. Hulse et al. (2001) noted that, in Pennsylvania, these lizards occupy fairly open habitat where rocks and logs provide cover.

Coal Skinks are typically active from early April through September or early October (Hulse et al. 2001). Little is known about their reproductive ecology in Maryland or neighboring states. Mating in Maryland probably occurs in early spring after emergence from hibernation. Females lay a clutch of 4-11 eggs in leaf debris between late April and late June (H. M. Smith 1946; CREARM 1973) and guard them for 4-5 weeks until hatching in late May to late July (H. M. Smith 1946). Information on reproduction and growth is lacking for Maryland, but Coal Skinks in Georgia are known to reach sexual maturity in their second year, at around 5 cm in snout-to-vent length (Hotchkin et al. 2001).

Reported prey include earthworms, beetles, grasshoppers, and spiders (Mitchell 1994). Predators include snakes, larger lizards, mammals (J. T. Collins 1974), and probably predatory birds (J. W. Gibbons et al. 2009). Minton and Minton (1948) reported ant predation on Coal Skink eggs. Coal Skinks most likely depend on their secretive behavior to avoid predation, often remaining under leaf litter and other cover objects when active, but when attacked, they can use tail autotomy to escape. Individuals in the southern part of the species' range reportedly take refuge by diving to the bottom of shallow water to escape predation, hiding under stones or debris (Martof et al. 1980; Conant and Collins 1998).

During the MARA project, Coal Skinks were documented three times within two quads and two blocks in Allegany County. These records are significant because, although documented in this county previously (H. S. Harris 1975), locality was unknown. And although no records were reported from Garrett County, Coal Skinks could be present within their historical range but overlooked due to their secretive behavior and potentially low densities. Intensive efforts to observe this species in suitable habitat and at historically documented sites in Maryland proved fruitless in the past, with habitat changes hypothesized to explain its apparent disappearance (McCauley 1945; Thompson 1984a).

The Coal Skink was listed as endangered in Maryland in 1972 (Thompson 1984a). With the discovery of new sites during the MARA project, intensive efforts should continue to survey for this species in Maryland at historical locations and in potential habitat. Quantitative study of known populations in Maryland could help determine immediate threats and the need for conservation actions.

TASHA FOREMAN AND GEORGE MIDDENDORF

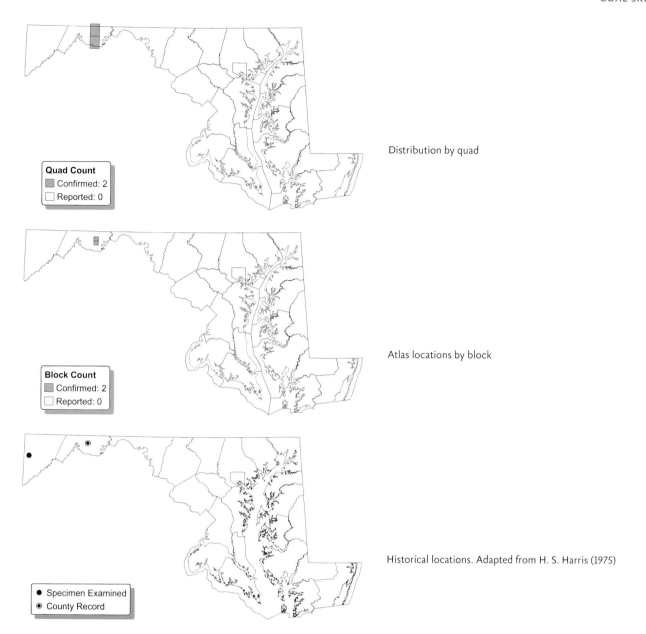

Distribution by quad

Quad Count
Confirmed: 2
Reported: 0

Atlas locations by block

Block Count
Confirmed: 2
Reported: 0

Historical locations. Adapted from H. S. Harris (1975)

Specimen Examined
County Record

Photograph by Lance H. Benedict

Common Five-lined Skink
Plestiodon fasciatus

The Common Five-lined Skink is Maryland's most common lizard. Its body is covered with smooth, cycloid scales, giving this lizard, like all skinks in Maryland, a shiny appearance. The species is variable in coloration, depending on age and sex. Juveniles are dark brown to black, with five white to yellowish longitudinal stripes and a bright blue tail. The stripes begin on the head and continue through the length of the body to halfway down the tail. The middorsal stripe is forked at the neck, forming two lines on the top of the head that rejoin at the snout. As individuals mature, the body and tail color turns a brownish gray, and the stripes fade. Older males may become completely patternless, whereas females retain faint stripes. Mature males have a wide, maroon-colored head that brightens to orange or deep red during the breeding season.

Although similar in appearance to the Broad-headed Skink, the Common Five-lined Skink has generally different facial scale patterns. Usually, Common Five-lined Skinks have four upper labial scales anterior to the subocular scale. However, this field mark is not always diagnostic. An examination of specimens from Virginia discovered that approximately 18% of Common Five-lined Skinks had five upper labial scales on either one or both sides of the head (Mitchell 1994). A more reliable field mark to distinguish this species from the Broad-headed Skink is the size and number of postlabial scales; Common Five-lined Skinks have two large postlabial scales abutting the anterior side of the ear canal. Common Five-lined Skinks are also considerably smaller than Broad-headed Skinks; adults attain a total

length of 12.5-22.2 cm. The blue tail coloration is usually lost at a snout-to-vent length of about 5.2-5.5 cm, compared with 7.6 cm in Broad-headed Skinks (Mitchell 1994).

The Common Five-lined Skink is found from New England to northern Florida and west to Wisconsin and eastern Kansas, Oklahoma, and Texas. In Maryland, this species has been recorded in all counties except Carroll (H. S. Harris 1975). It is common on the Atlantic Coastal Plain, uncommon on the Piedmont, and rare on the Appalachian Plateaus (McCauley 1945), with only two records reported from Garrett County (H. S. Harris 1975).

These lizards can be found in a variety of habitats, including urban settings (N. B. Green and Pauley 1987), but they prefer moist forests with woody debris such as fallen logs and stumps. They are often found among woodpiles, in rotting logs, and under bark or artificial cover at forest edges or in light gaps. Overwintering sites include rotting wood, tree stumps, mammal burrows, and piles of wood or sawdust (Tyning 1990). In developed areas, individual Common Five-lined Skinks may be associated with stone walls and brick buildings, which can radiate heat, harbor insects, and provide good places to hide.

In Maryland, this species may emerge in late March if the weather is warm, but most emerge in April and remain active until late September or early October, depending on the ambient temperature (McCauley 1945; Tyning 1990). Mating occurs in April and May, followed by egg laying in June and hatching in July and August, after an incubation period of 27-47 days, depending on temperature (Hulse et al. 2001). Only one clutch of pinkish white eggs—which become discolored due to contact with the substrate—are laid per season (Hulse et al. 2001). Clutch sizes in Virginia ranged from 5 to 14 eggs. Females attend the eggs throughout incubation, guarding them from predators and moving them around to keep them moist. Sexual maturity is reached at about 1-3 years (Tyning 1990), and some individuals can live 10 years or longer (McCauley 1945).

Common Five-lined Skinks primarily eat small invertebrates such as earthworms, millipedes, spiders, and insects. Natural predators are generally snakes and small mammals, but feral and domestic cats can kill a significant number (McCauley 1945). Tree climbing is an escape strategy and is most often done by males. Like other lizards, Common Five-lined Skinks employ caudal autotomy as an escape mechanism when captured by a predator. In a study by Vitt and Cooper (1986), tailed skinks successfully escaped predation more than half the time (56%), while skinks without tails escaped only 9% of the time. This species may also be parasitized by ticks, including the Black-legged Tick (*Ixodes scapularis*), a primary vector of Lyme disease. Giery and Ostfeld (2007) found that Common Five-lined Skinks may play an important role in Lyme disease ecology. Since Five-lined Skinks are incapable of transmitting Lyme disease to other ticks, their presence can reduce infection prevalence in tick populations by approximately 10% to 50% (Giery and Ostfeld 2007).

During the atlas project, Common Five-lined Skinks were documented in 357 blocks within 135 quads and 21 counties. The distributional pattern was similar to that reported by H. S. Harris (1975), except there were no sightings of Common Five-lined Skinks from the Appalachian Plateaus Province in the MARA project.

No conservation actions appear necessary for the Common Five-lined Skink at this time. This species is well adapted to disturbed habitats and areas near human development. Public education directed at the potential benefit of these lizards in reducing Lyme disease prevalence by their coexistence in local neighborhoods, as well as the adverse effects of predation by cats and dogs, would help this species continue to survive in the state's urbanizing landscapes.

GEORGE M. JETT

Distribution by quad

Quad Count
Confirmed: 110
Reported: 25

Atlas locations by block

Block Count
Confirmed: 228
Reported: 129

Historical locations. Adapted from H. S. Harris (1975).

Specimen Examined
Specimen Reported

Photograph by Michael Kirby

Broad-headed Skink
Plestiodon laticeps

The Broad-headed Skink is the largest native lizard with legs in eastern North America. This skink is stocky and cylindrically bodied, with glossy, smooth cycloid scales. As in the Common Five-lined Skink, the color and pattern of the Broad-headed Skink change with age and differ in males and females. Juveniles are boldly patterned, with five or seven yellowish stripes on a black background and a bright blue tail that fades at maturity. Mature males are uniformly olive brown with an orange-red head and swollen jaws, which become bright orange and enlarged during the breeding season. Mature females are brown but retain faint stripes, although the middorsal stripe may be indistinct to absent. The belly is usually cream-colored. Identification of young adults and juveniles is often difficult due to resemblance to the Common Five-lined Skink and requires careful inspection of scale patterns on the head. Unlike the Common Five-lined Skink, the Broad-headed Skink usually has five upper labial scales anterior to the subocular scale and lacks enlarged postlabial scales. Adults range from 16.5 cm to 32.4 cm in total length. Unlike in other species of skinks, Broad-headed males are much larger than females (Mitchell 1994).

The Broad-headed Skink ranges from southeastern Pennsylvania south to central Florida and west to eastern Texas and Kansas, with only a few isolated populations in the Appalachian Mountains. Maryland lies near the northern limit of this species' range, with Hulse et al. (2001) reporting only two locations in southeastern Pennsylvania and White and White (2007) indicating no records for Delaware. In Maryland, they are most common on the Atlantic Coastal Plain, where they have been reported in all counties except Caroline, Harford, and Kent (H. S. Harris 1975). Historical localities on the Piedmont and Blue Ridge were associated with the Potomac River in Montgomery, Frederick, and Washington counties. However, R. W. Miller (1979) reported a specimen from Carroll County, well north of the Potomac River.

This skink's habitat typically includes swamps, mature forest, and urban lots, especially when debris is present for cover. The species is considered arboreal because of its use of trees and tree cavities (P. W. Smith 1961; Mount 1975), unlike the other skinks in Maryland. It differs ecologically from the Common Five-Lined Skink by preferring more xeric habitats. Because Broad-headed Skinks spend much of their time in trees and, when terrestrial, spend time under leaf litter and debris, observing this lizard can be difficult.

Male Broad-headed Skinks are territorial: males fight with other males and defend females during the reproductive season (Vitt and Cooper 1985a) and may continue guarding their female after mating. Mating occurs in May and June. Females are thought to release a pheromone to attract a mate (Vitt and Cooper 1986). The female constructs a nest chamber in rotted wood and remains with the eggs during the incubation period to guard against predators and move the eggs around to keep them moist. Females lay only one clutch of 9-13 eggs per year. Incubation averages about 49 days in South Carolina (Vitt and Cooper 1985b), and hatchlings are first seen in mid-July. Sexual maturity for both sexes seems to be reached at the same size, about 7.5-8.5 cm in snout-to-vent length (Vitt and Cooper 1985b; Mitchell 1994). Broad-headed Skinks live longer than most other skinks; W. E. Cooper and Vitt (1997) reported an individual more than 8 years of age.

Broad-headed Skinks are primarily active foragers with a diverse diet, including snails, isopods, spiders, insects, small lizards, and lizard eggs (McCauley 1939a; Vitt and Cooper 1986). They use both chemosensory and visual cues to locate prey. Predators are snakes and mammals, including domestic cats (Mitchell and Beck 1992). Broad-headed Skinks should be handled with great care because they can be aggressive and give a strong bite—and when mishandled, tail autotomy is possible.

The Broad-headed Skink was documented in 86 blocks within 52 quads and 16 counties during the atlas period. The atlas distribution was similar to the historical distribution described by H. S. Harris (1975), with most records from the Atlantic Coastal Plain. However, none were reported from

Prince George's County, while several blocks were confirmed in Kent County. Locations were near the Potomac River, as reported by Harris (1975), but also north of the Potomac in Frederick and Montgomery counties, as well as one on the Piedmont in Baltimore County.

The wide distribution of the Broad-headed Skink increases its long-term population viability; however, a low overall density could make isolated populations vulnerable to local extinction. Conservation of this species is probably dependent upon protecting appropriate habitat such as large, open forests with mature and dead trees (Mitchell 1994) and the absence of urban predators such as feral and domestic cats (Loss et al. 2013).

DOUGLAS RUBY

Distribution by quad

Atlas locations by block

Historical locations. Adapted from H. S. Harris (1975)

Photograph by Michael Kirby

Little Brown Skink
Scincella lateralis

Of the 14 native skink species in the United States, the Little Brown Skink is the only member of the genus *Scincella* (de Quieroz et al. 2017). Its slender, elongate body has smooth, cycloid scales and small limbs. A transparent scale is present on the lower eyelid that allows limited vision through closed eyes. Two narrow, black stripes run dorsolaterally from the nose, through the eyes, and down through the length of the tail, separating the coppery brown dorsum from the creamy tan sides. The back and sides are often peppered with black specks, while the ventral surface is immaculate white to cream-colored. The species does not exhibit sexually dimorphic coloring, but females are often 10% larger overall, whereas males have larger heads and stronger jaws (Becker and Paulissen 2012). In both length and build, the Little Brown Skink is the smallest of Maryland's four skink species. Adult total length varies between 7.5 cm and 14.6 cm, and tail length ranges from 1.5 to 3 times snout-to-vent length (H. M. Smith 1946).

The Little Brown Skink is found in the southern half of the eastern United States, from southern New Jersey south to the Florida Keys and west to eastern Kansas and central Texas. In Maryland, this species has been documented in every county on the Atlantic Coastal Plain except Baltimore and Harford (McClellan et al. 1943; McCauley 1945; H. S. Harris 1975; White and White 2007). Historically, observations of Little Brown Skinks in Maryland were primarily restricted to the southern half of the Atlantic Coastal Plain (McClellan et al. 1943; McCauley 1945; H. S. Harris 1975). According to McCauley (1945), the species was not found north of Washington, DC, but Harris (1975) later confirmed its presence in northern Anne Arundel County, extending its range northward.

The species inhabits deciduous and mixed coniferous-deciduous woodlands and associated open-canopied habitats such as abandoned pastures and fields (Mitchell 1994), both near water and in drier, mesic habitats (McClellan et al. 1943). Occasionally, these lizards are also found in vegetated suburban habitats, including gardens and parks (Conant and Collins 1998). Unlike the Common Five-lined and Broad-headed Skinks, Little Brown Skinks rarely exhibit arboreal tendencies, instead hiding and foraging under or within decaying woody debris, leaf litter, and abandoned rodent burrows (Fitch and von Achen 1977).

Because Little Brown Skinks require internal body temperatures between 23 °C and 34 °C (D. G. Smith 1997), their activity is seasonally and diurnally constrained. In Maryland, these skinks emerge from hibernation in mid- to late spring and remain active until early to mid-fall (roughly March through October) (Mitchell 1994). During spring and fall, they are most active in the afternoon, while summer activity is restricted to mornings and evenings (Fitch and von Achen 1977). Even though males are more intrasexually and intersexually aggressive (Akin 1998; Becker and Paulissen 2012), only females exhibit territoriality (G. R. Brooks 1967). Adult males establish overlapping home ranges that often contain several small female territories (G. R. Brooks 1967).

Breeding activity commences soon after emergence from hibernation in the spring and typically extends through late summer. Females store sperm (Sever and Hopkins 2004) and can produce multiple clutches per season, averaging 4 eggs each (Fitch and von Achen 1977). Females lay their eggs under or within cover, sometimes communally (Mitchell 1994); they provide no parental care (Fitch and von Achen 1977). Roughly 1 cm in length (Mitchell 1994), eggs are cream colored and elliptical, and they hatch in about 6 weeks (McClellan et al. 1943). Little Brown Skinks reach sexual maturity within their first year, at around 4 cm in snout-to-vent length, and are short lived (~4 years) (G. R. Brooks 1967).

Little Brown Skinks feed primarily on small arthropods such as isopods, arachnids, and insects (Watson 2004) and are susceptible to a variety of ticks and intestinal parasites (G. R. Brooks 1967). Common predators include snakes (McClellan et al. 1943), other skinks (G. R. Brooks 1967), birds, mammals (G. R. Brooks 1967; Mitchell and Beck 1992), and even wolf spiders (Rubbo et al. 2001). Defense mechanisms against predation include immobility and crypsis, fleeing to a known refuge, and tail autotomy (D. G. Smith 1997). After autotomy, Little Brown Skinks often return to eat their tails, presumably to regain lost fat reserves—a behavior not often observed in other skinks (D. R. Clark 2005).

During the MARA project, Little Brown Skinks were documented in 60 blocks in 39 quads and 10 counties. They were observed within the nine historical counties of occurrence reported by H. S. Harris (1975) and for the first time in Caroline County. Little Brown Skinks were not documented in any of the counties on the upper Eastern Shore, even though their occurrence was reported there by White and White (2007).

Once generally considered uncommon in Maryland due to low observation rates (admittedly attributed to the skink's secretive habits [McCauley 1945]), the Little Brown Skink is currently considered a common and stable species in the state. Although the distribution appears relatively unchanged since H. S. Harris's (1975) publication, the atlas re-

sults do not address the status of populations. Furthermore, any site-specific differences from earlier observations in distribution could simply reflect variation in search effort (J. W. Gibbons et al. 1997). However, given the increased urbanization throughout much of the Little Brown Skink's range in Maryland, the MARA results are promising. Conservation efforts should focus on maintaining suitable forested habitats with healthy leaf-litter and humus layers to support the requisite cover and prey base for this species (Watson 2004).

TASHA FOREMAN

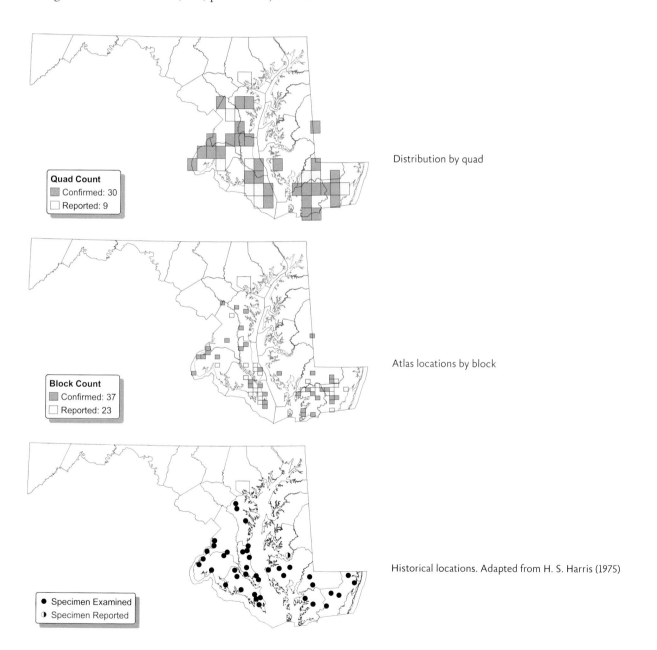

Distribution by quad

Quad Count
Confirmed: 30
Reported: 9

Atlas locations by block

Block Count
Confirmed: 37
Reported: 23

Historical locations. Adapted from H. S. Harris (1975)

Specimen Examined
Specimen Reported

Photograph by Ed Thompson

Six-lined Racerunner
Aspidoscelis sexlineata

The Six-lined Racerunner is well named for its habit of racing away in a quick burst of speed when disturbed. The lizard's six light cream, white, or yellow stripes lie on a dark brown to black background and extend from behind the eyes to the base of the tail, where the dorsal stripes fade and the lateral stripes continue onto the tail. The snout is short and triangular, while the jaws are broad and slightly wider than the neck. Granular, nonoverlapping dorsal scales contrast with eight rows of large, rectangular belly and tail scales. The skin often looks wrinkled or folded near the neck and along the body. Males may show light blue on the sides of the head, chin, throat, sides of the belly, and anterior of the forelimbs and hind limbs. This slender, moderately sized lizard attains a total body length of 15.2–26.7 cm, of which one-third is body and two-thirds is tail. Adult males are generally smaller than females. Juveniles resemble adults but possess a light blue tail (Mitchell 1994).

The Six-lined Racerunner ranges throughout most of the coastal states from Maryland south to Florida and west to mid-Texas and extends north to the southern Great Plains. In Maryland, the lizards are mostly found on the southern Atlantic Coastal Plain on the Western Shore and on the Ridge and Valley along the Potomac River (Klingel 1944; H. S. Harris 1975). Harris (1975) also reported a specimen from Frederick County, and Grogan (1985) observed an individual in Montgomery County, both near the Potomac River. Although it may be locally abundant, the species is not evenly distributed throughout these regions.

These lizards are associated with sunny, open, well-drained, sandy habitats, where shrubs and grass clumps are used for shade and cover (McCauley 1945; Grogan 1981). Completely terrestrial in habit, Six-lined Racerunners do not leave the ground or climb low objects (McClellan et al. 1943). This species is active from late spring through fall, with the pattern of seasonal activity influenced by local climate. In colder regions, the period of daily activity is shortened, while in warmer regions, the lizards may exhibit a bimodal activity pattern: active in the early morning and late afternoon and inactive at midday (Fitch 1958). At night or when temperatures are too cool or too hot, the lizards retreat into burrows. After overwintering in burrows, they emerge in spring, then retreat again in early fall (Etheridge et al. 1983). In Virginia, males were observed to emerge in late April, females in early May, and both remained active until mid-September (Mitchell 1994). Given its high body temperature needs, the species may emerge slightly later and retire a bit earlier in Maryland.

Reproductive data on the Six-lined Racerunner in Maryland are limited, but mating most likely occurs soon after emergence in spring (Etheridge et al. 1986). While the species is reported to produce 1–2 clutches per year (Mitchell 1994), the likelihood of multiple clutches probably depends on local climate and seasonality. Because Maryland represents the northernmost distribution, the species is likely to have only a single clutch per year in this state. Females nest solitarily in loose, sandy soil and sawdust piles and do not exhibit parental care (Mitchell 1994). Egg laying in Maryland was reported in mid-July (McClellan et al. 1943); clutch sizes range from 1 to 5 eggs, with incubation estimated at 60 days (E. E. Brown 1956; Mitchell 1994). Females become sexually mature at 1 year, and clutch size increases with the age and size of the female (D. R. Clark 1976; Trauth 1983).

Racerunners are active foragers, moving quickly from one location to another almost constantly in search of prey. Unlike other lizards in Maryland, the Six-lined Racerunner has a forked, snake-like tongue that it uses in olfaction while hunting for prey. These lizards consume many kinds of invertebrates, including grasshoppers, ants, flies, lepidopteran larvae and adults, beetle larvae and adults, true bugs, and spiders (McClellan et al. 1943; H. S. Harris 1975; Mitchell 1994). When they detect an item, they immediately attack, grasp the prey, and shake it into submission. If the prey is buried (e.g., beetle larva), these lizards dig furiously until they capture it. Six-lined Racerunners can, in turn, be eaten by predators with the speed and skill to capture them. E. E. Brown (1979) reported predation by a North American Racer. McCauley (1945) reported nematode parasites of the genus *Physaloptera* observed in a few stomachs.

During the MARA project, Six-lined Racerunners were observed in 19 blocks in 11 quads and five counties: Allegany, Anne Arundel, Calvert, Charles, and St. Mary's. Although the overall extent of the range remained similar to that reported by H. S. Harris (1975), fewer locations were detected on the Atlantic Coastal Plain during the atlas period. Given the spotty distribution, the species may occur elsewhere along the Potomac or other rivers extending into the Piedmont.

Populations are likely to be influenced by habitat preferences, particularly the availability of open, sandy habitats often associated with disturbance. Given the development and sprawl along the coastal region of Maryland in recent decades, it is encouraging that the general distribution of this lizard has not appreciably changed. The Six-lined Racerunner can most likely persist if appropriate habitat is conserved and anthropogenic impact is limited.

GEORGE MIDDENDORF, TASHA FOREMAN,
AND DOUGLAS RUBY

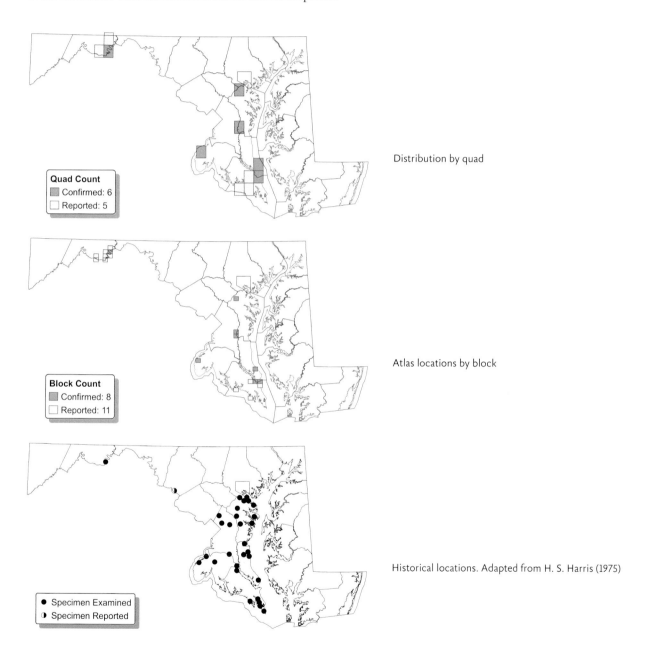

Distribution by quad

Quad Count
Confirmed: 6
Reported: 5

Atlas locations by block

Block Count
Confirmed: 8
Reported: 11

Historical locations. Adapted from H. S. Harris (1975)

Specimen Examined
Specimen Reported

Photograph by Robert T. Ferguson II

Scarletsnake
Cemophora coccinea

The nonvenomous Scarletsnake is one of three Maryland Batesian mimics of the venomous Harlequin Coralsnake (*Micrurus fulvius*). Batesian mimics have coloration and patterns closely matching the warning signals of neighboring harmful species (Davis Rabosky et al. 2016), which fools potential predators into associating their coloration with the harmful effects of the species they mimic. The successful replication of the notorious red, yellow, and black bands inspired the "red to yellow, kills a fellow; red to black, friend of Jack" variant rhymes.

The Scarletsnake is medium sized and boldly colored. Individuals have dorsal red blotches bordered by a black band and separated by a yellowish or white band. These bands do not encircle the body, leaving the white to yellow ventral surface without any marks. The head is red with a transverse black bar between or behind the eyes. The snout is rounded and projects over the lower jaw. The scales are smooth, and the anal plate is undivided. Sexual dimorphism is not readily apparent, but females are usually longer with shorter tails. Adults average 36–51 cm in total body length.

The distribution of the Scarletsnake ranges from southern New Jersey, Delaware, and eastern Maryland south into Florida and west to eastern Oklahoma and Texas, with scattered, disjunct populations along the northern and western extents of its range. For Maryland, only 20 accepted records of Scarletsnakes exist, most of which are from the Atlantic Coastal Plain (H. S. Harris 2013). H. S. Harris (1975) reported historical records from Baltimore City and Anne Arundel, Calvert, Prince George's, Talbot, and Wicomico counties. Records of this snake have since been reported from Balti-more, Charles, Howard, and Worcester counties (H. S. Harris 2013).

This nocturnal and semifossorial species is found in pine, hardwood, or mixed oak-pine woodlands (Palmer and Tregembo 1970; H. S. Harris 2013) and surrounding fields, grassy ecotones (Nelson and Gibbons 1972), and scrub habitats (Palmer and Tregembo 1970; Enge and Sullivan 2000). Associated substrate is often composed of well-drained, sandy soils (Palmer and Tregembo 1970). Although the Scarletsnake most likely utilizes rodent burrows and decayed tree root paths in compact soils (Palmer and Tregembo 1970), it also actively burrows into loose soils (Mitchell 1994; Ernst and Ernst 2003). Scarletsnakes have been found more than a meter deep in both soil types (Palmer and Tregembo 1970). Individuals can be observed under logs, large rocks, leaf litter, and other forms of cover during the day, but searching for them at night (Palmer and Tregembo 1970; Gibbons and Semlitsch 1991) or utilizing drift fences with pitfall traps (Nelson and Gibbons 1972; Gibbons and Semlitsch 1991; Enge and Sullivan 2000) has proven most effective.

In Virginia, Scarletsnakes are typically observed above ground in April through November, with peak activity during the summer (Mitchell 1994). Studies suggest peak activity periods are correlated more with the nesting activity of prey species (Dickson 1948; Enge and Sullivan 2000) than with temperature and precipitation (Minton and Bechtel 1958; Nelson and Gibbons 1972). Mating presumably occurs in the spring (Ernst and Ernst 2003), with egg laying throughout the summer (Palmer and Tregembo 1970; Herman 1983); hatching occurs in the fall (Woolcott 1959; Nelson and Gibbons 1972; Palmer and Braswell 1995). Females lay their eggs in a seemingly disparate array of locations, including burrows (Ernst and Ernst 2003), dry sandy soils (Gibbons and Semlitsch 1991), and moist humus (Woolcott 1959). Clutches contain 2–9 eggs (Woolcott 1959; Palmer and Tregembo 1970; Herman 1983; Palmer and Braswell 1995). Incubation lasts 70–80 days, with most hatching occurring in September (Ernst and Ernst 2003)—although Gibbons and Semlitsch (1991) observed a hatchling emerge from its egg in March, which might indicate that young can overwinter in the egg. Both sexes are thought to be sexually mature at approximately 37 cm in total length or 2–3 years of age (Ernst and Ernst 2003). Longevity is unknown in the wild, but a captive individual lived for 6 years (Ernst and Ernst 2003).

Scarletsnakes are known to eat small-bodied snakes, small lizards, and insects (Ernst and Ernst 2003), but they primarily feed on eggs (Palmer and Tregembo 1970; Trauth and McAllister 1995), including those of lizards (Trauth and McAllister 1995), turtles (Dickson 1948), and snakes, even those of other Scarletsnakes (Palmer and Tregembo 1970). They swallow small eggs whole (Trauth and McAllister 1995) and slit open larger eggs with their rear teeth to drink the contents (Dickson 1948; Minton and Bechtel 1958; Palmer and Tregembo 1970). Predators are likely to be similar to those of other snakes. Ernst and Ernst (2003) listed four observed predators: Southern Toad (*Anaxyrus terrestris*), Loggerhead Shrike (*Lanius ludovicianus*), Harlequin Coralsnake, and conspecifics.

During the MARA project, there was one record of Scarletsnake from Howard County, a new county record for Maryland (H. S. Harris 2013). Only two other records have been reported in Maryland since 2000, both observations made prior to the MARA effort and both from the same locality in Worcester County, with the most recent in 2005 (H. S. Harris 2013). The MARA results represent a significant reduction in distribution relative to the historical oc-

currence in Maryland (H. S. Harris 1975). However, due to the species' nocturnal and fossorial behaviors, it is difficult to conclude whether the observed reduction represents a decline in the number of Maryland populations or the Scarletsnake being overlooked due to its secretive lifestyle.

The Scarletsnake remains one of Maryland's most elusive snakes and, as a result, is included on the MDNR's (2016b) list of rare species. Further efforts to assess its status in Maryland are necessary to determine whether conservation efforts are warranted and, if so, to what extent. However, protection of suitable habitat from further development should be an immediate consideration (Hammerson 2007).

TASHA FOREMAN

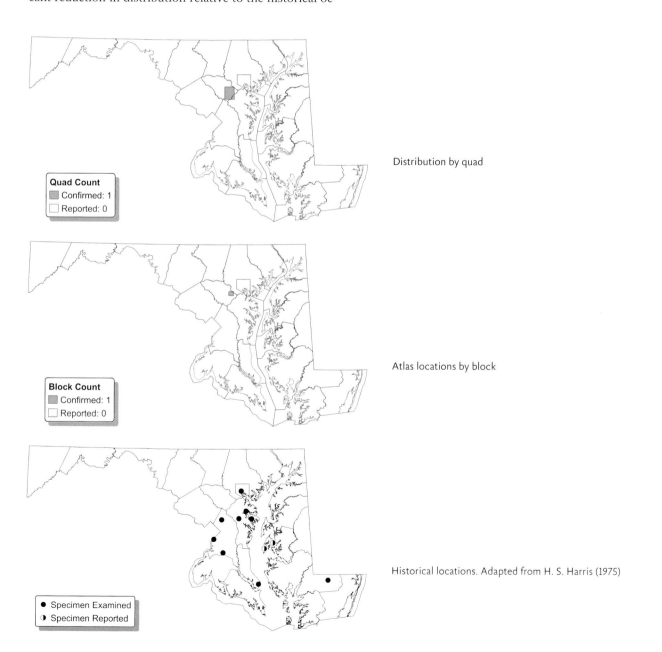

Distribution by quad

Quad Count
Confirmed: 1
Reported: 0

Atlas locations by block

Block Count
Confirmed: 1
Reported: 0

Historical locations. Adapted from H. S. Harris (1975)

• Specimen Examined
◗ Specimen Reported

Photograph by Andrew Adams

North American Racer
Coluber constrictor

The North American Racer is one of two "black snakes" found in Maryland. It goes by such colloquial names as Black-snake, Racer Black-snake, Racer, Black Racer, Blue Racer, Black Chaser, and Mouser (McCauley 1945). It is often mistaken for the similar Eastern Ratsnake due to its similar black color, but the Racer has some distinct morphological and behavioral differences. It can be distinguished from Eastern Ratsnakes by the presence of smooth scales, a bluish black ventral surface with white coloration restricted to the chin and throat, and a body round in cross section. Individuals from western Maryland are more likely to have a substantial amount of white on the chin and throat, while such coloration is rare on the Eastern Shore and intermediate in specimens from southern Maryland (McCauley 1945). The eyes are ridged above by a bony extension that gives the snake a "fierce aspect" (Gibbs et al. 2007). The anal plate is divided. Adult North American Racers attain a total length of 90-152 cm, second in size only to the Eastern Ratsnake in Maryland (Ditmars 1944). Juveniles are cryptically patterned, with a series of dark blotches on a light background. This coloration fades to black after three years, when the snakes are about 61 cm in length (H. A. Kelly et al 1936; Ditmars 1939).

The North American Racer is transcontinental in distribution. It occurs across much of eastern North America, excluding northernmost New England. Its range extends south into Florida and west to Texas and through the Midwest to the Pacific Coast. Its distribution is statewide in Maryland (H. S. Harris 1975). Racers occur infrequently on the Appalachian Plateaus but are particularly abundant on the Atlantic Coastal Plain (McCauley 1945).

This species occupies a variety of habitats in upland and wetland settings, including forests, swamps, marshes, roadsides, and agricultural areas. When basking on the ground in open areas, the snakes are rarely far from cover. They may also bask in the lower limbs of trees and shrubs (H. A. Kelly et al. 1936). As its name implies, this is a fast-moving snake, with a maximum ground speed just under 4 mi per hour (Leviton 1971). When threatened, Racers usually retreat in a series of rapid, darting movements and seek protective cover quickly, or may move upward into low vegetation. If pursued or cut off from escape, they will often turn in a defensive posture and may strike aggressively. They are also well known for rapidly vibrating the tail tip, which, when against dry leaves, produces a sound resembling the buzz of a rattlesnake. If captured, these snakes will bite readily and thrash frantically.

North American Racers often hibernate communally with other species, including Eastern Ratsnakes, Eastern Copperheads, and Timber Rattlesnakes (Gibbs et al. 2007). They emerge from hibernacula when spring temperatures allow, with mating occurring in early spring (H. A. Kelly et al. 1936). They are oviparous and, according to McCauley (1945), egg laying takes place from mid-June through mid-July in Maryland, with approximately 10-30 eggs laid under stones or logs. Communal nesting occurs occasionally (Beane et al. 2010). Numerous small granules cover the eggshell, giving it a rough texture and thus making the eggs easy to identify. Incubation lasts 65-70 days, and most hatching occurs in the month of August (McCauley 1945). Both sexes are mature at a total body length of approximately 85-90 cm (Hulse et al. 2001) and can live 10 or more years (Fitch 1963).

Rodents and other small mammals are their most abundant prey, but H. A. Kelly et al. (1936) and McCauley (1945) reported small quantities of other prey animals such as frogs, toads, and birds. North American Racers are also well-known predators of other reptiles, including venomous snakes. Klimstra (1959) reported consumption of insects and other arthropods by juveniles. North American Racers feed by day and hunt for prey in a slow and deliberate searching manner. Contrary to their scientific name, Racers are not constrictors. Instead, they pin prey to the ground with a loop of their body and then swallow the prey alive (Conant and Collins 1998).

During the MARA project, North American Racers were documented in 449 blocks in 181 quads in every county in Maryland. The MARA distribution was similar to the historical distribution in Maryland (H. S. Harris 1975), with most observations from the Atlantic Coastal Plain.

Killing by humans and by vehicular strikes most likely remain substantial sources of mortality. Nearly a third of the observed sightings (32%) during the atlas project were of remains. Aside from educating the public about the ecological and economic benefits of these snakes, and the sense- lessness of their killing, very little is needed in the way of conservation actions for this widespread species. The future of the North American Racer in Maryland appears to be secure at this time.

KYLE RAMBO

Quad Count
Confirmed: 155
Reported: 26

Distribution by quad

Block Count
Confirmed: 304
Reported: 145

Atlas locations by block

Historical locations. Adapted from H. S. Harris (1975)

Photograph by Matthew Kirby

Eastern Kingsnake
Lampropeltis getula

Considered the "king of snakes," the Eastern Kingsnake garnered its moniker from its habit of preying upon other snakes, including venomous species. This snake was previously classified within a transcontinental species group, but Pyron and Burbrink (2009) provided evidence that the Eastern Kingsnake is a distinct species. Regardless of its taxonomy, this is an impressive snake.

The Eastern Kingsnake has a black body with thin white to yellowish bands across the back that connect laterally, forming a chainlike pattern (Gibbons and Dorcas 2005). Its underside is irregularly checkered with black and white or yellow (White and White 2007). The scales are smooth, giving it a glossy appearance, and the anal plate is undivided. Adults typically reach 90–122 cm in total length. A large male collected in Harford County, Maryland, in 1933 measured 146 cm (H. A. Kelly et al. 1936).

The range of the Eastern Kingsnake extends from southern New Jersey south through Florida and west to southeastern Alabama along the Atlantic Coastal Plain and Piedmont. In Maryland, the Eastern Kingsnake occurs throughout the Atlantic Coastal Plain (McCauley 1945) and enters onto the eastern Piedmont via river valleys (H. S. Harris 1975).

A variety of habitat types are used by Eastern Kingsnakes, including hardwood and pine forests, meadows, agricultural areas, and the shorelines of wetlands, swamps, and streams (Gibbons and Dorcas 2005). The snakes choose sites with coarse woody debris, a thick leaf layer, and dense shrub layers (Wund et al. 2007; Steen et al. 2010). At a study site in Kent County, Maryland, this species was found more often in forested habitat with fewer trees and more ground cover than other forest types (McLeod and Gates 1998). McCauley (1945) found that in Maryland this species preferred upland habitats in the vicinity of water. Eastern Kingsnakes can also be found around old farm buildings and in trash and debris piles that provide ample cover (White and White 2007). They are completely terrestrial and rarely climb into vegetation (Mitchell 1994).

Like most reptiles in Maryland, Eastern Kingsnakes hibernate during the winter and emerge when the temperature warms. Mating occurs during the spring months (Ernst and Ernst 2003). Egg laying in the southeastern United States occurs during June and July (Gibbons and Dorcas 2005). In Virginia, known egg-laying dates were in June (Mitchell 1994). Clutch sizes range from 1 to 29 eggs, and eggs hatch in 39–83 days, depending on temperature (Ernst and Ernst 2003). Females are known to produce clutches of eggs fertilized by multiple males (Gibbons and Dorcas 2005).

Eastern Kingsnakes are most active during the daytime (Mitchell 1994; Gibbons and Dorcas 2005; Howze and Smith 2012), but during the heat of the summer they shift their activity to dusk and into the night (Ernst and Ernst 2003). When not active, Eastern Kingsnakes spend the majority of their time concealed under the cover of soil, leaf litter, or debris (Wund et al. 2007). They also hide in small mammal burrows and stumpholes (Steen et al. 2010).

When active, Eastern Kingsnakes are usually on the prowl for food. Kingsnakes are predators of other snakes, including venomous snakes. They are immune to the venom of pit vipers (Gibbons and Dorcas 2005). This species also feeds on a variety of other reptiles, small mammals, and eggs (Mitchell 1994). Posey (1973) observed a Kingsnake in Charles County feeding on turtle eggs. The snake pushed its head sideways to loosen the soil and uncover the eggs. Eggs are swallowed whole, and small prey are often swallowed alive (Gibbons and Dorcas 2005). Large prey are constricted and immobilized in the Kingsnake's coils. Prey snakes are often grabbed by the head and chewed and twisted before constriction begins (Ernst and Ernst 2003).

During the atlas period, Eastern Kingsnakes were found in 71 quads and 124 blocks. They were found scattered throughout the Atlantic Coastal Plain of Maryland. A few were found on the Piedmont just above the Fall Line. The MARA distribution was similar to the historical range in Maryland (H. S. Harris 1975), though this snake may have disappeared from some areas. No Eastern Kingsnakes were found in Cecil County during the atlas period. Only a few were found in Anne Arundel County compared with the numerous records reported by H. S. Harris (1975). This species also appears to be less common in the areas surrounding Washington, DC, than historically. Mitchell (1994) noted that in Virginia, Eastern Kingsnake populations had undoubtedly been lost through urbanization. The same may be true for Maryland.

The status of the Eastern Kingsnake in Maryland seems to be secure currently, though this species may not tolerate heavy urbanization. Many of the blocks where the snake was located correspond to some of the less developed areas of the Atlantic Coastal Plain. The future of this species in the state depends on the conservation of forested habitats in association with fields, meadows, and wetlands.

GLENN D. THERRES

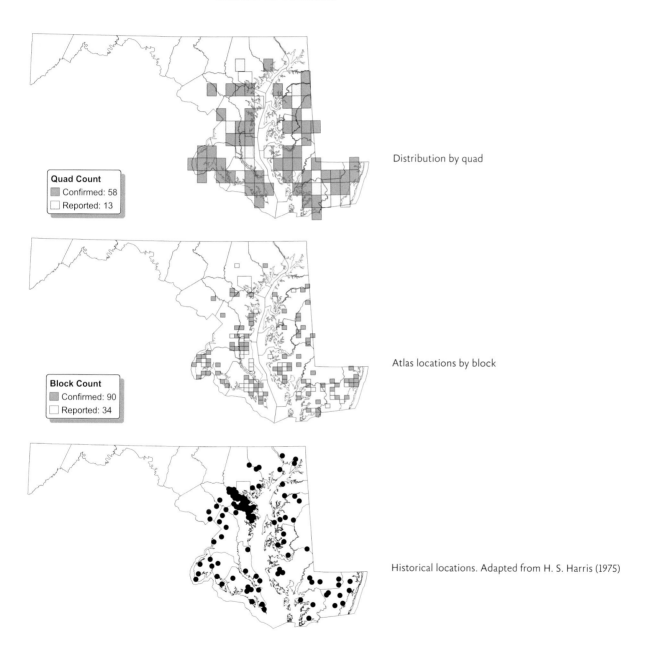

Quad Count
Confirmed: 58
Reported: 13

Distribution by quad

Block Count
Confirmed: 90
Reported: 34

Atlas locations by block

Historical locations. Adapted from H. S. Harris (1975)

Photograph by Lance H. Benedict

Northern Mole Kingsnake
Lampropeltis rhombomaculata

The Northern Mole Kingsnake is one of Maryland's most enigmatic snakes, as its subterranean habits and limited range in this state have resulted in few historical observations. Its body shape is similar to that of the ratsnakes, being slightly taller than wide, but its scales are smooth and the anal plate is undivided. Dorsal color is light to dark brown with scattered, dark-bordered, reddish brown blotches. The blotches extend along the back and are interspersed with smaller blotches along the sides, from the head to the tip of a relatively short tail (H. A. Kelly et al. 1936). The brown color fades to light yellowish along the lower sides and belly (Powell et al. 2016), with the belly having indistinct brown spots (Linzey and Clifford 1981). Adults typically range from 76 cm to 106.7 cm in total length, with the largest Maryland specimen (a female) measuring 109.8 cm (McCauley 1945).

The Northern Mole Kingsnake is distributed from southern Maryland south to northern Georgia and Alabama. It is absent from coastal areas in southern Virginia and northern North Carolina. The westernmost region of the range is disjunct in southern Mississippi and southeastern Louisiana. Isolated populations occur in eastern Tennessee and the panhandle of Florida. In Maryland, Northern Mole Kingsnakes are found on the western Atlantic Coastal Plain Province from south of Baltimore City to St. Mary's County (excluding Calvert County) and west to the eastern Piedmont in southern Montgomery County (H. S. Harris 1975). Most early records were from within and around Washington, DC, prompting McCauley (1945) to state that the District of Columbia served as the center of the species' range in Maryland.

Very little information has been published on the habitat preferences of the Northern Mole Kingsnake. McKelvy and Burbrink (2016) list dry pine and oak forests as its primary habitats. McCauley (1945) noted the habitat adjacent to the Maryland road on which he found a dead specimen as a cultivated field on one side and woods and dense thicket on the other. Howden (1946) referenced Northern Mole Kingsnake captures from the edges of fields bordered by woods and noted that most captures were in areas of sandy soil. The author (D. R. Smith) encountered the species in southern Prince George's County on dry ground within an open powerline cut surrounded by marsh, mixed pine-hardwood forest, and cultivated land. The Northern Mole Kingsnake is a fossorial species, spending the majority of its time in underground burrows of small mammals or in tunnels that it digs itself (McCauley 1945). These snakes also use rock crevices and hollow logs or stumps and can be found beneath surface objects (Ernst and Ernst 2003). While seldom encountered, Northern Mole Kingsnakes can at times be active above ground, particularly after heavy rains (Mitchell 1994). The secretive nature of this snake has led to a conception by some that the species is not so much rare as underrepresented (H. A. Kelly et al. 1936; McCauley 1945; Howden 1946).

Northern Mole Kingsnakes are generally active from April through early November in the northern part of their range. Ernst and Ernst (2003) found these snakes to be active during the early morning hours during spring and fall, but that activity shifted to twilight and night with the onset of warmer weather. Not much is known about the reproductive habits of the Northern Mole Kingsnake in Maryland. In Virginia, courtship and mating occur during the spring (Mitchell 1994). The onset of courtship is a series of tongue flicks and body jerks, with copulation lasting up to an hour (Ernst and Ernst 2003). Northern Mole Kingsnakes are oviparous. Egg laying typically occurs during June and July in rodent burrows, sawdust piles, or loose soil of recently plowed fields (Ernst and Ernst 2003). Clutch size was found to range from 7 to 18 eggs in Virginia (Mitchell 1994), while the only reported ovipositing female Northern Mole Kingsnake in Maryland produced 15 eggs (Howden 1946). Tyron and Carl (1980) reported an incubation period between 49 and 54 days for 17 eggs in Georgia.

Northern Mole Kingsnakes are constrictors, seizing prey with the mouth then wrapping several coils of the body around the prey. While rodents appear to make up the majority of the diet, other prey items include snakes, lizards, frogs, and toads (Linzey and Clifford 1981).

Four Northern Mole Kingsnakes were found during the atlas period. These snakes were found in only three blocks within just two quads in Prince George's and Charles counties. While these locations fall within the historical range of the species, the MARA distribution suggests a substantial range contraction since H. S. Harris's (1975) publication. Prior to the atlas period, the most recent report of Northern Mole Kingsnake was from Anne Arundel County in October 1993 (H. S. Harris 2005).

Loss of habitat, particularly in and around Washington, DC, and along the Baltimore-Washington corridor, has been a factor in the reduction of Northern Mole Kingsnake records over the past 50 years. However, available habitat re-

mains over large areas of Anne Arundel, Prince George's, Charles, and St. Mary's counties, suggesting that the species may still be present but simply went underreported during the atlas period because of its secretive habit. Future studies targeting the Northern Mole Kingsnake may be necessary to better assess its population status and fully understand the conservation needs of this species.

DAVID R. SMITH

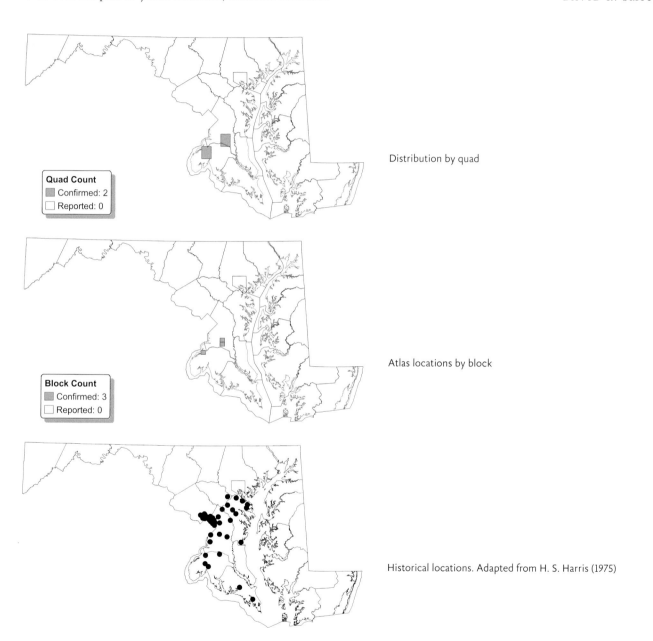

Distribution by quad

Atlas locations by block

Historical locations. Adapted from H. S. Harris (1975)

Photograph by Andrew Adams

Eastern Milksnake
Lampropeltis triangulum

Named after a regional myth and frequently harassed due to a case of mistaken identity (see below), the Eastern Milksnake endures in Maryland because of its ability to survive in forest, field, basement, and barn. The Eastern Milksnake has a series of three (or occasionally five) rows of reddish brown, black-bordered dorsal blotches set on a lighter brown, tan, or grayish body. The middorsal blotches number 18-52 from the head to the base of the tail. The first middorsal blotch has two projections forward that meet on top of the head and enclose a V- or Y-shaped area of body color at the back of the head. The underbelly is irregularly black and white checkered. The scales are smooth and the anal plate is undivided. The Eastern Milksnake is moderately sized, with adults typically ranging from 61 cm to 90 cm in total length, with a maximum length of 132 cm. Juvenile Eastern Milksnakes are similar but more vibrantly colored, with a cream body and red blotches.

Historically, milksnakes with particular morphological features were identified as the "Coastal Plain" Milksnake. The taxonomy of this snake has been contentious (Grogan and Forester 1998); it was proposed as a subspecies (*L. t. temporalis*) of the Eastern Milksnake (Conant 1943a) and later as an intergrade between the Eastern Milksnake (*L. t. triangulum*) and Scarlet Kingsnake (*L. t. elapsoides*) subspecies (K. L. Williams 1978, 1988). However, recent genetic analyses of the Milksnake complex concluded that this snake is merely a color form of the Eastern Milksnake not warranting subspecies status (Ruane et al. 2014). Nevertheless, the "Coastal Plain" moniker continues to be useful in referring to populations on the Atlantic Coastal Plain that do seem unique in their pattern, prey, and preferred habitat. Coastal Plain Milksnakes exhibit several distinguishing characteristics: a single row of 18-32 dorsal blotches, with head markings separated from the first blotch by a light "collar"; the red or brown portion of the anterior blotches reaching the edge of the ventral scales; black coloration on the belly, associated with borders of dorsal blotches or concentrated midway between adjacent blotches (Conant 1943a); dorsal blotches on the tail completely encircling the tail to form red rings (William Grogan, pers. comm.); and absence of the V or Y shape on the posterior of the head. For considerable distances on either side of the Fall Line, a wide range of intermediate forms exists (Conant 1943a).

The range of the Eastern Milksnake extends from southern Maine, parts of Ontario and Quebec, and southeastern Minnesota south to northern Georgia and west to northern Alabama, northeastern Louisiana, and northern Arkansas. It is largely absent from the Atlantic Coastal Plain in the eastern United States. In Maryland, Eastern Milksnakes are fairly common west of the Fall Line, reaching their highest abundance on the Appalachian Plateaus Province (McCauley 1945). East of the Fall Line, Eastern Milksnakes are uncommon to rare, with many individuals matching the description of the Coastal Plain form (Grogan and Forester 1998). Eastern Milksnakes seem to be absent from the central Eastern Shore, with no records reported from Caroline, Queen Anne's, and Talbot counties (H. S. Harris 1975; Grogan 1985; Grogan and Forester 1998).

West of the Fall Line in Maryland, Eastern Milksnakes inhabit deciduous and mixed forests, rocky hillsides, and meadows. They readily make use of surface objects for concealment, both natural (e.g., flat rocks and logs) and artificial (e.g., discarded building materials, scrap metal, piled bricks and stone). The Coastal Plain form inhabits sandy, mixed deciduous-coniferous forests, where it is primarily fossorial and rarely encountered. It is mostly found by chance around houses within this type of habitat, either in foundations or under various surface objects.

Eastern Milksnakes are generally active from April through October. During the day, particularly in warm summer weather, they become very difficult to find, emerging from beneath their cover only at dusk or during the night in search of food (McCauley 1945). Rodents are the primary prey of this species, but other snakes are also consumed. The Coastal Plain form readily preys on lizards (Grogan and Forester 1998). Mating occurs in the spring; females lay 5-20 eggs in June or July, and hatching takes place in August and September (Mitchell 1994). R. W. Miller and Grall (1978) captured a pair of Coastal Plain Milksnakes on 13 May; subsequently, they observed mating on 14 May, egg laying in mid-June, and hatching in early August.

Legend has it that the Eastern Milksnake was so named because of a prevailing belief that the snake would creep into barns and consume milk from dairy cows—straight from the udder (Hutchins et al. 2003). Though this is biologically impossible, the moniker stuck. It is likely that the myth took shape because Eastern Milksnakes are frequently found in barns, which provide cool, dark hiding spots and access to abundant rodents. An Eastern Milksnake may consume about 150 rodents each season (H. A. Kelly et al. 1936).

The Eastern Milksnake may suffer more from being secretive than being uncommon. Seldom seen by the general public, this nonvenomous snake may not be immediately recognizable. Due to a color palette and pattern similar to those of the Eastern Copperhead, many harmless Eastern Milksnakes may suffer the same harassment and extermination meted out to the more feared venomous snake. Eastern Milksnakes are usually sluggish to escape when encountered and, if unable to elude their harasser, may strike repeatedly, though quite harmlessly, as a means of defense (McCauley 1945). The regional folklore that gave the Eastern Milksnake its name and its persistent misidentification as a venomous species are reflected in some of its nicknames, such as Cowsucker and House Moccasin (McCauley 1945).

Eastern Milksnakes were documented in 84 quads and 165 blocks during the atlas period. They were found throughout western Maryland from the Appalachian Plateaus through the Piedmont, but in only a few atlas blocks on the Atlantic Coastal Plain. The overall MARA distribution was similar to the historical distribution in Maryland (H. S. Harris 1975); however, no Eastern Milksnakes were found in the Baltimore-Washington corridor, where much development has occurred. They were also not reported from historical locations in Anne Arundel, Calvert, Kent, Prince George's, and Wicomico counties. Coastal Plain Milksnakes were probably present in many more blocks on the Atlantic Coastal Plain, but their more secretive nature makes them more difficult to find.

The main threat to the Eastern Milksnake is the development of upland forest habitat and adjacent open-canopied areas. While this species is currently secure in Maryland, development is rapidly occurring along the eastern Piedmont. The Coastal Plain form is most threatened by development in southern St. Mary's County. During the atlas period, Coastal Plain specimens were still turning up in new developments surrounded by forest, but as development density increases, the outlook for the snakes there is unclear. It will be crucial to maintain significant tracts of upland habitat for this species to persist.

BRIAN C. GOODMAN AND LANCE H. BENEDICT

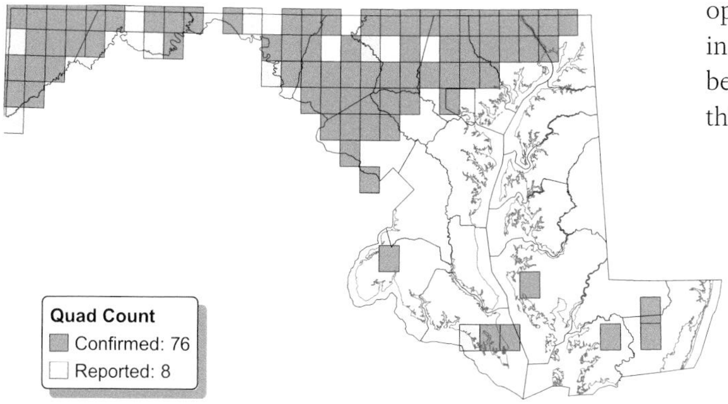

Quad Count
Confirmed: 76
Reported: 8

Distribution by quad

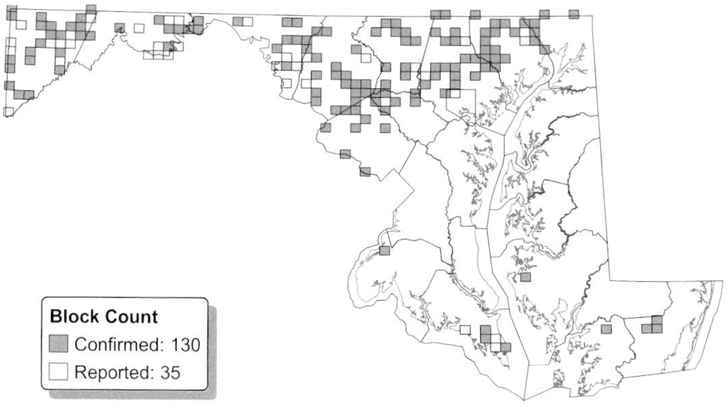

Block Count
Confirmed: 130
Reported: 35

Atlas locations by block

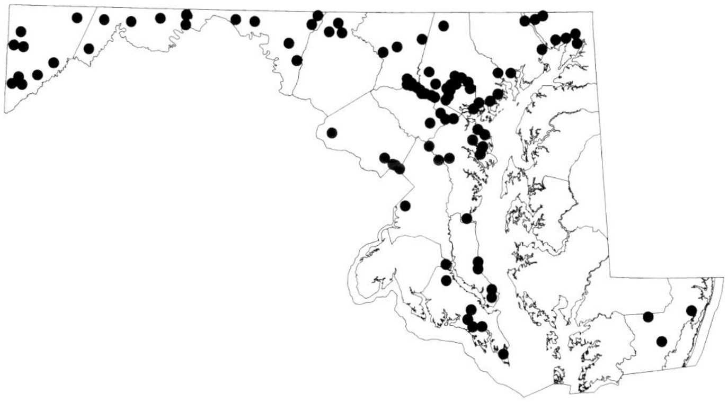

Historical locations. Adapted from H. S. Harris (1975)

Photograph by NPS/J. Chase

Rough Greensnake
Opheodrys aestivus

Distinctly slender for its length, the Rough Greensnake is well suited to its arboreal habits. This species is characterized by a bright green body with large eyes and a long, prehensile tail comprising approximately 32% to 42% of its total length (Mitchell 1994). Ventral and labial coloration is a paler, light shade of green, yellow, or white. Dorsal scales are keeled, and the anal plate is divided. Because the species is often encountered dead on the road, it is worth noting that the bright green fades to blue soon after death (Mitchell 1994). The sexes are similar, but females average slightly larger than males. Adults range from 56 cm to 81 cm in total length, with a record length of 116 cm reported. The only similar species in Maryland is the Smooth Greensnake, which has smooth dorsal scales and is smaller and more terrestrial.

The Rough Greensnake occurs in eastern North America from southern New Jersey south to Florida and west to Kansas, Oklahoma, Texas, and northeastern Mexico. In Maryland, Rough Greensnakes are mainly restricted to the Atlantic Coastal Plain and the eastern Piedmont, although a specimen was reported from the Ridge and Valley along the Potomac River in Washington County (McCauley 1945; H. S. Harris 1975). McCauley (1945) suggested that river and stream valleys account for westward range extension beyond the Atlantic Coastal Plain.

Harmless and beautiful, this colubrid is encountered less often than most naturalists would prefer. The Rough Greensnake is diurnal and primarily arboreal. These snakes inhabit open deciduous or mixed forests, old fields, marshes, and the wrack zone of barrier islands (McCauley 1945; Mitchell

1994; White and White 2007). Their cryptic color, shape, and behavior make them difficult to find in the shrubs, vines, and small trees where they spend much of their time. Encounters are often by chance, when the snake is crossing a road or path, or when an outdoor cat brings one home.

Werler and McCallion (1951) reported that Rough Greensnakes are active from March through November in Virginia Beach, Virginia, whereas Clifford (1976) reported activity from May to September in Amelia County, Virginia. Mitchell's (1994) review of museum records for Virginia provided an activity range of April through October. These snakes are especially active in late summer and early fall, when they are often seen crossing roads (N. B. Green and Pauley 1987; White and White 2007). Mating occurs in the spring but may also take place in the fall (Richmond 1956). Females lay 2-14 eggs in June or July in rotting logs or under rocks, boards, or other debris. McCauley (1945) noted 6 small eggs in the oviducts of a Talbot County specimen that was preserved in May and 4 larger eggs in a Worcester County specimen from June. Young hatch in August or September and reach maturity in 1-2 years (Ernst and Barbour 1989; White and White 2007).

Rough Greensnakes feed on invertebrates, although McCauley (1945) recounted a thwarted attempt to feed on a Green Treefrog in captivity; young lizards have also been reported as potential prey (Groves 1941). Major prey types are moth and butterfly larvae, crickets and grasshoppers, and spiders and harvestmen (E. E. Brown 1979; M. V. Plummer 1991). Less commonly recorded prey types are land snails, wood roaches, and sphecid wasps (Mitchell 1994). In six dissected stomachs from Maryland specimens, McCauley (1945) found mainly grasshoppers and caterpillars but did find one snail as well. Known predators of Rough Greensnakes include Loggerhead Shrikes (*Lanius ludovicianus*), several species of hawks, and domestic cats (Mitchell and Beck 1992; Mitchell 1994). Rough Greensnakes remain motionless for extended periods of time, with neck and head extended, then, periodically, may sway, waving the head laterally. Whether these movements are meant to simulate twigs waving in the breeze or have some other purpose is not clear (McCauley 1945). These snakes tend to thrash when first picked up and may open the mouth to reveal a purple black lining, but they rarely bite (N. B. Green and Pauley 1987; White and White 2007).

During the MARA project, Rough Greensnakes were found in 115 quads and 248 blocks. Similar to H. S. Harris's (1975) description of geographic coverage, the atlas project found Rough Greensnake on the Atlantic Coastal Plain and eastern Piedmont in Maryland. No records were reported from the Ridge and Valley. An individual confirmed along the Susquehanna River in Cecil County was very near to where Harris (1975) marked an unconfirmed sighting of Smooth Greensnake.

Mitchell (1994) expressed concern for the species in Virginia and identified herbicide and pesticide use, increased road mortality, and domestic and feral cats as problematic. For now, the Rough Greensnake remains widely distributed on Maryland's Atlantic Coastal Plain. However, given its cryptic nature, demographic trends within the species will be detectable only with more focused monitoring.

RONALD L. GUTBERLET, JR.

Quad Count
Confirmed: 102
Reported: 13

Distribution by quad

Block Count
Confirmed: 183
Reported: 65

Atlas locations by block

Historical locations. Adapted from H. S. Harris (1975)

Photograph by Lance H. Benedict

Smooth Greensnake
Opheodrys vernalis

The Smooth Greensnake, sometimes referred to as the Grass Snake, was first described in 1827 and named *Coluber vernalis* (Harlan 1827). Subsequently, it has undergone a number of name changes, including being classified in its own genus, *Liochlorophis* (Oldham and Smith 1991). However, since this classification created a monotypic genus as a sister group to *Opheodrys*, the use of *Liochlorophis* is no longer considered valid (Crother et al. 2017).

One of two green-bodied snakes in Maryland, the Smooth Greensnake resembles the closely related Rough Greensnake. Both of the slender colubrids are unpatterned, vivid green above and pale greenish white to yellow below, but the Smooth Greensnake has smooth body scales rather than keeled scales. The bright green color of adults changes to blue after death (Mitchell 1994). Adults average 30.3–51 cm in total length, with females being slightly longer. Tail length averages 29% of total body length in females, compared with 34% in males (Ernst and Ernst 2003). During their first year, young Smooth Greensnakes are darker, olive green above and pale blue gray below.

The Smooth Greensnake is also separated from the Rough Greensnake by its distribution. While the latter occurs in the southeastern United States, the Smooth Greensnake is primarily a northern species. It is distributed from Nova Scotia south through New England and the Appalachian Mountains to southwestern Virginia and eastern West Virginia. Its range extends west through southern Canada and the Great Lakes region to Saskatchewan and North Dakota. Scattered, disjunct populations occur along the southern limits of its range and westward through the central Great Plains and southern Rocky Mountains, and an extremely disjunct population is found in eastern Texas. In Maryland, Smooth Greensnakes inhabit the western provinces, from the western Piedmont in Frederick County through the Appalachian Plateaus. A historical record from Conowingo on the eastern Piedmont was an unconfirmed report (H. S. Harris 1975) and more likely represented a misidentified Rough Greensnake.

Smooth Greensnakes live mostly on the ground, where they are camouflaged among the grass and tall weeds of fields, pastures, farmlands, shrublands, and other sunny, open habitats. They also occupy branches of low bushes, assisted in climbing by a prehensile tail. This species is considered much less arboreal than its southern congener (McCauley 1945). In Virginia, Smooth Greensnakes are more commonly found in high-elevation fields and balds but may be found in open, mixed hardwood forests (Mitchell 1994). Ernst and Ernst (2003) consider this snake a denizen of primarily mesic habitats such as wet meadows, borders of marshes and bogs, and open woodlands. Prey items are almost exclusively arthropods, including spiders, centipedes, caterpillars, grasshoppers, ants, and beetles (adults and larvae), though other species of prey such as earthworms, slugs, snails, crayfish, and salamanders have also been reported (Ernst and Ernst 2003).

Hibernation occurs underground in mammal burrows, rocky areas, or ant mounds and is often communal with other reptiles (Ernst and Ernst 2003). These snakes are diurnal and generally active from April through October; males may be most active in June and September, while females are more active in May and late July through September (Hulse et al. 2001; Ernst and Ernst 2003). Mating has been reported in early April in Pennsylvania (Hulse et al. 2001). The snake lays an average of 7 eggs per clutch, usually from June through August, in sites such as rotting logs and stumps, piles of sawdust or rotting vegetation, and mammal burrows. Nest sites may be shared by multiple females, and communal nests may contain as many as 31 eggs. Embryonic development begins before the eggs are laid; thus, the incubation period averages only 15 days. Prolonged egg retention within the female may be an adaptation to colder climates (McCauley 1945; Hulse et al. 2001; Ernst and Ernst 2003).

Twenty-one sightings of Smooth Greensnakes were submitted to the atlas project, from 16 blocks in 11 quads. Two of these were in Frederick County, one in Allegany County, and the remainder in Garrett County. Although the Garrett County distribution appears very similar to that described by H. S. Harris (1975), there were fewer records from Allegany County, and no records from Washington County, which may indicate that this species is declining in Maryland.

Hulse et al. (2001) reported that this snake seems to be

declining in Pennsylvania. The Virginia Wildlife Action Plan gives Smooth Greensnake a rating of Tier III (high conservation need) due to declining populations, low population levels, or range restriction. In Maryland and nearby states, this higher-elevation species may be stressed by warming temperatures resulting from climate change. Many grassland and shrubland birds are declining regionally due to factors such as succession and habitat loss (Brennan and Kuvlesky 2005; Sauer et al. 2013), and these habitat changes may adversely affect the Smooth Greensnake as well. Given the possible population decline and the number of threats this snake faces, it was included in Maryland's Wildlife Action Plan (MDNR 2015) as a species of greatest conservation need. Additional surveys are needed to better define the conservation status of Smooth Greensnakes in Maryland.

LYNN M. DAVIDSON

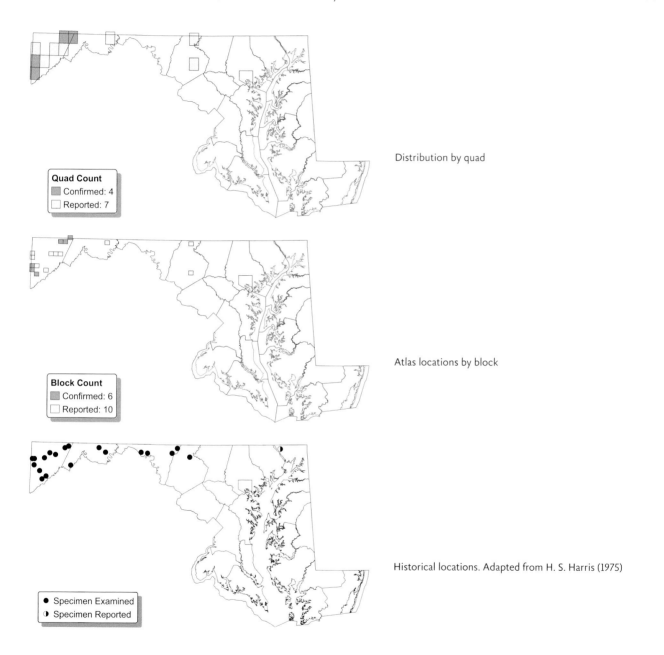

Quad Count
Confirmed: 4
Reported: 7

Distribution by quad

Block Count
Confirmed: 6
Reported: 10

Atlas locations by block

Historical locations. Adapted from H. S. Harris (1975)

● Specimen Examined
◑ Specimen Reported

Photograph by Robert T. Ferguson II

Eastern Ratsnake
Pantherophis alleghaniensis

Frequently referred to as the Black Ratsnake, the Eastern Ratsnake is one of the most common snakes in Maryland. The glossy black dorsal scales are weakly keeled, and a faint spotted pattern often overlies the dorsal blackish ground color. Light areas between the dorsal scales are often evident. The body has straight sides (not rounded) and, together with the flat bottom, this gives the snake a shape like a loaf of bread in cross section. The chin and throat are white, and the ventral surface is grayish but often shows a black and white checkerboard pattern in the anterior half of the body. The anal plate is divided. The Eastern Ratsnake is the largest snake in Maryland. Adults range in total length from 106.7 cm to 183 cm, but a record specimen exceeded 256 cm. Young Ratsnakes look considerably different from adults, and variation with age can make identification difficult. Juveniles are pale gray and boldly patterned, with dark gray to brown blotches on the dorsal surface and tail. A black stripe runs from the eye to the posterior angle of the mouth. Young Ratsnakes are sometimes mistaken for Eastern Copperheads or Eastern Pinesnakes (*Pituophis melanoleucus*).

Taxonomic revision based on phylogenetic analyses (Burbrink 2001) defined the range of the Eastern Ratsnake as eastern North America east of the Appalachian Mountains and Apalachicola River, from western New Hampshire south to Florida. However, species status in the northeastern portion of the range and throughout the Appalachian region and all of Maryland remains uncertain due to potential hybridization with the Gray Ratsnake (*Pantherophis spiloides*) and an absence of sampling locations from this region

(Burbrink 2001). Owing to these taxonomic uncertainties and the similar appearance of ratsnakes throughout Maryland, all ratsnakes were considered Eastern Ratsnakes for the purposes of the MARA project—even though the Gray Ratsnake may also be present in Maryland (Burbrink 2001), particularly west of the Blue Ridge. Ratsnakes have been recorded in every county in Maryland (H. S. Harris 1975).

Habitats used by Eastern Ratsnakes include agricultural fields, old fields, mixed deciduous forests, meadows, marsh edges, urban woodlots, abandoned barns and buildings, and backyards. An excellent climber, this snake is often seen in trees and shrubs, where it may be found searching for prey or basking on branches. This behavior also means that the snakes will occasionally enter houses and may even take up residence, undetected, in attics and basements. Normally shy, Eastern Ratsnakes usually avoid confrontation. If faced with danger, they often remain motionless or vibrate their tail, or they may rear back in a defensive posture. If picked up, they will wrap their body around an arm and spread smelly musk and feces on their handler.

Eastern Ratsnakes in Maryland emerge from overwintering sites as early as mid-February or March and remain active until late November. Common hibernacula are hollow trees or other enclosed spaces, often with an eastern exposure. The period of greatest activity is May to September. Mating takes place from mid-April to early June, and eggs are laid from late May to July beneath logs, rocks, or debris, or in hollow logs, tree stumps, tree holes, or abandoned burrows (Ernst and Ernst 2003). Mean clutch size is 15 eggs, and mean incubation period is about 62 days, with most hatching taking place in August and September. Home range size in a study of 32 radio-tracked snakes on the Eastern Shore was about 10 ha (Durner and Gates 1993). Male snakes grow faster, attain a greater size, and live longer than females (Stickel et al. 1980; Blouin-Demers et al. 2002). Ratsnakes reach sexual maturity at about 5 years and in the wild may live to 20 years or more (Stickel et al. 1980; Blouin-Demers et al. 2002).

In a long-term study at the Patuxent Wildlife Research Center, Stickel et al. (1980) determined that the diet consisted of 13 bird species, 10 mammal species, and one reptile (Common Gartersnake). The most common prey was the Meadow Vole (*Microtus pennsylvanicus*), followed by the Pine Vole (*M. pinetorum*) and the White-footed Mouse (*Peromyscus leucopus*). Ratsnakes are also common predators on Eastern Bluebird (*Sialia sialis*) eggs and chicks, which they obtain by climbing into small, human-made nest boxes. Eggs are swallowed whole and broken up in the esophagus by contractions of the upper body.

During the atlas project, Eastern Ratsnakes were the most frequently documented snake, with 1,332 sightings; they were observed in almost every quad in Maryland: 225 quads and 827 blocks. The MARA distribution encompassed the entire state and remained unchanged from H. S. Harris's (1975) distribution.

Snakes play an important role in the ecosystem by preying on small rodents and songbirds, thereby serving as an essential control on the population growth of prey species. On farms and ranches, Eastern Ratsnakes benefit people by controlling the small rodents that feed on grain. Unfortunately, persecution by the uninformed or by those who are afraid of snakes leads to unnecessary mortality. Many of these snakes are crushed by vehicles, and juveniles may be killed when mistaken for an Eastern Copperhead. Informing the public about the biology and value of snakes can help people develop a respect and appreciation for these fascinating reptiles. This species has no special protections in Maryland, and populations appear to be stable at this time.

CHRISTOPHER SWARTH

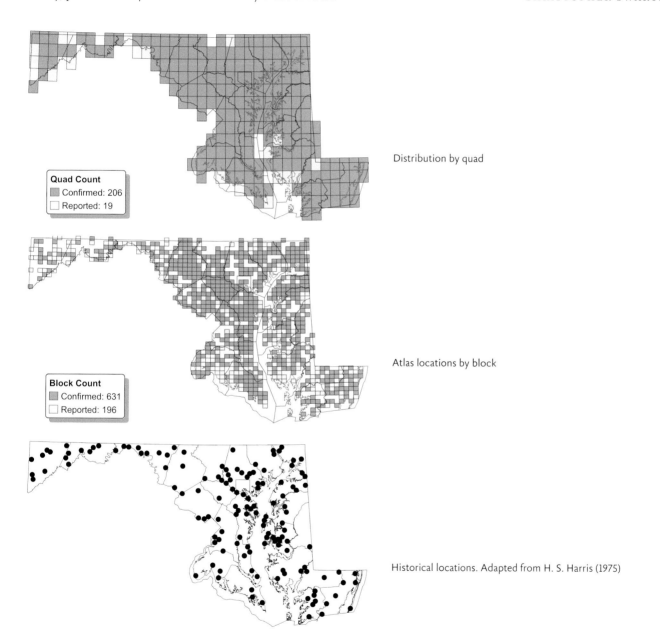

Quad Count
Confirmed: 206
Reported: 19

Distribution by quad

Block Count
Confirmed: 631
Reported: 196

Atlas locations by block

Historical locations. Adapted from H. S. Harris (1975)

Photograph by Lance H. Benedict

Red Cornsnake
Pantherophis guttatus

Captive-bred examples of the Red Cornsnake, in all its many color mutations, represent perhaps the most popular snake in the pet trade. This species is also native to Maryland, and though the state's Red Cornsnakes are less vibrantly colored than their relatives from the Deep South, this is certainly one of Maryland's more colorful snakes.

In Maryland, this snake begins life with a gray background color, which becomes beige, brown, or dull orange with age. The change begins anteriorly and gradually progresses toward the tail. The dorsum is marked with a series of large red-brown to orange-brown blotches, outlined in black and generally rectangular in shape. The first blotch has projections from the corners that extend onto the head, forming a spearhead pattern, one of the keys to identification. Between the eyes is an additional band, the same color as the blotches, that continues behind the eyes diagonally downward toward the neck. Laterally, there are smaller pale orange blotches, which, along with the spearhead and divided anal plate, help to distinguish this species from the similar Eastern Milksnake. The belly has a checkerboard pattern of black and white, with two black longitudinal stripes on the underside of the tail. The scales of this snake are very weakly keeled along the spine and smooth laterally. In cross section, the body is shaped like a loaf of bread (i.e., flat across the belly). Red Cornsnakes typically reach 76–122 cm in total length, with a maximum recorded length of 183 cm. Juveniles are very similar to juvenile Eastern Ratsnakes in appearance, but the spearhead pattern on the head sets them apart.

The Red Cornsnake is widely distributed east of the Mississippi River from southern New Jersey south to the Florida Keys and west to western Tennessee, Mississippi, and southeastern Louisiana. Isolated populations are also found in Kentucky. In Maryland, the Red Cornsnake is a rare snake with a patchy distribution across the state. Historically, these snakes were spottily distributed on the Atlantic Coastal Plain as far north as Anne Arundel County on the Western Shore and Talbot County on the Eastern Shore; however, records were lacking from Worcester County (H. S. Harris 1975). Piedmont records are scant, with two reported from southern Montgomery County (McCauley 1945; H. S. Harris 1975); Harris (1975) believed a report from Baltimore to be a released pet. An isolated population occurs on the Ridge and Valley Province in Allegany County (Norden 1971; Grogan and White 1973).

In Maryland, the Red Cornsnake is primarily a denizen of dry, open-canopied, mixed pine-hardwood forest, where it is mostly fossorial and rarely encountered. On the Atlantic Coastal Plain, the snakes are found in sandy upland areas, whereas on the Ridge and Valley, they occupy steep slopes with a lengthy photoperiod. Most activity seems to be at night, when individuals are sometimes found crossing roads. Otherwise, these snakes are sometimes found around farm buildings or under artificial cover objects, where they may rest in preparation for shedding or forage for rodents.

Despite the familiarity of herpetologists with this species from the pet trade, little has been written concerning its natural history in Maryland. At this northern latitude, Red Cornsnakes hibernate during the winter months in suitable hibernacula such as rodent burrows, rock crevices, and building foundations (Ernst and Ernst 2003). They emerge from hibernation in April and begin to forage. Rodents make up the bulk of their diet, but fledgling birds and lizards are also taken opportunistically (Mitchell 1994). Courtship and mating typically occur in May (Ernst and Ernst 2003). The Red Cornsnake is oviparous. Egg laying in Virginia occurs in June and July, with a single clutch of 7–20 eggs laid annually (Mitchell 1994). In Maryland, a female captured in Wicomico County laid 12 eggs, 8 of which hatched in September (Grogan 1994).

The Red Cornsnake is a relatively docile snake that rarely bites when handled. When cornered, however, it will sometimes vibrate its tail and assume a defensive posture, raising the front half of its body off the ground in a compact S shape and gaping its mouth. Although a completely harmless snake, many are mistaken for Eastern Copperheads and killed on sight.

Red Cornsnakes were found in only 11 quads and 17 blocks during the MARA project. Four of these quads are contiguous on the Ridge and Valley Province. The remaining quads are mostly in southern St. Mary's and Calvert counties, the stronghold of the Atlantic Coastal Plain population in Maryland. Historical locations in Anne Arundel, Caroline,

Charles, Dorchester, Montgomery, Prince George's, Somerset, Talbot, and Wicomico counties were not reconfirmed during the atlas period. Prior to the atlas period, Red Cornsnakes were most recently seen in Anne Arundel County, but they are not as common as they once were (Herbert Harris, Jr., pers. comm.). The last known records from Wicomico County are from the 1990s (Grogan 1994).

Red Cornsnake populations in Maryland seem to have declined and become isolated to a few remaining, rela-tively small areas on the Atlantic Coastal Plain and Ridge and Valley provinces. While the population on the Ridge and Valley seems relatively secure, the Atlantic Coastal Plain population may be under pressure from habitat loss due to development. Further population assessment of the Red Cornsnake may be necessary to fully understand its conservation needs.

LANCE H. BENEDICT

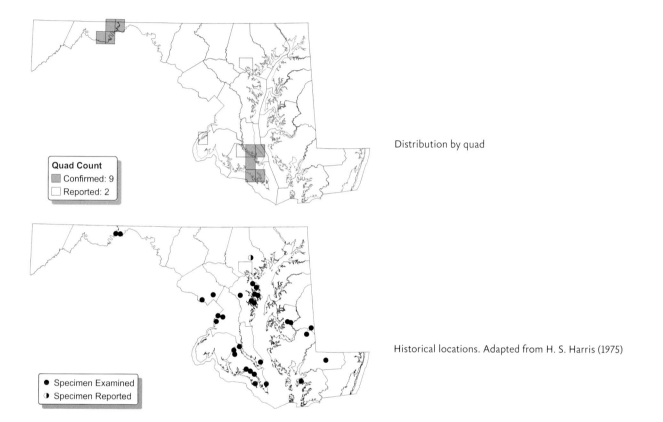

Distribution by quad

Historical locations. Adapted from H. S. Harris (1975)

Photograph by Nathan H. Nazdrowicz

Common Wormsnake
Carphophis amoenus

The Common Wormsnake is aptly named, as it resembles an earthworm with eyes. The snake has a plain brown dorsum with smooth scales. The belly and the first one or two scale rows of the sides are distinctly pinkish in color. The head is uncharacteristic of most snakes—conical, not distinct from the neck, and with very small, black eyes. The tail has a sharp spine on the tip (McCauley 1945). This is one of Maryland's smallest snakes; adults are 19-28 cm in total length, with females typically larger and heavier than males (Willson and Dorcas 2004).

The Common Wormsnake is widely distributed throughout the eastern United States. It occurs as far north as Rhode Island and southwestern Massachusetts, south to central Georgia and west to the Mississippi River. A few isolated populations in eastern Arkansas and southwestern Illinois mark the northwestern extent of its distribution. In Maryland, Common Wormsnakes occur throughout most of the state, except for the Appalachian Plateaus (H. S. Harris 1975). McCauley (1945) reported the Common Wormsnake as most common on the Atlantic Coastal Plain in Maryland. The snakes are fairly common on the Delmarva Peninsula but absent on the barrier islands (White and White 2007).

This species is a denizen of forested habitats. The snakes typically occur in moist hardwood forests (Gibbons and Dorcas 2005) but can be found in pine and mixed pine-hardwood forests and open areas with soft, loamy soil that is suitable for burrowing (Hulse et al. 2001; White and White 2007). Common Wormsnakes can tolerate a wide range of soil moisture, pH, and temperature (J. M. Orr 2006). The species can also be fairly common in rocky, forested areas (Hulse et al. 2001; Gibbons and Dorcas 2005) and near the edges of wetlands and streams (McCauley 1945). In general, Common Wormsnakes favor areas of high canopy cover and deep leaf litter (Sutton et al. 2010). They are fossorial and are rarely encountered in the open. Like earthworms, the primary prey of Common Wormsnakes, these small snakes can be found under logs, rocks, and other debris on the forest floor (Gibbons and Dorcas 2005). Rotting logs are particularly favorable. The author (Therres) has found that logs with Bess Beetles (*Passalus cornutus*) are good microhabitat for Common Wormsnakes; McCauley (1945) reported that Wormsnakes often use burrows created by these beetles under the bark of rotting logs. These snakes can also be found under cover objects in old fields and powerline rights of way (Willson and Dorcas 2004), and people occasionally find them in yards while raking leaves (H. A. Kelly et al. 1936). Most Common Wormsnakes stay within an average home range area of 253 m^2 (Barbour et al. 1969).

After emerging from hibernation, this species is generally active from March through November in the northern regions of its range (Ernst and Ernst 2003). The snakes hibernate underground or in rotting logs and stumps. In Maryland, one was found hibernating about 72 cm deep in the soil (Grizzell 1949). Most daily activity is nocturnal (Ernst and Ernst 2003). Diurnal activity peaks during the late afternoon and early evening hours, and active periods typically do not last longer than 12 hours (Croshaw 2008).

Courtship and mating generally take place in the spring. In Maryland, courtship occurs during May, with oviposition in June and July (M. F. Groves 1946). Females lay 4-5 smooth, elongate eggs (Gibbons and Dorcas 2005) in rotting logs or under rocks (Ernst and Ernst 2003). Egg laying in Virginia occurs in June and July (Mitchell 1994). In Maryland, egg laying may occur as early as mid-May; a gravid female with 5 eggs was collected in Worcester County on 13 May (McCauley 1945). The incubation period is not known, but eggs typically hatch in late summer (Ernst and Ernst 2003).

Common Wormsnakes were documented in 290 blocks within 130 quads during the atlas period. They probably were present in many more blocks, but the fossorial habit of Wormsnakes makes them challenging to find—often requiring repeated visits to the same location. At one woodlot in Queen Anne's County, it took four visits by the author to find one, and in Talbot County, it took three visits before one was found under a log overturned on the previous two visits. During the atlas period, Common Wormsnakes were found throughout the Atlantic Coastal Plain, with a spotty distribution on the Piedmont. They were found in only nine atlas blocks on the Ridge and Valley Province. The MARA distribution was similar to the historical distribution in Maryland (H. S. Harris 1975). However, they were also found in western Montgomery County, where no historical records existed. None were found in Carroll County, where Harris (1975) reported one record of unknown origin.

Given the development and urban sprawl that has occurred in Maryland over the past few decades, it is en-

couraging that the distribution of this small woodland snake has not changed. It appears that the Common Wormsnake can persist in the Maryland landscape, provided some forest habitat is conserved.

GLENN D. THERRES

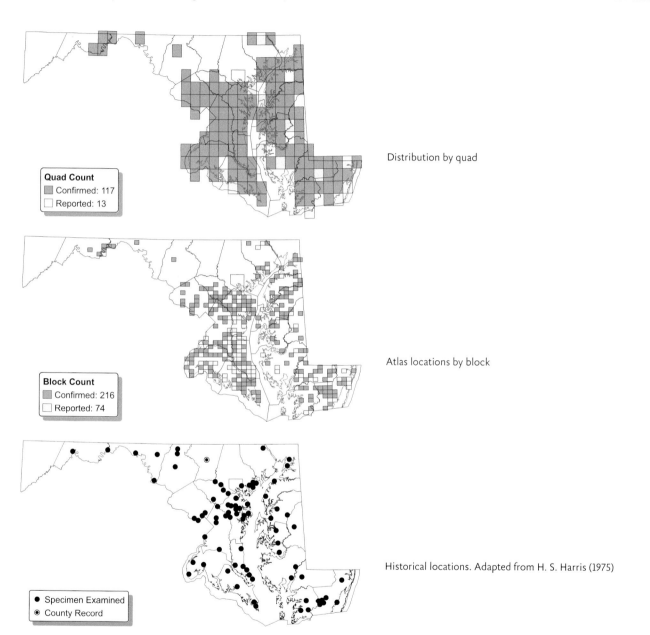

Quad Count
Confirmed: 117
Reported: 13

Distribution by quad

Block Count
Confirmed: 216
Reported: 74

Atlas locations by block

● Specimen Examined
◉ County Record

Historical locations. Adapted from H. S. Harris (1975)

Photograph by Michael Kirby

Ring-necked Snake
Diadophis punctatus

The Ring-necked Snake is one of only seven transcontinental snakes in North America (Fontanella et al. 2008) and varies in size and coloration across its range. In Maryland, the Ring-necked Snake is small and weakly aposematically colored. Dorsal coloration varies from light olive tan to black. A thin yellow to pale orange ring is present around the neck, although the ring may be dorsally broken in populations on Maryland's Eastern Shore. Ventral coloration ranges from yellow to pale orange and is variably marked in Maryland, ranging from a complete to incomplete single medial row of black, crescent-shaped marks running from below the neck to above the vent in eastern populations to unmarked in western populations. The head is flat, blunt nosed, and dorsally darker than the rest of the dorsum, and the tail is short with a sharp point. Sexual dimorphism is limited, with females slightly larger and entirely smooth scaled, and males having weakly keeled dorsal scales near their cloacal region. Total adult length ranges from 25.4 cm to 38 cm for individuals in eastern North America.

The Ring-necked Snake has a somewhat contiguous range in its eastern continental extent from southeastern Canada south through Florida, west to Mexico and southeastern Arizona, then diagonally back to the southern shores of Lake Superior, including most of the Midwest prairie region. The westernmost extent of its continental range is discontinuous, with isolated populations in the northwestern and southwestern United States (Ernst and Ernst 2003). In Maryland, the range is presumably statewide, with Ring-

necked Snakes recorded in all physiographic provinces and counties except Wicomico (H. S. Harris 1975).

Habitat is range specific (Ernst and Ernst 2003), but in Maryland, Ring-necked Snakes are primarily found in mature forests with loose moist soils, thick leaf litter, and woody debris (McCauley 1941, 1945; Willson and Dorcas 2004). Ring-necked Snakes are both fossorial and predominantly nocturnal (Blanchard et al. 1979), and they are most often found by searching under rotting logs, large rocks, loose tree bark, brush piles, and leaf litter.

Their annual and daily activity is typically dictated by regional temperature (Henderson 1970; Blanchard et al. 1979) and humidity (Elick and Sealander 1972; Seigel and Fitch 1985). In temperate climates such as Maryland's, individuals hibernate over the winter, entering their hibernacula in late fall and exiting in the spring (Ernst and Ernst 2003). Ring-necked Snakes do not bask, and their optimal temperatures for activity range from 25 °C to 29 °C (Henderson 1970).

Breeding occurs in the fall (Blanchard et al. 1979; Willson and Dorcas 2004), with fertilization occurring over the winter or in the spring (Ernst and Ernst 2003). Females lay their eggs, sometimes communally (Blanchard et al. 1979), in moist rotting logs, leaf litter, or soil in June and July (McCauley 1945; D. R. Clark et al. 1997). Eggs are leathery, thin shelled, and, like the adults, susceptible to dehydration (Fitch and Fitch 1967). Clutch size ranges from 1 to 6 eggs in Maryland (D. R. Clark et al. 1997), is positively correlated with female body size (McCauley 1945; Blanchard et al. 1979; Seigel and Fitch 1985; D. R. Clark et al. 1997), and has been positively correlated with annual precipitation (Seigel and Fitch 1985). Eggs typically hatch in late August or September after an average incubation period of 52 days (McCauley 1945; D. R. Clark et al. 1997; Ernst and Ernst 2003). In Michigan, males usually reach sexual maturity in around 3–4 years and females at 4–5 years (Blanchard et al. 1979). Size at maturity varies with range, but the general consensus is that individuals more than 22 cm in length are most likely mature (Ernst and Ernst 2003). Ring-necked Snakes are long lived, with some known to exceed 22 years in the wild (Blanchard et al. 1979).

Ring-necked Snakes rely primarily on chemical cues to locate prey (Henderson 1970; Lancaster and Wise 1996) and use a mild venom to subdue prey once captured. Their venom has some similarities to that of vipers but poses no threat to anything larger than the snake's prey (Hill and Mackessy 2000). Ernst and Ernst (2003) provide a list of prey species across the entire continental range. Maryland populations appear to feed primarily on Eastern Red-backed Salamanders and earthworms (Blanchard et al. 1979). Ernst and Ernst (2003) also list predators, which include lizards, snakes, birds, rodents, and larger mammals. Ring-necked Snakes with strong aposematic coloration perform a display in response to predators that involves tightly coiling and upturning the tail to reveal the bright orange to red ventral surface (Gehlbach 1970). In Maryland, Ring-necked Snakes rarely exhibit this display and more often resort to erratic thrashing, chewing, salivating, and defecating (McCauley 1945).

During the MARA project, Ring-necked Snakes were documented in 426 blocks within 169 quads and in all 23 counties and Baltimore City. The MARA results were comparable to the historical distribution (H. S. Harris 1975). Records from Wicomico County, not documented by Harris (1975), revealed a wide-ranging distribution, probably resulting from the increased effort to document the species.

The widespread distribution of the Ring-necked Snake in Maryland suggests that no species-specific conservation actions are necessary at this time. However, the apparent increased prevalence of snake fungal diseases is cause for concern (Allender et al. 2015). Until the impact of this newly emerging disease is better understood, continued protection of forested habitats from development would best benefit this snake, as well as numerous other species in Maryland.

TASHA FOREMAN

Distribution by quad

Quad Count
Confirmed: 149
Reported: 20

Atlas locations by block

Block Count
Confirmed: 327
Reported: 99

Historical locations. Adapted from H. S. Harris (1975)

Photograph by Matthew Kirby

Rainbow Snake
Farancia erytrogramma

The Rainbow Snake is not only one of the most colorful of Maryland's snakes, it is also perhaps the rarest and most enigmatic, with only 18 documented sightings from 1937 to date (Saunders and Benedict 2008; Lance H. Benedict, unpubl. data). Unlikely to be confused with any other Maryland snake, this harmless, state-endangered colubrid has a glossy, iridescent black back—hence the "rainbow" moniker—with three longitudinal red stripes. Laterally, the first two scale rows adjacent to the belly are half yellow, half red. The belly is red or pinkish with two rows of large black spots. A discontinuous central row of black belly spots may also be present. The anal plate is usually divided. The top of the head is black with an intricate pattern of red markings. The throat is bright yellow, and the upper labial scales are also bright yellow, each with a prominent black blotch. The eyes are impenetrably dark, and the head is indistinct from the body. The snake's tail comes to a sharp point. Adults are typically 68.8-122 cm in total length, with the largest specimen in Maryland measuring 115 cm (Saunders and Benedict 2008).

The Rainbow Snake is distributed on the Atlantic Coastal Plain from Charles County, Maryland, to south-central Florida and in the Gulf States westward to southeastern Louisiana. In Maryland, Rainbow Snakes are found in Charles County near the Potomac River (H. S. Harris 1975; Saunders and Benedict 2008). The first three Maryland specimens were unearthed along Chicamuxen Creek during road construction in 1937 (McCauley 1939b, 1945). In the period 1960-1988, three additional snakes were found near Trunk Gut, southwest of Newburg (J. E. Cooper 1960a; R. W. Miller

and Zyla 1992). Since 2005, 11 Rainbow Snakes have been documented in western Charles County, from Mallows Bay south to Thorn Gut Marsh (Saunders and Benedict 2008; L. Benedict, unpubl. data). A single, mutilated specimen was found in Montgomery County in 1987, but it was neither photographed nor preserved, and its origin remains uncertain (Saunders and Benedict 2008).

This species is primarily aquatic, utilizing freshwater streams and marshes as well as brackish rivers and marshes. The snakes do venture from the water to bask, shed, lay eggs, consume prey, and overwinter (Richmond 1945). The Rainbow Snake is nocturnal and fossorial out of the water; therefore, it is rarely encountered even in areas where it is known to occur. In fact, the first live Rainbow Snake was not photographed in Maryland until 2005 (Saunders and Benedict 2008). At night, these snakes have been found on roads (Saunders and Benedict 2008) and in small streams (Matthew Kirby, pers. comm.). During the day, they have mostly been found buried (Richmond 1945; Harrell et al. 2009) and, occasionally, basking in the spring and fall (Raymond Jarboe, pers. comm.).

Little is known about the natural history of the Rainbow Snake in Maryland. Rainbow Snakes are active in the state from April through October (L. Benedict, unpubl. data). Elsewhere in their range, they are active year-round (Mitchell 1994). However, there is evidence suggesting that individuals will sometimes hibernate. For instance, Rainbow Snakes have been plowed from sandy fields in March (Richmond 1945). In Maryland, Rainbow Snakes most likely overwinter. Females lay 20-40 eggs in a subsurface cavity in July. The female may remain with the eggs (Mitchell 1994). Hatching often occurs in September, with young snakes shedding a few days after birth and overwintering in the soil in or near the nest (Richmond 1945). Young snakes enter aquatic habitat in March or April (J. W. Gibbons et al. 1977).

Adult Rainbow Snakes prey almost exclusively on the American Eel (*Anguilla rostrata*), while juveniles are known to eat tadpoles and possibly stream salamanders (Neill 1964). Rainbow Snakes are extremely gentle and passive. Considered harmless, this snake seldom, if ever, bites when handled. Immediately after capture, the snake will thrash around for several seconds, sometimes attempting to poke the handler with its harmless tail tip. Thereafter, it hardly moves at all. The Rainbow Snake does not, however, make a good captive. As a state-listed endangered species, as stipulated by the MDNR, Rainbow Snakes can be held only in accordance with a scientific collection permit or an endangered species permit issued by the state agency.

Rainbow Snakes were documented in three blocks within two quads during the atlas period. These blocks lie in western Charles County, essentially overlapping with the area investigated by Saunders and Benedict (2008), but differ from localities documented by H. S. Harris (1975). The historical locations of Chicamuxen Creek in northwestern Charles County and Trunk Gut in southern Charles County were not reconfirmed during the atlas project, but given the secretive nature of this snake, the lack of sightings should not be interpreted as evidence of extirpation from these areas. Nonetheless, no specimens have been found in Chicamuxen Creek since 1937 (Forester and Miller 1992; Seth Berry, pers. comm.). The most recent specimen from the Trunk Gut area was found in 1988 (R. W. Miller and Zyla 1992).

With the range of the Rainbow Snake in Maryland apparently limited to streams and wetlands near the Potomac River in Charles County, conservation of the species will require preservation of wetlands and adjacent forest habitat in this region. Water quality and aquatic corridors must also be maintained for the American Eel, the Rainbow Snake's primary prey. Given the rarity of this snake and what little is known of its natural history in Maryland, all observations should be reported to the MDNR.

LANCE H. BENEDICT

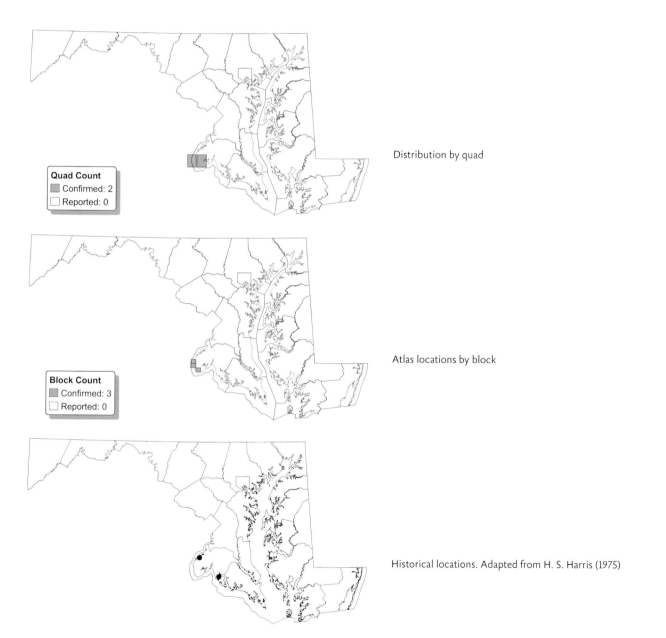

Distribution by quad

Quad Count
Confirmed: 2
Reported: 0

Atlas locations by block

Block Count
Confirmed: 3
Reported: 0

Historical locations. Adapted from H. S. Harris (1975)

Photograph by Ed Thompson

Eastern Hog-nosed Snake
Heterodon platirhinos

The name for the Eastern Hog-nosed Snake is derived from the distinctive, slightly upturned rostral scale at the tip of its snout. This is a medium-sized, thick-bodied snake with keeled dorsal scales and a divided anal plate. Its color pattern is highly variable, ranging from dark blotches on a light-colored body to entirely black (Ernst and Ernst 2003). For patterned individuals, the background color is varying combinations of gray brown, olive, orange, and yellow, with 20–30 large brown blotches on the back and smaller alternating blotches on the sides (White and White 2007). The head is the same color as the body, with a transverse dark bar that connects the eyes and continues to the corner of the mouth (Ernst and Ernst 2003). The underside is cream to yellow and sometimes mottled with gray. Adult total length ranges from 51 cm to 84 cm, and females are the larger sex.

The Eastern Hog-nosed Snake is widely distributed throughout eastern North America. It occurs from southwestern Ontario and southern New Hampshire to the Florida Keys and as far west as western Texas to central Minnesota. In Maryland, the species has a statewide distribution but is most common on the Atlantic Coastal Plain, while rare on the Appalachian Plateaus (McCauley 1945; H. S. Harris 1975). The first known Eastern Hog-nosed Snake from Maryland was found on 3 June 1878 in Laurel, Prince George's County (Cope 1900).

These snakes are found in a variety of habitats, from wooded hillsides to grassy and cultivated fields, including relatively dry areas with sandy or sandy-loam soils (Ernst and Ernst 2003). They also may occupy disturbed habitats such as edges of forest clear-cuts or small patches of remain-

ing woodlands in developments. Hog-nosed Snakes are exclusively diurnal and may travel long distances during the day in search of food or to find a mate. At night or during the hottest parts of the day, they burrow into the soil, using their snout to push forward while moving their head laterally from side to side (Ernst and Ernst 2003).

In Maryland, Eastern Hog-nosed Snakes emerge in early April and are active until mid-November, depending on the temperature. Individuals hibernate alone, and males locate receptive females in spring after emergence. Range-wide, most breeding occurs during the spring from March through May, but breeding may also take place in September and October in some areas (Plummer and Mills 1996). Nesting season extends from early May through August, with most oviposition occurring in June to July (Ernst and Ernst 2003). Nesting may occur in shallow cavities in loose or sandy soil or under rocks (Edgren 1955; Behler and King 1979), where females lay 4–61 elongate, thin-shelled eggs. Hatchlings emerge after 39–65 days, depending on the temperature and other environmental factors such as rain events (Behler and King 1979). Eastern Hog-nosed Snakes may live as long as 7 years (Snider and Bowler 1992).

Eastern Hog-nosed Snakes primarily prey on amphibians, particularly toads (F. M. Uhler et al. 1939; Platt 1969), and may also occasionally prey on insects (Platt 1969). In Virginia, the diet was reported to consist of 40% toads, 30% frogs, 11% salamanders, and 19% small mammals (Uhler et al. 1939). The saliva of Hog-nosed Snakes is mildly venomous, and enlarged teeth on the rear upper jaw may be used to convey venom to subdue prey (Behler and King 1979). These snakes are considered harmless to humans, however, with very rare bites resulting in only localized swelling (Behler and King 1979; T. R. Johnson 1987; Hulse et al. 2001). When threatened, Eastern Hog-nosed Snakes exhibit extraordinary defensive behaviors. At first, they flatten the head and neck and hiss loudly with a gaping mouth. Continued provocation often results in false strikes with the mouth closed. If this strategy does not deter the would-be attacker, the snake will writhe about on the ground, regurgitate any undigested meal, and defecate, followed ultimately by rolling over onto its back with a gaping mouth, feigning death. If the snake is righted, it will roll over again (Hulse et al. 2001).

During the MARA project, Eastern Hog-nosed Snakes were recorded from 93 quads and 162 blocks. They were documented in 22 counties, with none found in Harford County. As described earlier by H. S. Harris (1975), the majority of the records were from the eastern Piedmont and Atlantic Coastal Plain.

Loss of habitat due to housing developments, automobile traffic, and intentional killings are likely to be the major concerns for continued survival of the species in Maryland. Both the appearance and behavior of the Eastern Hog-nosed Snake increase the risk that it will be misidentified as a venomous species. Its overall color and pattern, in addition to its habit of flattening its head and neck and hissing when threatened, can make people think that the snake is a threat and react accordingly. Educating the public may help deter the unnecessary slaughter of this harmless and beneficial species.

GEORGE M. JETT

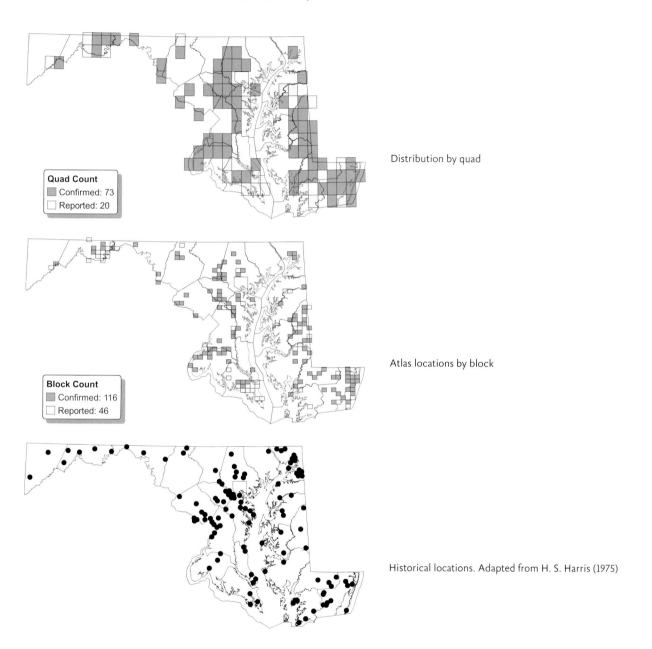

Distribution by quad

Quad Count
Confirmed: 73
Reported: 20

Atlas locations by block

Block Count
Confirmed: 116
Reported: 46

Historical locations. Adapted from H. S. Harris (1975)

Photograph by Matthew Kirby

Plain-bellied Watersnake
Nerodia erythrogaster

The Plain-bellied Watersnake is appropriately named, given that in the eastern portion of its range, including Maryland, this large snake is plain-colored with a conspicuously reddish orange but unmarked belly. Because of its coloration, the Plain-bellied Watersnake was known as the Red-bellied Watersnake during part of the atlas period, but its common name was changed in 2012 (Crother et al. 2012). Dorsally, adults are uniformly dark to coppery brown with strongly keeled scales. The upper labial scales are often reddish with dark brown vertical marks. The anal plate is divided. This watersnake typically attains a total adult length ranging from 76 cm to 122 cm, with females slightly larger than males (Mitchell 1994). Juveniles have dark brown blotches over a lighter, reddish brown background, similar to the Common Watersnake, but the belly is a plain cream to light orange.

The Plain-bellied Watersnake occurs from the southern Delmarva Peninsula south to northern Florida and west through the Gulf States. It is primarily an Atlantic Coastal Plain species in the mid-Atlantic region but occurs on the Piedmont further south. Its range extends up the Mississippi River valley to southern Illinois and west to Kansas, Oklahoma, Texas, and northern Mexico. Its distribution in Maryland is limited, occurring only on the lower Eastern Shore in Dorchester, Wicomico, and Worcester counties. Historical records were reported from Blackwater NWR and along the upper reaches of the Pocomoke River (H. S. Harris 1975). The species was first documented in Maryland in 1941 near Willards, Wicomico County (Conant 1943b). Grogan (1985) reported three additional records from Dorchester County and suspected that this snake was commonly distributed in the southern half of the county. A Plain-bellied Watersnake was reported from Anne Arundel County (J. E. Cooper 1969), but that record was determined to be questionable (H. S. Harris 1975).

Within its southeastern range, the Plain-bellied Watersnake is commonly associated with rivers and floodplains, large and small lakes and ponds, and other natural wetlands (Gibbons and Dorcas 2005). White and White (2007) described its habitat on the Delmarva Peninsula as headwater swamps, freshwater and brackish portions of tidal swamps and marshes, slow-moving streams, and ditches. Though an aquatic species, Plain-bellied Watersnakes are rarely seen in water and may be far from it, particularly in hot, dry weather (White and White 2007). They can be found basking on logs and vegetation along the edges of bodies of water. When resting near water, this species tends to select a spot directly on the ground rather than in branches (Linzey and Clifford 1981). The snakes can also be found under surface objects such as boards, logs, and litter (Mitchell 1994).

Plain-bellied Watersnakes emerge from hibernation in late spring. Mating occurs in April through mid-June (Ernst and Ernst 2003). Their reproduction and development have not been documented on the Delmarva Peninsula (White and White 2007). Plain-bellied Watersnakes are viviparous; females give live birth to as many as 50 young in late summer or early fall (Linzey and Clifford 1981). In Virginia, births were reported to occur during the first three weeks in September (Mitchell 1994).

Prey species include frogs and fish, though crayfish and salamanders are also eaten (Linzey and Clifford 1981). All prey are held by the watersnake's long, recurved teeth and swallowed alive (Mitchell 1994).

During the atlas period, Plain-bellied Watersnakes were detected in 11 quads and 28 blocks. They were documented in the same three lower Eastern Shore counties reported by H. S. Harris (1975): Dorchester, Wicomico, and Worcester. In Worcester County they were documented in association with the Coastal Bays watershed in addition to the Pocomoke River. Although not mapped, one Plain-bellied Watersnake was reported to the MARA project without voucher evidence from St. Mary's County in 2010. It was determined that this report was more likely to be a misidentified Common Watersnake, as there is a population of uniformly patterned Common Watersnakes on the Virginia side of the Potomac River (Mitchell 1994).

Though limited in its distribution in Maryland, this species may have expanded its range slightly since the late 1900s. Because this is a semiaquatic species, the conservation of water quality and wetlands on the lower Eastern Shore of Maryland is essential to sustain the Plain-bellied Watersnake. Fortunately, much of the occupied range of this species consists of hydric soils and wetlands, which limits development within this range to upland locations.

The species is found even in developed areas such as Berlin and West Ocean City due to the abundance of ditches, creeks, impoundments, and wetlands. White and White (2007) expressed concern that the Plain-bellied Watersnake may be declining on the Delmarva Peninsula due to habitat loss. The results of the atlas study, however, suggest that the range of this species on the lower Eastern Shore has not declined. Given Maryland's strong wetland protection laws and the wet conditions of the forests within the species' range, the Plain-bellied Watersnake's continued existence appears secure.

GLENN D. THERRES

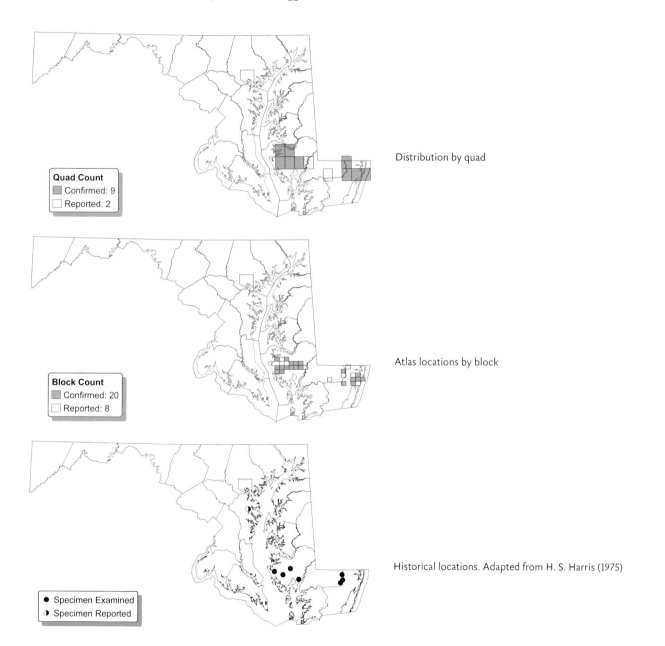

Quad Count
Confirmed: 9
Reported: 2

Distribution by quad

Block Count
Confirmed: 20
Reported: 8

Atlas locations by block

Specimen Examined
Specimen Reported

Historical locations. Adapted from H. S. Harris (1975)

Photograph by Andrew Adams

Common Watersnake
Nerodia sipedon

The Common Watersnake is one of Maryland's most widespread and abundant snakes. It is also the one aquatic snake that many Marylanders mistakenly call "water moccasin," even though the range of the true venomous "water moccasin," the Northern Cottonmouth (*Agkistrodon piscivorus*), does not extend north of southeastern Virginia.

The Common Watersnake has a stout, roundish body, with a moderate-sized head somewhat distinct from the neck. The tail is relatively short. Scales are strongly keeled, and the anal plate is divided. These snakes are extremely variable in color and pattern. The dorsal color of adults can be brown, tan, gray, or reddish, usually with dark crossbands on the neck and anterior half of the body, which change to dark, alternating dorsal and lateral blotches on the posterior half. The belly is typically cream colored with reddish and black spots, often in the shape of half-moons, on each ventral scale. The upper labial scales on the head are lighter than the background body color and are marked with dark vertical bars. Adults tend to darken with age and the dorsal markings become obscured, with some individuals appearing uniformly dark brown to black. Adult females are larger and heavier than adult males (Ernst and Ernst 2003). Adults typically measure 56-106.7 cm in total length, with individuals in Maryland documented to be as long as 122 cm (McCauley 1945). Juveniles are similar to patterned adults, but the contrast is stronger and the blotches are dark brown or black.

The range of the Common Watersnake extends over much of central and eastern North America. From eastern Maine and southeastern Canada, the species is distributed south to the panhandle of Florida and southeastern Louisi-

ana. South of North Carolina, it is absent from the Atlantic Coastal Plain. Nebraska and eastern Colorado mark the western extent of its distribution. In Maryland, Common Watersnakes are found in abundance throughout the state (McCauley 1945; H. S. Harris 1975).

Common Watersnakes inhabit freshwater swamps, rivers, and streams in Maryland. They are also common in stormwater management ponds in more developed areas. These snakes can also occasionally be found in brackish tidal waters, though according to Ernst and Ernst (2003), the species requires fresh water. Within these aquatic habitats, Common Watersnakes can be found basking in sun-exposed areas such as on the roots or branches of trees or shrubs, stumps, cypress knees, rocks, or bare or grassy banks (McCauley 1945). When not basking, these watersnakes may hide under cover objects such as boards, tin, rocks, or other debris. When disturbed, Common Watersnakes typically escape directly into the nearby waterbody, where they may seek shelter under banks of streams, under rocks, or in mud. Common Watersnakes can be extremely vicious when cornered or captured, striking repeatedly and biting aggressively. They will also void the contents of their anal glands while wildly thrashing the tail, releasing a characteristically noxious-smelling substance.

The primary activity period throughout its range is April through August (Ernst and Ernst 2003). On the lower Eastern Shore of Maryland, the Common Watersnake may be active all year, with short bouts of inactivity during extremely cold weather. In western Maryland, by contrast, individuals may not emerge from hibernation until later in the spring. Courtship and mating typically occur during the day in April through June; some mating also seems to occur in the fall (Ernst and Ernst 2003). Copulation mostly takes place on land but can also occur on low branches of trees and shrubs and in shallow water; it may last for 2-3 hours (Ernst and Ernst 2003). The Common Watersnake is a viviparous species. The birth of 30 or more young usually occurs in late August or early September (H. A. Kelly et al. 1936). Females are mature after about 3 years, and males mature in about 2 years.

Common Watersnakes hunt primarily during the day within or near their aquatic habitats, chasing down prey, capturing it in their mouths, and swallowing it head first; however, frogs are typically swallowed feet first. Common prey species include fish, frogs and toads, salamanders, crayfish, and various aquatic worms, snails, and insects (Ernst and Ernst 2003). These watersnakes also eat carrion. Because of their propensity to eat fish, Common Watersnakes can cause significant economic damage in commercial fish farms. They are very active foragers, swimming slowly through the water, probing into every nook or other likely spot where prey could be hidden. They can detect prey by sight and odor.

During the atlas project, Common Watersnakes were found throughout the state, with the species documented in 219 quads and 683 blocks. This distribution was similar to that reported by H. S. Harris (1975).

The Common Watersnake continues to be one of Maryland's most abundant and widespread snake species. Its abil-ity to persist in just about any aquatic environment that contains fish or other prey—from pristine streams to degraded stormwater management ponds—suggests that its population within the state is secure.

DAVID R. SMITH

Distribution by quad

Quad Count
Confirmed: 192
Reported: 27

Atlas locations by block

Block Count
Confirmed: 463
Reported: 220

Historical locations. Adapted from H. S. Harris (1975)

Photograph by Matthew Kirby

Queensnake
Regina septemvittata

Among snakes worldwide, the Queensnake has one of the most specialized diets, feeding almost exclusively on recently molted crayfish (Wood 1949; Godley et al. 1984). To contend with the low relative abundance of freshly molted crayfish in aquatic systems (R. A. Stein 1977), Queensnakes have a chemosensory ability to detect hormones released by the freshly molted crayfish (Jackrel and Reinert 2011).

The dorsal surface of the Queensnake is an olive brown to gray color, with one middorsal and two dorsolateral faint black stripes (H. A. Kelly et al. 1936; McCauley 1945). The ventral surface is yellow to cream with four brown, longitudinal stripes that border the midline and the sides of the ventral surface (Hulse et al. 2001). The ventral coloration extends onto the lower lateral scales and upper labial scales on the head, forming a lateral stripe. The scales are keeled, and the anal plate is divided. This species is relatively small and slender when compared with other aquatic snakes of North America. Adults grow to a total length of 38–61 cm, with females outgrowing males. Juvenile Queensnakes have more prominent patterning, with the dorsal stripes fading with age.

The distribution of the Queensnake extends throughout the eastern and central United States. From southern Ontario, southeastern Pennsylvania, and western Wisconsin, its range extends south to Mississippi, with an isolated population centered in Arkansas. Although widely distributed, populations are often disjunct, separated by habitat fragmentation and a lack of riparian vegetation and suitable prey populations (Ernst and Ernst 2003). In Maryland, Queensnakes are found throughout the Piedmont, Blue Ridge, and Appalachian Plateaus, but a conspicuous distributional gap occurs on the Ridge and Valley Province, with populations found only on the periphery (H. S. Harris 1975). The range also extends onto the northern Atlantic Coastal Plain on the Western Shore and Eastern Shore (McCauley 1945; H. S. Harris 1975), but these snakes are absent from the Atlantic Coastal Plain in the southeastern United States. The southernmost observations on the Atlantic Coastal Plain in Maryland are in Calvert County (Lee 1973c).

Queensnakes occupy small- to medium-sized streams with slow to moderate currents, a rocky substrate, and abundant crayfish populations (Branson and Baker 1974; Hulse et al. 2001). Although riparian vegetation provides basking opportunities, vegetation must remain sparse enough to allow ample direct sunlight (Branson and Baker 1974). Queensnakes can be found basking in overhanging branches, hiding in crayfish burrows, or resting under stones in and along the stream (Gibbons and Dorcas 2004). Though they will often dive into water when approached while basking, Queensnakes rarely bite and instead expel feces and musk as a defense.

Seasonal activity for Queensnakes begins in March or April, with increasing activity following the onset of warm spring weather (Hulse et al. 2001). Daily foraging occurs in the early morning and late afternoon to evening (Branson and Baker 1974). Activity decreases with hot, dry weather in the summer, when the snakes take refuge under rocks in shallows. An additional, small peak in activity typically occurs in the fall (J. T. Wood 1949). Overwintering is communal for this species, with aggregations occurring in Muskrat (*Ondatra zibethicus*) and crayfish burrows or in streamside rock crevices (Ernst and Ernst 2003; Gibbons and Dorcas 2004).

Breeding occurs in the spring, with females giving birth to live young around mid-August through September (Hulse et al. 2001; Ernst and Ernst 2003). Litter size varies with geographic location and size of the female, ranging between 5 and 17 young (Trauth 1991; Gibbons and Dorcas 2004). Queensnakes reach sexual maturity at 2 years, but females typically reproduce for the first time at 3 years (Branson and Baker 1974).

During the MARA project, Queensnakes were observed in 81 blocks within 56 quads. The overall MARA distribution showed similarities to the distribution reported by H. S. Harris (1975), with an absence from the Ridge and Valley regions of Allegany and Washington counties and sparse observations from the Atlantic Coastal Plain. However, Queensnakes were found to be much more common than previously documented in the upper Piedmont. Additional occurrences were also documented on the Atlantic Coastal Plain, extending the distribution farther south into Charles and Queen Anne's counties. Although the MARA results increased the known Maryland distribution of the

Queensnake, this was probably an artifact of increased survey efforts rather than a range expansion.

Although not considered threatened or endangered in Maryland, the Queensnake, like other aquatic reptiles, is particularly sensitive to anthropogenic and environmental disturbances such as urban development, habitat fragmen-

tation, and destruction of riparian and aquatic habitats (Gibbons et al. 2000; Reading et al. 2010). The Queensnake's permeable skin may cause it to be more sensitive and vulnerable to exposure to numerous aquatic contaminants than other reptiles (Stokes and Dunson 1982).

ANDREW P. LANDSMAN

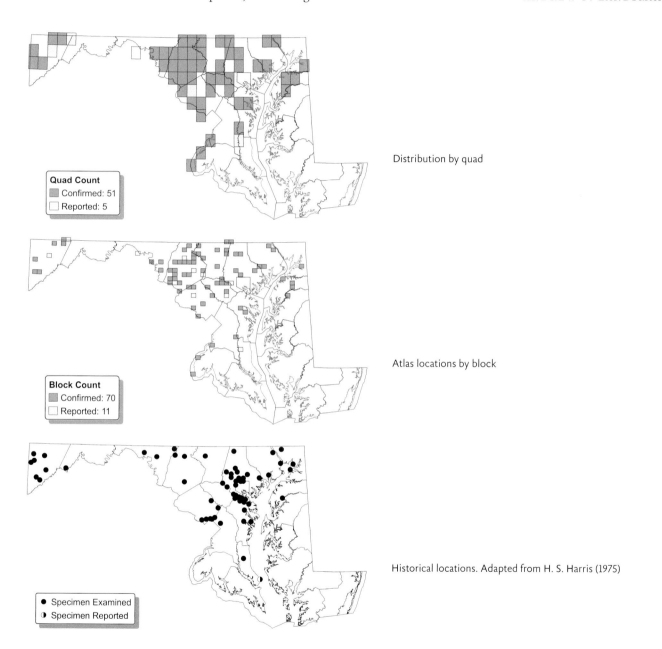

Quad Count
Confirmed: 51
Reported: 5

Distribution by quad

Block Count
Confirmed: 70
Reported: 11

Atlas locations by block

• Specimen Examined
◗ Specimen Reported

Historical locations. Adapted from H. S. Harris (1975)

Photograph by Michael Kirby

Dekay's Brownsnake
Storeria dekayi

The scientific name of the Dekay's Brownsnake honors two prominent zoologists: David Humphreys Storer and James Ellsworth De Kay. Dekay's Brownsnake is rather plain, with keeled scales, and it can superficially resemble earthsnakes with their brown coloration and lack of conspicuous markings. The overall color can vary from reddish brown to yellowish brown, gray, tan, or dark brown, with two rows of small, dark brown spots down the back. A light middorsal stripe is usually evident between these spots. Boldly marked individuals have brown lines connecting the spots or have additional dark spots along the sides, forming a faint checkered pattern. On the sides of the head, behind the eyes, there is a dark, nearly vertical streak, which is a diagnostic characteristic for this species. The belly color varies from a pale yellow to pinkish, cream, or brown color and is nearly unmarked. Adults average 23-33 cm in total length. Juveniles have a light yellowish, usually incomplete, collar on the neck.

Dekay's Brownsnake is found throughout most of eastern North America from southern Canada south to the Gulf Coast and west to the Great Plains. In Maryland, it has been recorded in every county (H. S. Harris 1975). McCauley (1945) considered this species most common on the Atlantic Coastal Plain; fewer records exist in the remainder of the state (H. S. Harris 1975). Records are also absent from the Blue Ridge region in the neighboring states of Pennsylvania (Hulse et al. 2001) and Virginia (Mitchell and Reay 1999), suggesting that Dekay's Brownsnakes may be rare to absent within this region of Maryland as well.

This snake lives in a wide variety of terrestrial habitats,

both moist and dry, including woodlands, meadows, wetland edges, upland forests, suburban yards, and even urban parks and abandoned city lots (Wright and Wright 1957; Linzey and Clifford 1981; White and White 2007). They are most often found in areas with sufficient ground cover, such as logs, rocks, fallen leaves, and trash. McCauley (1945) discovered two inside a beetle burrow in a rotten log. Although one of Maryland's more widely distributed snakes, they are encountered infrequently due to their secretive, inconspicuous nature. After more than 20 years of gardening and other yardwork, the author (Davidson) first discovered this species as a regular, common resident of her backyard through the deployment of cover boards.

Dekay's Brownsnake is generally active from late March or early April into November (Wright and Wright 1957). This species is diurnal during early spring and fall, but it becomes crepuscular or nocturnal in the summer months (Ernst and Ernst 2003). Mating usually occurs in the spring, from late March through May; males find reproductive females by following their pheromone trails (Ernst and Ernst 2003). Dekay's Brownsnakes are viviparous, with gestation lasting about 2.5 months (King 1993), giving birth usually in July and August. Litters range from 3 to 41 young (Wright and Wright 1957; M. A. Morris 1974), although the average litter size is 13 (Ernst and Ernst 2003). Littleford (1945) reported an August litter of 27 young from a female found in Calvert County. Maturity is reached at 2-3 years of age (Ernst and Ernst 2003).

During winter months, Dekay's Brownsnakes may intermittently move from hibernacula to the surface in warm weather (Ernst and Ernst 2003). Otherwise, they hibernate below the frost line, often communally. Grogan (1975) reported a Maryland burrow containing four individuals, and four hibernacula on Long Island, New York, collectively contained more than 200 individuals (Clausen 1936). Brownsnakes have also been known to hibernate with other snake species (Ernst and Ernst 2003). They exhibit high site fidelity, returning to the same hibernaculum each year (Ernst and Ernst 2003).

Earthworms and slugs form the bulk of their diet, but they may consume other prey such as snails, soft-bodied insects, and amphibian eggs (Wright and Wright 1957; McCauley 1945). Rossman and Myer (1990) described the snake's method of extracting snails from their shells as the application of torsion to the soft body, after the shell was wedged against an immovable object (Ernst and Ernst 2003).

Dekay's Brownsnakes were found in 297 blocks within 135 quads during the atlas period. They most likely occurred in many additional areas, but the secretive nature of this species makes it difficult to locate. The atlas project documented the species across Maryland, with most records on the Atlantic Coastal Plain and adjacent Piedmont, scattered records on the Appalachian Plateaus and Ridge and Valley provinces, and no records in the Blue Ridge region. This distribution of records was very similar to that reported by H. S. Harris (1975).

Perhaps because of its generalist habitat selection and reclusive nature and the abundance of its prey, the Dekay's Brownsnake appears relatively tolerant of urbanization and the numerous threats to which most species succumb. Unless the trend toward neat, tidy lawns becomes even more of a national obsession, this species is unlikely to require conservation measures for continued existence in Maryland anytime in the near future.

LYNN M. DAVIDSON

Distribution by quad

Quad Count
Confirmed: 127
Reported: 8

Atlas locations by block

Block Count
Confirmed: 261
Reported: 36

Historical locations. Adapted from H. S. Harris (1975)

Photograph by Michael Kirby

Red-bellied Snake
Storeria occipitomaculata

The harmless Red-bellied Snake is a small, secretive snake with a reddish belly, keeled scales, and a divided anal plate. The species is variable in color, with dorsal coloration being any shade of brown or gray and occasionally black. The belly is typically some shade of red but can be yellow, orange, or black. On the neck, there are usually three light-colored spots that are sometimes fused. The dorsum has two dark longitudinal stripes that may border a narrow middorsal stripe that is lighter in color than the body. This snake usually has a light spot at the rear of the upper labials, just below and behind the eye. Adults can reach a total length of 42.2 cm but typically average 20.3–25.4 cm, with the largest specimen reported in Maryland reaching 35.6 cm (McCauley 1945). As with most species of snakes, adult females are larger than adult males. Juveniles are patterned similarly to adults, but are often darker above and duller below, with the neck spots more conspicuous and fused to form an indistinct ring.

The Red-bellied Snake ranges from Nova Scotia westward to southeastern Saskatchewan and south to central Florida through eastern Texas. Red-bellied Snakes have a patchy distribution in Maryland, but H. S. Harris (1975) considered them most common on the Appalachian Plateaus Province. Historically, no records existed for Caroline, Carroll, Dorchester, Harford, Kent, Montgomery, Prince George's,

or Talbot counties (H. S. Harris 1975). Subsequently, this species was documented in Dorchester County (Grogan 2008).

Habitats frequented by Red-bellied Snakes vary widely from moist to xeric woodlands, fields, and bogs or the edges of marshes and swamps (Ernst and Ernst 2003). The species generally remains hidden beneath cover objects such as rocks, logs, and other debris (McCauley 1945; Ernst and Ernst 2003). During the spring and fall, Red-bellied Snakes can be active in the late morning or afternoon, but during the summer they are mostly active at night or at dawn and dusk. These small snakes have the ability to traverse considerable distances within a relatively short time. For example, one individual moved roughly 400 m in a single day (Blanchard 1937). Red-bellied Snakes may, on occasion, climb low shrubs and tangles of weeds (P. W. Smith 1961; Minton 1972).

In Maryland, Red-bellied Snakes are generally active from March through October, though in western Maryland they may not emerge until May and may seek hibernacula in late September. They often hibernate in groups and with other species of snakes (Carpenter 1953; Harding 1997). Courtship and mating may occur in spring, summer, or fall. The Red-bellied Snake is viviparous, giving birth to live young. Births typically occur during July or August across most of the range (Linzey and Clifford 1981; Ernst and Ernst 2003). Young are born singly and are initially encased in a thin membrane that ruptures soon after birth. Females produce a single litter each year of 4–9 young but may give birth to as many as 20 young (Ernst and Ernst 2003). Both males and females are likely to be sexually mature by year 2. While the longevity of Red-bellied Snakes in the wild is uncertain, one female lived for more than 4 years in captivity (Snider and Bowler 1992).

Little is known about the feeding behavior of the Red-bellied Snake, though foraging probably occurs mostly at night. Prey is predominantly slugs, but earthworms, soft-bodied insects, and snails are also eaten (McCauley 1945; Ernst and Ernst 2003). Defensive strategies of Red-bellied Snakes include a flattening of the head, curling of the upper lip to expose the maxillary teeth, and expulsion of musk from the anal gland. Death-feigning behavior has also been documented (R. Jordan 1970; Watermolen 1991). Red-bellied Snakes are eaten by many predators, including other snakes, large fish, crows, hawks, and falcons (Ernst and Ernst 2003).

During the atlas project, Red-bellied Snakes were documented in 59 blocks within 35 quads. The species likely occurs in many more atlas blocks, but its secretive nature makes it very difficult to detect. Red-bellied Snakes were most common on the Ridge and Valley and Appalachian Plateaus provinces. The MARA distribution was similar to the historical distribution in Maryland (H. S. Harris 1975), although none were detected in Baltimore, Cecil, Howard, or Queen Anne's counties.

The conservation status of this small, reclusive snake is difficult to ascertain. However, its current distribution appears to be similar to its historical distribution, suggesting that it has persisted despite increased development pressure.

DAVID R. SMITH

Distribution by quad

Quad Count
Confirmed: 30
Reported: 5

Atlas locations by block

Block Count
Confirmed: 40
Reported: 19

Historical locations Adapted from H. S. Harris (1975)

Photograph by Lance H. Benedict

Eastern Ribbonsnake
Thamnophis saurita

With three yellowish longitudinal stripes on a dark body, the Eastern Ribbonsnake may be mistaken for the much more abundant Common Gartersnake. A closer inspection, however, reveals a more slender snake, with distinct lateral stripes on the third and fourth scale rows, a vertical, light-colored line in front of the eye (H. A. Kelly et al. 1936; Mc-Cauley 1945; White and White 2007), and no markings on the upper labial scales (Gibbons and Dorcas 2005). Eastern Ribbonsnakes have a yellow, cream, or tan dorsal stripe extending from the neck almost to the tip of the tail. In general, the coloration of the top of the head and dorsal background is dark brown to black and extends below the lateral stripes, slightly onto the belly. The majority of the belly is a solid yellowish to light greenish color and is unmarked (Ernst and Ernst 2003; White and White 2007). The scales are keeled, and the anal plate is undivided. Adults reach a total body length of 45.7-66 cm.

The range of the Eastern Ribbonsnake covers eastern North America to the Mississippi River (Ernst and Ernst 2003). It occurs from southern Nova Scotia and New England west through Michigan and south along the Atlantic and Gulf Coast states to Florida and Mississippi, with scattered, disjunct populations within the interior states. According to H. S. Harris (1975), the distribution of the Eastern Ribbonsnake in Maryland was apparently statewide, but the species was extremely uncommon on the Appalachian Plateaus. This species has yet to be documented in Wicomico County (White and White 2007), and there are only a few Garrett County reports (R. W. Miller 1979).

The Eastern Ribbonsnake is semiaquatic. Individuals occur in or near freshwater habitats such as nontidal ponds, streams, and wetlands; however, they are absent from the coastal barrier islands (White and White 2007). They may be encountered both in the open and under cover objects (McCauley 1945). It is not uncommon to observe individuals several meters above the ground or water, in bushes or shrubby trees (McCauley 1945; Mitchell 1994; White and White 2007). In Nova Scotia, Bell et al. (2007) found Ribbonsnakes only within 5 m of water and within home ranges on land that rarely exceeded 50 m^2, from late May to September. Escape tactics include swimming across the water surface and into dense herbaceous shoreline vegetation (McCauley 1945; Scribner and Weatherhead 1995; White and White 2007). Ernst and Ernst (2003) reported the use of Muskrat (*Ondatra zibethicus*) bank burrows as retreats.

Hibernation occurs in and near swampy areas (H. A. Kelly et al. 1936). Carpenter (1953) documented Eastern Ribbonsnakes hibernating with other species of snakes and amphibians in ant mounds and Meadow Vole (*Microtus pennsylvanicus*) tunnels. Some hibernacula are underwater, which may increase survivorship by decreasing desiccation (J. Todd et al. 2009). On warm winter days, Eastern Ribbonsnakes may become active, emerging briefly to bask (Mitchell 1994). In the spring and summer, the snakes are active primarily during the day but sometimes forage at night when frogs are breeding (H. A. Kelly et al. 1936; Ernst et al. 1997; Ernst and Ernst 2003).

In Maryland, mating of Eastern Ribbonsnakes is assumed to take place in early spring (McCauley 1945). This species is viviparous, and females give birth in summer to 7-15 young (H. A. Kelly et al. 1936; McCauley 1945), but most commonly 10 or 11 young (McCauley 1945). Overall for the species, the number of young varies from 3 to 26 (Ernst and Ernst 2003), with larger litter sizes in southern populations (Rossman et al. 1996). Maturity is reached in 2-3 years (Harding 1997).

Eastern Ribbonsnakes actively search for their prey. They feed primarily on amphibians, mostly frogs (Ernst and Ernst 2003), but also toads, salamanders, and fish (P. D. Evans 1942; Rossman 1963). In Michigan, Carpenter (1952) found that Eastern Ribbonsnakes ate mainly (90%) amphibians, along with fish and caterpillars.

During the MARA project, Eastern Ribbonsnakes were found in 75 blocks within 56 quads and in every county in the state except Allegany, Garrett, Washington, and Wicomico. The MARA distribution was similar to the historical distribution in Maryland (H. S. Harris 1975) and reflects the rarity or absence of Eastern Ribbonsnakes in western Maryland.

Ernst and Ernst (2003) stated that populations of Eastern Ribbonsnakes were declining over much of their range due to habitat destruction, and Bell et al. (2007) surmised that their small activity range would make these snakes vulnerable to local extinction. The Eastern Ribbonsnake continues to be a somewhat elusive inhabitant of Maryland's watery habitats and may be declining or disappearing from parts of the state.

DAVID E. WALBECK

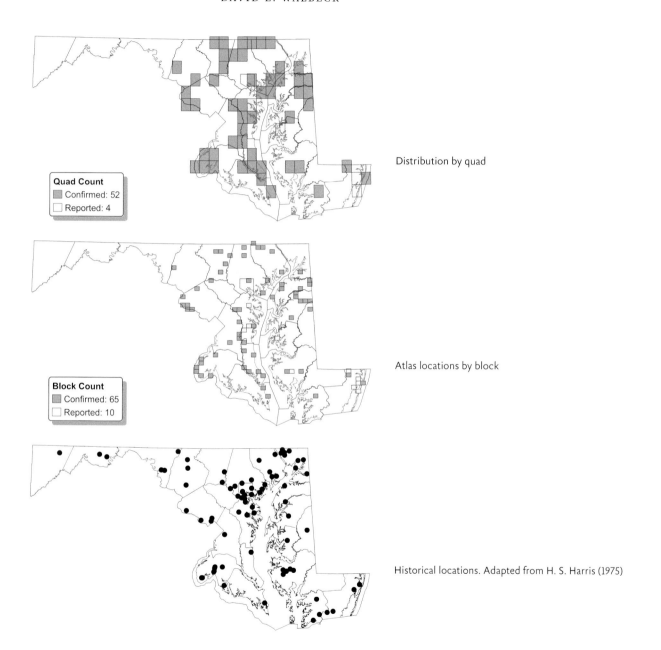

Quad Count
Confirmed: 52
Reported: 4

Distribution by quad

Block Count
Confirmed: 65
Reported: 10

Atlas locations by block

Historical locations. Adapted from H. S. Harris (1975)

Photograph by Kevin M. Stohlgren

Common Gartersnake
Thamnophis sirtalis

The Common Gartersnake is the most widely distributed snake in North America, and its range extends farther north than that of any other North American reptile. This species has a light-colored dorsal stripe, two scale rows wide, from the neck to the tip of the tail. The stripe is usually yellowish but can be faint to absent. Below the dorsal stripe is a double row of alternating black blotches that form a checkerboard pattern over a green to brown background. The lower sides are lighter in color, usually forming yellowish lateral stripes on the second and third rows of scales, but frequently these stripes are less distinct (H. A. Kelly et al. 1936). The ventral scales are yellowish to bluish green to light gray, with a lateral row of small black spots on each side. The upper labial scales are yellowish with dark vertical lines. The scales are keeled, and the anal plate is undivided. The similar Eastern Ribbonsnake is more slender, has more distinct lateral stripes, and has no dark lines on the upper labial scales. Adult Common Gartersnakes measure 45.7-66 cm in total body length.

The Common Gartersnake is distributed throughout much of Canada and the United States. The range extends as far north as the southern Hudson Bay and southern Northwest Territories of Canada. The species is primarily absent from the American Southwest except for California and isolated populations in eastern Texas and New Mexico. In Maryland, Common Gartersnakes are found throughout almost the entire state (H. S. Harris 1975).

These snakes occupy all habitats in Maryland with the exception of salt marshes and coastal barrier islands (White and White 2007). They are semiaquatic and normally stay near standing fresh water and wetlands, although some inhabit brackish areas (Batts 1961). Gartersnakes can be found both in the open and under rock, wood, and bark cover (McCauley 1945). When underwater, their pupil diameter constricts to overcome the change in refractive index (Fontenot 2008).

Spring emergence of Common Gartersnakes in Maryland is usually in April, and activity can extend into November (H. A. Kelly et al. 1936). Gartersnakes may occasionally be found above ground on warm winter days (Ernst and Ernst 2003); for instance, the author (Walbeck) has observed Common Gartersnakes in deciduous woods in December in Baltimore County and in open grassy habitat in January in Calvert County. Most activity in the spring and fall is diurnal, but crepuscular and nocturnal activity occurs during the spring and summer, when frogs are breeding (Ernst and Ernst 2003). Hibernation occurs in soft soil, under old stumps, and in rocky areas at a depth below the frost line (H. A. Kelly et al. 1936). Some hibernacula are communally shared with other individuals and other snake species (Ernst and Ernst 2003). Gartersnakes sometimes hibernate underwater (McCauley 1945; Carpenter 1953), which can increase survival rates by reducing dehydration and energy use (Costanzo 1989). The author found a Common Gartersnake under a tire that was half submerged along an intermittent stream in Anne Arundel County in late November. The snake was presumably in hibernation but flailed around wildly once exposed. In another example, McCauley (1945) quoted the diary of Father McClellan, who found a hibernating Common Gartersnake in running spring water in late January.

Common Gartersnakes in Maryland mate in May and June (McCauley 1945), but breeding does not occur every year (Larsen et al. 1993). Females move to summer habitat with abundant cover to give birth, and gravid females rarely feed late in gestation (Larsen et al. 1993). Common Gartersnakes bear 8-50 young in August and September (H. A. Kelly et al. 1936; McCauley 1945). Litters can be twice as large on the Atlantic Coastal Plain of Maryland compared with western Maryland (McCauley 1945). Neonates may hibernate closer to the summer habitat than adults, thus reducing the risks of migrating to distant adult hibernacula and increasing the time spent closer to more potential food sources (Larsen et al. 1993). Maturity occurs at 1-2 years for males and 2-3 years for females (Rossman et al. 1996).

Common Gartersnakes have a varied and opportunistic diet, preying on earthworms, frogs, toads, salamanders, and small fish, with metamorphic frogs and toads being a main food source, when available (Arnold and Wassersug 1978; Kephart and Arnold 1982). In Michigan, Carpenter (1952) found that Gartersnakes ate 80% earthworms and 15% amphibians, along with small mammals, young birds, fish, caterpillars, and leeches.

During the MARA project, Common Gartersnakes were found in 709 blocks within 207 quads and in every county in the state. The MARA distribution mirrored the historical distribution in Maryland (H. S. Harris 1975; Grogan 1981), although the species was mostly absent from the tidewater areas of southern Dorchester and western Somerset counties.

The Common Gartersnake continues to be one of Maryland's most abundant snakes. Its ability to occupy urban environments, so long as habitats with prey items are sufficiently available, suggests that its status within the state is secure.

DAVID E. WALBECK

Quad Count
Confirmed: 188
Reported: 19

Distribution by quad

Block Count
Confirmed: 532
Reported: 177

Atlas locations by block

Historical locations. Adapted from H. S. Harris (1975)

Eastern Smooth Earthsnake subspecies. Photograph by Robert T. Ferguson II

Mountain Earthsnake subspecies. Photograph by Ed Thompson

Smooth Earthsnake
Virginia valeriae

The Smooth Earthsnake currently consists of two subspecies in Maryland: the Eastern Smooth Earthsnake (*V. v. valeriae*) and the Mountain Earthsnake (*V. v. pulchra*). Both subspecies are rarely encountered in Maryland, owing in large part to their mostly fossorial existence. Additionally, the snakes are small and drab when compared with other species, making them easy to overlook or misidentify. On closer inspection, however, both are elegant animals. Dorsal coloration ranges from light gray to various shades of brown, often with small back spots that may be arranged in two to four parallel rows extending from head to tail, sometimes giving the appearance of dark longitudinal stripes on either side of the midbody. A dark but faint band may be present from the nostril through the eye. Occasionally, a light longitudinal stripe extends from behind the head to the tail between the parallel rows of spots; this is most often observed in Mountain Earthsnakes. The Eastern Smooth Earthsnake's dorsal scales are almost always smooth, though some individuals may have weakly keeled scales. Mountain Earthsnakes almost always have distinct, weakly keeled scales throughout most of the dorsal scale rows. An important distinction between the subspecies is that the Eastern Smooth Earthsnake typically has 15 dorsal scale rows at midbody compared with 17

in the Mountain Earthsnake. Distribution, however, may be the best aid in distinguishing these subspecies, as they have complementary ranges, meaning that their ranges do not overlap. Ventral coloration is typically white, gray, or beige, and all Smooth Earthsnakes have a divided anal plate. These snakes have a thick appearance and a small tail. Adults are typically 18–25.4 cm in total length, though the record length reported for this species is 39.3 cm.

The Eastern Smooth Earthsnake ranges from central New Jersey to southern Ohio in the north, and south to Florida and Alabama. The Mountain Earthsnake is found only in Pennsylvania, Maryland, Virginia, and West Virginia on the Appalachian Plateaus. H. S. Harris (1975) reported the Eastern Smooth Earthsnake largely from the Piedmont and Atlantic Coastal Plain provinces of Maryland, and extending into the Great Valley region of the Ridge and Valley Province via river valleys. He also described a lack of records from the central portion of Maryland's Eastern Shore. Harris (1975) reported Mountain Earthsnakes from only a few sites in western Garrett County near the West Virginia border, and Thompson (1984a) documented an additional five localities in the same general area west of the Youghiogheny River.

Both subspecies can be found within or near deciduous, mixed coniferous-deciduous, or coniferous woodlands. While the Eastern Smooth Earthsnake is a denizen of flat or low hills, the Mountain Earthsnake is an inhabitant of mountainous terrain. Individuals are often encountered in open habitats near woodlands, including meadows and roadside openings, and even around human dwellings. Mountain Earthsnakes, particularly gravid females, are often encountered under cover objects in forest openings, including disturbed habitats such as utility rights of way and sloping roadside banks. Individuals of both subspecies are typically found under surface cover objects such as rocks and logs, but they will also readily utilize human-made objects and junk piles. This surface activity typically occurs during the spring or late summer and fall, or after heavy rains. Occasionally, specimens are found active on roads on rainy nights.

Due to a fossorial lifestyle and infrequent capture, little is known about Earthsnakes. The activity season of this species is April through October, but in southern populations of Eastern Smooth Earthsnakes, individuals may also be active on the surface during warm periods in winter months (B. D. Todd et al. 2008). B. D. Todd et al. (2008) reported that male capture rates of this species increase in August, suggesting a late summer and fall mating season, with females birthing the following year. Females give birth to 4-14 young in August or September. Birthing locations are largely unknown, though the author (Ruhe) found a female Eastern Smooth Earthsnake and litter of six neonates under an old rug in Burlington County, New Jersey, in early September. The author also found a female Mountain Earthsnake and four neonates under a rock in Clinton County, Pennsylvania, in mid-September.

Both subspecies in Maryland feed exclusively on earthworms, but White and White (2007) report that this species may also feed on other soft-bodied invertebrates. Predators are unknown but presumed to be other snakes, mammals, and birds. Large, carnivorous invertebrates may also attack this species, particularly the young. Earthsnakes are harmless and seldom bite. The mouth is quite small, and the rare bite is painless. Like other species of snakes, Earthsnakes will musk when captured, and while harmless, the smell can be quite unpleasant.

The MARA project documented the presence of Eastern Smooth Earthsnakes in 90 blocks within 56 quads. MARA locations generally mirrored those reported by H. S. Harris (1975), but this subspecies was not found in Harford County. It was encountered in Allegany County, however, representing a new county of occurrence. An individual found in Caroline County also represented a new county record.

Mountain Earthsnakes were confirmed in five blocks in four quads, all on the Appalachian Plateaus of Garrett County. This subspecies was found in the historical areas of collection (H. S. Harris 1975; Thompson 1984a), with new localities documented east of the Youghiogheny River.

No direct conservation efforts specifically target the Eastern Smooth Earthsnake in Maryland; the Mountain Earthsnake is listed as state endangered (MDNR 2016b). Both subspecies may benefit from the maintenance of meadows and other forest openings. Taxonomists have classified the Smooth Earthsnake group with and without subspecies and as multiple species at various times in the twentieth and twenty-first centuries. There is a strong possibility that the currently recognized subspecies may indeed warrant elevation to species status, based on anatomical differences, complementary isolated ranges, and differences in habitat. Modern phylogenetic studies may unravel this mystery.

BRANDON M. RUHE

Quad Count
Eastern Smooth Earthsnake
■ Confirmed: 52
□ Reported: 4
Mountain Earthsnake
▨ Confirmed: 4
▨ Reported: 0

Distribution by quad

Block Count
Eastern Smooth Earthsnake
■ Confirmed: 82
□ Reported: 8
Mountain Earthsnake
▨ Confirmed: 5
▨ Reported: 0

Atlas locations by block

Historical locations. Adapted from H. S. Harris (1975)

● Eastern subspecies
▲ Mountain subspecies

Photograph by Matthew Kirby

Eastern Copperhead
Agkistrodon contortrix

The Eastern Copperhead is one of only two venomous snakes in Maryland and perhaps the best camouflaged. Adult Eastern Copperheads are light grayish brown or tan with wide dorsal blotches, chestnut brown in outline with a lighter interior. The blotches narrow toward the spine, forming an hourglass pattern. Typically, some blotches are offset or lack an opposite half. In many individuals, a small, dark brown spot is present at the center of some dorsal blotches and midway between them. Scales are strongly keeled. The triangular head, distinct from the neck, is primarily gold with coppery flecking and two tiny, dark brown dots centered on top. A lighter region on the side of the head, weakly outlined in white, encloses the jaw line. The pupil is vertically elliptical. The belly is cream, with lateral, dark brown spots below and between the dorsal blotches, and the anal plate is undivided. Adult Eastern Copperheads are heavy-bodied snakes, typically 61-90 cm in length, with a maximum recorded length of 135 cm. Males are somewhat larger than females. Juveniles have greater contrast in pattern but less color and have a bright, sulfur yellow tail tip.

Other patterned snakes, such as Common Watersnakes, Eastern Milksnakes, and juvenile Eastern Ratsnakes and North American Racers, are often harmless recipients of mistaken identity. Contrary to popular belief, the shape of the head is not the ideal trait by which to identify an Eastern Copperhead; the vertical pupil is the best means of confirming a suspected Eastern Copperhead.

The Eastern Copperhead is widely distributed from Massachusetts to northern Florida and west to eastern Texas, Oklahoma, and Kansas. In Maryland, the species has a pecu-

liar distribution. It is absent from the Atlantic Coastal Plain of the Eastern Shore from southern Cecil County south through Caroline and Talbot counties (H. S. Harris 1975); otherwise the distribution is essentially statewide. On the Appalachian Plateaus, these snakes are limited to the lower elevations along river courses (McCauley 1945). Harris (1975) considered specimens from near Cambridge in Dorchester County as introduced, but this idea has since been rejected (Grogan 1994).

This species is found in a variety of habitats. On the Atlantic Coastal Plain, Eastern Copperheads are found in mixed pine-hardwood and pine forests, as well as on high ground around swamps and marshes (Ernst and Ernst 2003) On the Piedmont, they mostly occupy rocky stream corridors and surrounding oak-hickory forests. In the mountains, they are found on rocky slopes, often above streams. In all areas, they will make use of dilapidated buildings, rock walls, and brush piles as suitable sites for shedding or gestating (Mitchell 1994). The Eastern Copperhead is a terrestrial snake that lies coiled in ambush for unsuspecting prey (Hulse et al. 2001). It is diurnal in the spring and fall but very active nocturnally during the summer (Mitchell 1994), when these snakes are often seen crossing roads after dark.

In the mountains, Eastern Copperheads tend to hibernate communally with conspecifics as well as several other snake species, including Timber Rattlesnakes (Ernst and Ernst 2003). These hibernacula, in suitable rock formations, are used year after year, as are gestation sites in open rocky areas (Hulse et al. 2001). On the Piedmont and Atlantic Coastal Plain, Copperheads frequently hibernate singly or in small numbers, often utilizing more ephemeral sites such as hollow roots and tree stumps. Gestation sites are in clearings or fields with suitable hiding spots (Ernst and Ernst 2003). The activity season is generally April to October, its exact length influenced by elevation (Mitchell 1994). Courtship and mating may occur in spring or fall (Ernst and Ernst 2003). Courtship may be accompanied by male "combat" in which two males raise the anterior third of their bodies and intertwine, one attempting to show dominance over the other (Hulse et al. 2001). Eastern Copperheads are viviparous. Reproduction is biennial, with most birthing in September. Average litter size is seven young, each newborn initially enclosed in a membranous, transparent sac (Ernst and Ernst 2003). A wide variety of prey is taken, including invertebrates, amphibians, reptiles, and small birds and mammals. Juveniles eat a higher proportion of invertebrates than adults (Mitchell 1994).

Eastern Copperheads are frequently killed because of the assumed threat they pose to people and their pets; however, Mitchell (1994) was not able to ascertain any fatalities caused by this snake in Virginia. In 2013, the director of the Maryland Poison Control Center estimated that 20–35 Eastern Copperhead bites occur annually in Maryland (Zumer

2013). Adults are unlikely to experience serious medical consequences, but anyone bitten by an Eastern Copperhead should seek medical attention. A full envenomation from a large snake could be more serious for a small child and has dire consequences for a small dog (Wolfe-Arnovits 2000).

Eastern Copperheads were documented in 229 blocks within 106 quads during the atlas period. Overall, the atlas distribution was similar to the historical distribution in Maryland (H. S. Harris 1975). They were found throughout the southern Atlantic Coastal Plain, with a spotty distribution on the Piedmont. They were common in the mountains of the Blue Ridge and Ridge and Valley provinces, while absent from the Great Valley. The distribution was very spotty on the Appalachian Plateaus.

Given the extensive loss of habitat due to development since H. S. Harris's (1975) publication, it is remarkable that this species remains widespread in Maryland. However, in many blocks on the Piedmont and Atlantic Coastal Plain, the species persists mostly on conservation lands or in narrow forest corridors along stream floodplains. The conservation and maintenance of large forested tracts, as well as forested riparian corridors, is critical to the conservation of this species.

LANCE H. BENEDICT

Distribution by quad

Quad Count
Confirmed: 95
Reported: 11

Atlas locations by block

Block Count
Confirmed: 174
Reported: 55

Historical locations. Adapted from H. S. Harris (1975)

● Specimen Examined
◑ Specimen Reported

Photograph by Lance H. Benedict

Timber Rattlesnake
Crotalus horridus

The Timber Rattlesnake is Maryland's largest venomous serpent, a highly evolved predator, and a majestic symbol of our wilderness areas. For those who have seen one in the wild, and perhaps felt the shiver down the spine that accompanies an unexpected rattling at close range, it is both a fearsome and wondrous sight.

Timber Rattlesnakes are variably colored with two color phases, yellow and black, determined by head color. The yellow phase is typically mustard or golden yellow. The black (or dark) phase is medium gray to dark brown. Both phases have dark brown to black dorsal and lateral blotches that meet to form chevrons outlined in cream or pale yellow. A middorsal stripe, if present, is tan or taupe in yellow-phase snakes and dark brown in black-phase snakes. The tail is black, ending in the rattle. Generally, the belly is cream, peppered with black. Scales are strongly keeled, and the anal plate is undivided. The triangular head, distinct from the body, is covered with tiny scales. A heat-sensing pit is located midway between the eye, which has a vertically elliptical pupil, and the nostril. These are heavy-bodied snakes; adults are 90-152 cm in total length but seem bigger because of their girth. Males are somewhat larger than females. Juveniles are light gray with dark gray blotches and gradually acquire their adult coloration during their first two years. Newborns have only a "button" after the postnatal shed. The first free rattle segment is acquired in the summer after birth and with each shed thereafter. As shedding may occur twice in some years, the number of rattle segments does not equal age.

The Timber Rattlesnake is widely distributed from Ver-

mont to northern Florida and west to eastern Texas, Oklahoma, and Kansas. In the Mississippi Valley, it is distributed as far north as southern Minnesota. It has been extirpated from Delaware, Maine, Michigan, Ontario, Quebec, and Rhode Island (W. H. Martin et al. 2008). In Maryland, Timber Rattlesnakes currently inhabit mountainous regions on the Blue Ridge Province west through the Appalachian Plateaus (H. S. Harris 1975). Evidence suggests that the species was statewide in colonial times and was present on the Eastern Shore approximately 100 years ago (McCauley 1945; Mitchell 1994; White and White 2007). On the Piedmont, the species still exists, albeit in declining numbers, in southern Frederick County and may persist in northern Baltimore County (J. E. Cooper and Groves 1959), although the latest reports stem from the 1990s (William H. Martin, pers. comm.).

Land usage, activity season, birthing period, and prey species vary across the range of the species (W. H. Martin et al. 2008). In the mid-Atlantic, most insights into the habits of this snake have come from W. H. Martin (1992, 1993, 2002), who has been studying its life history since 1973. Timber Rattlesnakes are dependent on ancestral den sites on steep rocky slopes, where they hibernate for more than half the year (W. H. Martin 1992) and gestate in open rocky areas (W. H. Martin 1993). Gestation and den sites are used year after year, anchoring the local population. Surrounding these anchor points is a foraging area, typically deciduous forest, which allows summer dispersal for hunting and mating (W. H. Martin et al. 2008). Genetic diversity is maintained through the wide-ranging movement of males (up to 7 km), which ensures that up to a third of breeding is with females from a different den (R. W. Clark et al. 2008). Roads and development can severely reduce the inter-den interactions and genetic diversity of local populations (R. W. Clark et al. 2010; R. W. Clark et al. 2011).

The activity season is generally from mid-April to mid-October, varying with elevation and yearly weather events (W. H. Martin 1992). Movement to summer foraging habitat occurs in May. White-footed Mice (*Peromyscus leucopus*), Deer Mice (*P. maniculatus*), voles (*Clethrionomys* and *Microtus*), and Eastern Chipmunks (*Tamias striatus*) are the main prey items, with occasional squirrels (*Sciurus*) and birds taken by large individuals (W. H. Martin et al. 2008). Females typically reproduce every 3 years (Ernst and Ernst 2003). Mating occurs in the previous year from late July to early September (W. H. Martin 1993). Gravid females arrive at gestation sites in late May to early June and remain until giving birth, from mid-August through September. Timber Rattlesnakes are viviparous, with an average litter size of about 7 or 8 young (Mitchell 1994). The mother remains with the newborns for 7-10 days, after which time they all disperse to forage prior to hibernation (W. H. Martin 1992). Yearly survivorship is 50% to 70% in the first year and up to 90% as adults (W. H. Martin 2002). Age at first reproduction is 5-11 years. Reproduction repeats at 2- to 5-year intervals, depending on elevation and food supply (W. H. Martin 1993). Maximum age has recently been documented at more than 40 years (William Brown, unpubl. data).

The Timber Rattlesnake is generally a passive snake, typically resting in a coiled position during the daytime and fleeing into a crevice when approached. Bites occurring without intentional human provocation are rare but potentially life threatening, as the venom destroys platelets and interferes with coagulation (Keyler 2008). Roughly half of bites do not result in significant envenomation (Keyler 2008), but anyone bitten should be taken to a hospital for evaluation and potential antivenom treatment. Historical first aid treatments such as tourniquets, incisions, and suction are highly discouraged (D. L. Hardy 1992).

Timber Rattlesnakes were documented in 30 quads and 59 blocks during the atlas period, from the western Piedmont west through the Appalachian Plateaus. They were probably present in several more quads in Garrett County but remained undetected. Overall, the atlas distribution was similar to the twentieth-century distribution in Maryland (H. S. Harris 1975). On the Piedmont, their presence was confirmed in southern Frederick County but not in northern Baltimore County.

This species is currently secure in Maryland, as much of its remaining habitat is on state and federal lands. However, the periphery of these lands is constantly encroached upon by roads and development. New threats include wind power facilities, encroachment of forest canopy on gestation sites, and snake fungal disease. As a slow-maturing and infrequently reproducing species, the Timber Rattlesnake is vulnerable to anthropogenic disturbances. Human persecution is, hopefully, waning as people come to appreciate this snake's importance to the ecosystem and its intrinsic value as a symbol of our wilderness areas. Under MDNR regulations, Timber Rattlesnakes may be held only in accordance with a scientific collection permit or endangered species permit issued by the state agency.

LANCE H. BENEDICT

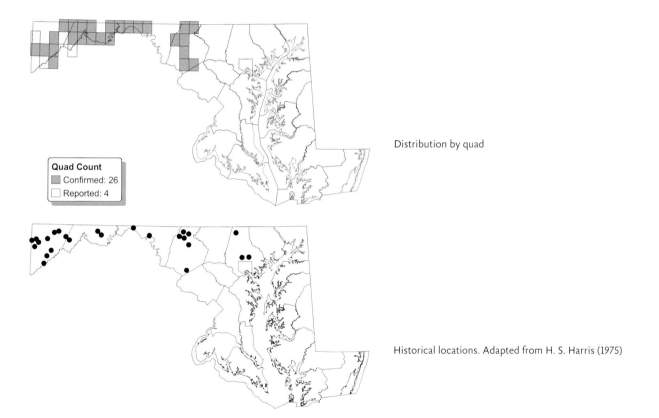

Distribution by quad

Quad Count
Confirmed: 26
Reported: 4

Historical locations. Adapted from H. S. Harris (1975)

Appendix A

Number of species in each block and quad, as documented during the Maryland Amphibian and Reptile Atlas project

Numbers in parentheses indicate blocks and quads with less than 40.47 ha (100 ac) of upland. Dashes indicate blocks with no land or water in Maryland.

Quad	Blocks						Quad Total
	NW	NE	CW	CE	SW	SE	
Aberdeen	19	26	15	12	18	24	35
Accident	12	8	13	13	12	16	24
Alexandria	—	—	—	7	—	9	12
Anacostia	5	15	11	8	13	10	28
Annapolis	16	12	19	(0)	13	(1)	29
Artemas	11	8	31	15	34	41	44
Assawoman Bay	—	—	10	3	17	4	20
Avilton	13	20	17	22	21	23	33
Baltimore East	9	13	9	10	10	6	20
Baltimore West	10	9	7	14	18	7	23
Barren Island	(0)	5	(0)	(0)	(0)	(1)	6
Barton	26	20	26	16	14	10	33
Bay View	11	6	13	11	17	13	20
Bel Air	19	24	21	17	19	16	30
Bellegrove	5	12	28	10	37	29	42
Beltsville	23	24	14	17	18	22	32
Benedict	10	36	18	25	20	27	39
Berlin	23	29	22	30	26	28	35
Betterton	17	13	14	12	13	17	26
Big Pool	—	25	—	—	—	—	25
Bittinger	19	25	25	26	15	24	34
Blackwater River	20	17	30	26	13	(2)	34
Bloodsworth Island	(1)	4	(0)	(2)	(1)	(1)	7
Blue Ridge Summit	12	11	14	22	22	18	30
Bowie	22	22	25	18	17	17	39
Boxiron	25	3	11	(0)	(1)	(0)	25
Brandywine	27	15	22	14	24	32	44
Bristol	28	27	38	22	43	26	49
Broomes Island	35	38	26	20	26	32	47
Buckeystown	14	15	12	22	14	31	37
Burgess	(0)	(0)	(0)	(0)	—	(0)	(0)
Burrsville	7	—	10	—	7	—	12
Cambridge	12	17	18	14	19	13	26
Catoctin Furnace	22	27	30	19	33	20	40
Cecilton	12	15	13	12	9	13	26
Centreville	20	16	18	22	23	17	31
Charles Town	—	8	—	(0)	—	—	8
Charlotte Hall	20	26	9	15	19	21	38
Cherry Run	3	14	13	28	4	24	35

Quad	Blocks						Quad Total
	NW	NE	CW	CE	SW	SE	
Chicamacomico River	21	19	17	24	9	14	31
Church Creek	16	17	18	19	17	16	28
Church Hill	27	18	19	19	19	18	36
Claiborne	19	10	(0)	15	16	15	25
Clarksville	28	28	25	26	25	30	36
Clear Spring	16	12	30	11	13	11	34
Cockeysville	25	25	17	17	15	26	34
Colonial Beach North	20	23	(2)	15	(0)	12	27
Colonial Beach South	(0)	(0)	(0)	(0)	—	—	(0)
Conowingo Dam	8	9	25	19	22	20	34
Cove Point	25	(0)	42	9	29	37	49
Cresaptown	5	1	2	(0)	2	—	9
Crisfield	6	2	(0)	(0)	(0)	(0)	7
Cumberland	4	3	5	7	1	13	21
Curtis Bay	16	7	19	18	12	16	30
Dahlgren	(0)	(0)	—	(0)	—	(0)	(0)
Damascus	15	19	12	18	12	11	28
Davis	18	14	10	—	—	—	24
Deal Island	(1)	5	4	17	4	(3)	23
Deale	10	18	15	13	19	13	31
Deer Park	18	13	16	12	14	19	31
Delmar	—	—	19	20	25	19	32
Delta	7	10	17	28	18	25	36
Denton	13	20	14	14	20	11	28
Dividing Creek	18	23	18	20	13	23	36
Earleville	21	13	14	13	6	11	30
East New Market	15	15	17	19	16	16	29
East of North Beach	(0)	(0)	(0)	(0)	(0)	(0)	(0)
East of Point Lookout	(0)	(0)	(0)	(0)	(0)	(0)	(0)
Easton	24	13	12	15	14	20	31
Eden	23	29	25	16	19	19	35
Edgewood	29	25	26	33	22	15	39
Elkton	22	17	20	22	25	22	34
Ellicott City	21	7	28	23	25	25	36
Emmitsburg	15	11	27	14	12	15	30
Evitts Creek	6	19	12	16	13	6	26
Ewell	(0)	(3)	(0)	(0)	—	—	(3)
Falls Church	41	20	—	6	—	—	41
Fawn Grove	17	10	27	20	9	21	33
Federalsburg	10	24	10	19	13	13	29
Finksburg	20	20	19	19	17	19	29
Flintstone	4	6	17	13	17	20	33
Fort Belvoir	—	—	—	(0)	—	(0)	(0)
Fowling Creek	14	17	21	28	17	19	33
Frederick	24	19	18	18	16	21	31
Friendsville	9	11	15	23	11	22	25
Frostburg	21	6	23	5	15	9	33
Funkstown	7	14	8	14	9	19	24
Gaithersburg	14	15	16	14	13	21	25
Galena	11	11	13	8	17	15	24
Germantown	11	17	18	24	15	24	31
Gibson Island	23	14	17	13	23	20	32
Girdletree	25	16	15	18	20	14	32
Golden Hill	20	21	23	26	19	23	34
Goldsboro	14	18	16	25	19	17	33

Quad	Blocks						Quad Total
	NW	NE	CW	CE	SW	SE	
Grantsville	11	14	11	18	13	19	25
Great Cacapon	(0)	14	—	—	—	—	14
Great Fox Island	(0)	(1)	(0)	(0)	(0)	(1)	(2)
Gunpowder Neck	20	0	9	0	7	0	21
Hagerstown	2	3	11	8	11	5	16
Hallwood	4	—	—	—	—	—	4
Hampstead	21	14	11	11	13	11	26
Hancock	7	1	17	18	21	—	30
Hanesville	0	7	(0)	25	6	21	27
Harpers Ferry	16	14	23	21	—	—	30
Havre de Grace	18	23	24	20	19	(2)	35
Heathsville	(0)	(0)	—	(0)	—	—	(0)
Hebron	13	(0)	15	13	17	15	31
Hedgesville	23	11	—	10	—	—	27
Hereford	22	30	16	10	10	12	34
Hickman	7	—	6	—	30	—	30
Hobbs	25	12	22	8	17	8	31
Hollywood	37	25	36	38	39	31	49
Honga	14	18	6	10	3	13	21
Horseshoe Point	3	(0)	2	(0)	(0)	7	11
Hudson	(0)	7	(0)	19	(0)	11	23
Hughesville	12	12	16	22	11	14	30
Indian Head	(4)	26	42	39	40	27	47
Jarrettsville	11	11	15	14	15	18	21
Kedges Straits	(1)	(0)	(1)	(1)	(0)	(2)	(3)
Keedysville	12	13	20	17	16	16	29
Kensington	20	18	10	24	16	18	31
Kent Island	21	21	20	13	19	6	27
Kenton	—	—	(0)	—	2	—	2
Keyser	13	7	4	—	—	—	19
King George	23	4	(0)	—	—	—	23
Kingston	11	15	15	11	15	16	25
Kinsale	(0)	(0)	—	(0)	—	(0)	(0)
Kitzmiller	14	11	11	23	18	2	30
La Plata	13	10	18	16	12	10	31
Langford Creek	13	13	11	17	11	16	28
Lanham	12	21	13	13	14	21	31
Laurel	25	27	26	32	12	25	41
Leesburg	—	8	—	—	—	—	8
Leonardtown	24	18	20	34	19	35	42
Libertytown	17	13	17	15	16	21	27
Lineboro	14	15	16	16	16	18	28
Littlestown	17	11	14	17	12	16	25
Lonaconing	10	8	19	19	14	6	28
Love Point	(0)	(0)	(0)	(0)	12	15	17
Lower Marlboro	32	33	25	26	18	36	42
Manchester	10	10	15	15	22	13	26
Mardela Springs	16	21	14	19	13	22	30
Marion	10	17	7	13	11	11	24
Marydel	4	—	19	—	10	—	21
Mason-Dixon	7	6	8	9	10	8	15
Mathias Point	30	13	23	10	23	(0)	36
McHenry	11	13	12	11	15	21	27
Mechanicsville	21	24	18	17	22	29	38
Middle River	11	12	10	15	9	12	28

Quad	Blocks						Quad Total
	NW	NE	CW	CE	SW	SE	
Millington	16	25	18	34	28	32	43
Monie	17	17	14	15	6	5	26
Mount Storm	16	—	—	—	—	—	16
Mount Vernon	(1)	11	19	26	19	10	33
Myersville	19	14	22	14	25	18	32
Nanjemoy	23	24	22	20	27	19	42
Nanticoke	5	7	6	10	(0)	13	18
New Freedom	10	10	16	21	10	13	28
New Windsor	12	11	13	18	11	12	25
Newark West	14	13	20	16	19	17	26
Ninepin Branch	18	16	20	20	18	18	31
Norrisville	7	10	6	22	11	15	27
North Beach	23	19	25	24	30	24	42
North East	23	13	18	21	19	18	34
North of Cove Point	(1)	(0)	(0)	(0)	(0)	(0)	(1)
Oakland	22	15	22	15	18	17	31
Ocean City	24	—	23	—	5	—	28
Odenton	16	27	36	17	28	25	44
Oldtown	28	37	17	31	26	28	44
Oxford	17	17	18	10	(0)	11	25
Passapatanzy	—	(0)	—	(0)	—	—	(0)
Patterson Creek	9	5	19	19	—	23	29
Paw Paw	32	40	36	28	31	(0)	45
Perryman	21	14	26	12	7	0	28
Phoenix	14	12	11	12	12	19	25
Piney Point	34	33	0	28	(0)	23	46
Piscataway	11	17	14	13	28	14	36
Pittsville	—	—	15	13	15	14	23
Pocomoke City	15	19	23	19	13	14	27
Point Lookout	25	(0)	23	(0)	(0)	(0)	31
Point No Point	(1)	(0)	43	(0)	35	(1)	47
Point of Rocks	19	16	21	22	—	22	30
Poolesville	25	15	23	10	12	14	33
Popes Creek	23	17	15	14	16	14	35
Port Tobacco	31	24	27	18	18	23	40
Preston	18	11	12	15	10	14	25
Price	17	21	18	14	16	19	29
Prince Frederick	30	36	22	29	25	36	43
Princess Anne	21	13	13	11	12	17	29
Public Landing	19	16	17	18	16	19	28
Quantico	—	—	—	(0)	—	22	22
Queenstown	18	21	22	21	14	23	32
Reisterstown	12	12	15	17	21	19	29
Relay	24	20	23	11	19	12	39
Rhodesdale	13	16	13	14	18	15	29
Richland Point	(0)	(0)	(0)	(0)	(0)	(0)	(0)
Ridgely	27	23	23	25	11	11	35
Rising Sun	12	12	15	15	12	16	25
Rock Hall	16	12	26	19	15	18	31
Rock Point	16	28	17	19	14	20	35
Rockville	16	25	17	18	28	12	34
Round Bay	28	26	23	24	26	27	46
Saint Clements Island	14	21	(0)	(0)	(0)	(0)	24
Saint George Island	12	15	(0)	(0)	(0)	(0)	20
Saint Marys City	29	17	21	35	15	34	44

Quad	Blocks						Quad Total
	NW	NE	CW	CE	SW	SE	
Salisbury	18	23	24	17	18	15	40
Sandy Spring	16	27	15	21	14	17	33
Sang Run	16	13	20	16	24	15	31
Savage	29	28	25	27	33	25	43
Saxis	13	15	(0)	(0)	—	—	23
Seaford West	16	—	8	—	7	—	17
Selbyville	—	—	24	16	22	23	30
Seneca	13	13	32	26	—	21	35
Sharptown	14	—	15	—	14	—	24
Shepherdstown	11	14	—	13	—	(0)	20
Smith Point	(0)	(0)	(0)	(0)	(0)	(0)	(0)
Smithsburg	8	6	7	11	9	22	24
Snow Hill	26	23	30	26	27	22	41
Solomons Island	27	32	35	35	37	46	50
South River	26	22	18	23	22	37	41
Sparrows Point	1	18	0	(0)	26	(0)	28
Spesutie	15	(3)	15	3	(0)	12	27
Sterling	21	26	1	23	—	—	30
Stratford Hall	(0)	12	(0)	(0)	(0)	(0)	12
Sudlersville	28	15	15	16	19	20	35
Swan Point	(5)	(0)	(0)	11	(0)	7	14
Sykesville	22	31	20	23	23	24	37
Table Rock	19	15	14	14	16	16	25
Taneytown	12	12	17	12	15	16	23
Taylors Island	15	21	(0)	14	(0)	16	25
Terrapin Sand Point	(1)	(0)	(0)	(0)	(0)	(0)	(1)
Tilghman	15	17	15	(0)	(0)	(1)	20
Tingles Island	17	23	(1)	13	(1)	10	25
Towson	17	13	12	18	9	16	25
Trappe	13	17	15	17	15	15	29
Union Bridge	14	15	13	17	13	14	23
Upper Marlboro	12	16	9	13	15	20	31
Urbana	16	16	15	13	18	26	28
Vienna	—	(2)	—	—	—	—	(2)
Walkersville	12	19	19	18	22	17	31
Wango	22	23	30	13	24	21	39
Washington East	23	22	15	13	(0)	15	32
Washington West	7	12	12	—	—	—	18
Waterford	—	—	—	6	—	11	13
Westernport	14	10	6	(0)	—	—	17
Westminster	16	14	15	11	15	18	28
Wetipquin	12	12	18	20	18	15	30
Whaleysville	—	—	16	16	18	17	26
White Marsh	10	23	19	21	12	11	31
Whittington Point	10	3	6	—	4	—	13
Widewater	(0)	29	(0)	25	(0)	16	36
Williamsport	21	8	15	8	12	11	27
Winfield	18	15	12	13	22	14	25
Wingate	16	(2)	14	7	9	5	20
Woodbine	25	21	23	21	24	25	33
Woodsboro	15	16	14	21	16	19	29
Wye Mills	21	26	22	15	21	17	34

Appendix B

Number of search-hours in each block and quad, as documented during the Maryland Amphibian and Reptile Atlas project

Numbers in parentheses indicate blocks and quads with less than 40.47 ha (100 ac) of upland. Dashes indicate blocks with no land or water in Maryland.

Quad	Blocks						Quad Total
	NW	NE	CW	CE	SW	SE	
Aberdeen	22.13	32.96	1.47	0.95	10.49	56.28	124.28
Accident	6.50	4.50	5.00	4.50	5.00	6.00	31.50
Alexandria	—	—	—	6.25	—	4.20	10.45
Anacostia	6.00	3.25	6.00	4.73	2.55	1.75	24.28
Annapolis	19.50	13.00	14.73	(0)	6.75	(0)	53.98
Artemas	10.75	2.00	26.50	5.00	39.78	72.00	156.03
Assawoman Bay	—	—	4.42	0.00	10.09	0.00	14.51
Avilton	11.25	8.75	6.50	25.25	9.50	20.00	81.25
Baltimore East	0.25	6.95	4.50	5.15	4.16	2.41	23.42
Baltimore West	5.45	0.70	6.50	2.75	3.25	5.75	24.40
Barren Island	(0)	1.37	(0)	(0)	(0)	(0)	1.37
Barton	11.00	19.25	18.50	7.75	5.50	3.75	65.75
Bay View	4.51	1.18	2.46	2.99	8.01	3.48	22.63
Bel Air	3.76	34.70	20.09	6.78	26.30	5.83	97.46
Bellegrove	1.50	7.00	19.00	4.75	34.00	21.25	87.50
Beltsville	4.50	8.75	5.50	5.00	2.75	3.75	30.25
Benedict	5.75	124.44	11.73	43.23	15.25	41.25	241.65
Berlin	2.00	4.33	4.55	7.43	13.03	10.43	41.77
Betterton	3.71	2.66	3.46	7.18	5.91	9.48	32.40
Big Pool	—	18.00	—	—	—	—	18.00
Bittinger	5.50	20.50	12.60	12.25	1.50	15.00	67.35
Blackwater River	7.40	4.33	37.67	7.10	3.54	(2.08)	62.12
Bloodsworth Island	(0.58)	1.98	(0)	(0)	(0)	(0)	2.56
Blue Ridge Summit	4.25	8.00	6.25	12.50	27.25	16.50	74.75
Bowie	16.25	7.00	15.75	8.75	3.75	4.50	56.00
Boxiron	16.93	0.25	3.92	(0)	(0.67)	(0)	21.77
Brandywine	13.98	6.00	2.75	1.25	11.25	18.00	53.23
Bristol	27.25	152.25	28.23	8.50	67.25	86.18	369.66
Broomes Island	28.25	83.25	21.50	14.50	77.50	232.90	457.90
Buckeystown	14.00	10.00	5.50	12.50	15.50	39.25	96.75
Burgess	(0)	(0)	(0)	(0)	—	(0)	(0)
Burrsville	0.77	—	2.00	—	1.88	—	4.65
Cambridge	0.85	5.05	5.00	6.25	5.85	4.70	27.70
Catoctin Furnace	11.25	11.25	27.00	8.00	15.95	14.25	87.70
Cecilton	3.10	1.72	1.83	1.67	2.45	1.52	12.29
Centreville	9.01	9.15	8.96	14.93	14.09	8.03	64.17
Charles Town	—	12.00	—	(0)	—	—	12.00
Charlotte Hall	15.16	17.50	3.30	11.48	3.98	3.98	55.40
Cherry Run	1.50	8.00	5.25	24.00	3.50	10.50	52.75

Quad	Blocks						Quad Total
	NW	NE	CW	CE	SW	SE	
Chicamacomico River	9.44	8.52	6.07	9.59	0.63	7.21	41.46
Church Creek	3.57	5.42	3.50	6.34	8.89	11.13	38.85
Church Hill	6.77	4.73	9.31	8.61	10.08	5.64	45.14
Claiborne	9.00	5.98	(0)	5.75	6.00	6.25	32.98
Clarksville	8.98	25.75	22.50	15.25	12.25	15.75	100.48
Clear Spring	8.75	5.25	20.75	4.75	5.75	3.25	48.50
Cockeysville	19.50	17.48	0.45	7.95	2.20	12.20	59.78
Colonial Beach North	13.50	35.00	(0.25)	6.25	(0)	7.00	62.00
Colonial Beach South	(0)	(0)	(0)	(0)	—	—	(0)
Conowingo Dam	0.87	1.97	28.90	5.70	9.48	10.16	57.08
Cove Point	32.96	(0)	123.50	3.00	36.50	73.50	269.46
Cresaptown	5.00	1.00	0.00	(0)	0.50	—	6.50
Crisfield	2.63	1.21	(0)	(0)	(0)	(0)	3.84
Cumberland	2.00	2.00	0.75	2.25	0.50	4.00	11.50
Curtis Bay	6.00	3.00	5.25	3.25	7.25	13.48	38.23
Dahlgren	(0)	(0)	—	(0)	—	(0)	(0)
Damascus	5.25	26.25	8.25	14.73	4.00	6.55	65.03
Davis	12.00	6.00	3.25	—	—	—	21.25
Deal Island	(1.00)	1.05	0.36	13.35	4.07	(4.00)	23.83
Deale	5.50	5.75	9.25	15.00	7.25	6.00	48.75
Deer Park	3.25	3.00	2.50	5.25	2.25	10.92	27.17
Delmar	—	—	9.26	9.24	12.79	5.17	36.46
Delta	1.45	4.74	8.70	27.30	6.45	17.13	65.77
Denton	3.71	6.81	1.63	2.72	3.07	3.15	21.09
Dividing Creek	10.27	14.19	6.15	13.36	3.99	5.09	53.05
Earleville	7.87	3.90	4.00	3.98	1.86	1.91	23.52
East New Market	6.63	6.00	4.43	10.42	7.79	5.34	40.61
East of North Beach	(0)	(0)	(0)	(0)	(0)	(0)	(0)
East of Point Lookout	(0)	(0)	(0)	(0)	(0)	(0)	(0)
Easton	2.40	1.95	4.20	2.42	0.55	15.26	26.78
Eden	6.76	30.14	12.76	3.79	7.31	9.91	70.67
Edgewood	40.73	24.65	6.78	173.97	12.12	3.60	261.85
Elkton	7.47	4.89	7.98	7.11	10.65	7.12	45.22
Ellicott City	15.75	1.00	12.98	21.25	25.25	14.33	90.56
Emmitsburg	10.50	5.50	63.50	10.75	7.75	12.25	110.25
Evitts Creek	1.00	13.25	15.75	14.00	4.50	1.70	50.20
Ewell	(0)	(0)	(0)	(0)	—	—	(0)
Falls Church	39.00	10.25	—	1.50	—	—	50.75
Fawn Grove	5.20	3.83	23.20	29.70	1.26	25.31	88.50
Federalsburg	4.25	8.97	4.18	3.53	4.91	6.17	32.01
Finksburg	22.00	19.00	10.48	11.50	18.00	14.00	94.98
Flintstone	5.00	1.50	5.00	4.25	11.00	13.33	40.08
Fort Belvoir	—	—	—	(0)	—	(0)	(0)
Fowling Creek	2.74	1.35	1.58	5.00	2.75	2.38	15.80
Frederick	20.25	10.75	10.50	13.75	12.25	14.00	81.50
Friendsville	4.50	4.75	5.25	17.50	2.75	21.50	56.25
Frostburg	18.75	3.50	31.75	3.50	2.75	1.00	61.25
Funkstown	6.25	22.25	4.25	10.25	8.75	40.00	91.75
Gaithersburg	4.98	1.25	2.00	1.25	4.50	8.65	22.63
Galena	5.10	5.00	4.55	3.88	3.33	7.41	29.27
Germantown	4.50	3.50	6.00	4.50	6.00	7.75	32.25
Gibson Island	16.00	14.50	5.25	11.00	15.75	18.25	80.75
Girdletree	6.54	6.67	3.11	9.78	7.02	4.54	37.66
Golden Hill	8.10	15.78	18.18	17.73	11.68	6.79	78.26
Goldsboro	7.16	6.83	8.99	162.82	28.51	2.79	217.10

Quad	Blocks						Quad Total
	NW	NE	CW	CE	SW	SE	
Grantsville	4.75	8.75	4.00	10.50	13.50	8.00	49.50
Great Cacapon	(0)	9.50	—	—	—	—	9.50
Great Fox Island	(0)	(0)	(0)	(0)	(0)	(0)	(0)
Gunpowder Neck	11.00	0.00	0.00	0.00	0.00	0.00	11.00
Hagerstown	2.25	2.50	19.50	25.75	25.75	15.00	90.75
Hallwood	0.50	—	—	—	—	—	0.50
Hampstead	16.25	6.75	5.48	4.50	3.50	4.00	40.48
Hancock	6.25	0.50	7.00	4.25	14.50	—	32.50
Hanesville	0.00	2.39	(0)	10.65	0.99	10.33	24.36
Harpers Ferry	15.50	6.50	9.50	12.00	—	—	43.50
Havre de Grace	4.82	5.18	9.46	4.74	19.99	(0.21)	44.40
Heathsville	(0)	(0)	—	(0)	—	—	(0)
Hebron	3.40	(0)	25.30	3.30	4.30	0.82	37.12
Hedgesville	17.35	12.00	—	10.00	—	—	39.35
Hereford	14.00	13.00	18.25	7.25	5.75	9.25	67.50
Hickman	2.38	—	2.25	—	22.88	—	27.51
Hobbs	10.41	3.60	7.73	2.39	5.01	2.33	31.47
Hollywood	8.21	12.50	41.25	8.73	33.81	19.73	124.23
Honga	4.86	7.33	2.63	1.96	2.20	0.65	19.63
Horseshoe Point	1.50	(0)	0.75	(0)	(0)	1.50	3.75
Hudson	(0)	4.89	(0)	9.08	(0)	2.55	16.52
Hughesville	8.88	12.00	3.35	5.00	7.79	10.50	47.52
Indian Head	(4.00)	15.00	35.25	17.88	54.58	5.75	132.46
Jarrettsville	1.44	8.23	2.92	8.43	12.72	11.51	45.25
Kedges Straits	(0)	(0)	(0)	(0)	(0)	(0)	(0)
Keedysville	7.00	9.00	26.25	17.50	7.00	10.00	76.75
Kensington	8.25	4.58	4.25	7.75	4.25	6.50	35.58
Kent Island	13.25	5.75	32.75	13.94	9.50	4.00	79.19
Kenton	—	—	(0)	—	0.38	—	0.38
Keyser	5.50	4.50	1.00	—	—	—	11.00
King George	21.25	4.75	(0)	—	—	—	26.00
Kingston	2.34	4.31	2.72	2.35	4.36	5.03	21.11
Kinsale	(0)	(0)	—	(0)	—	(0)	(0)
Kitzmiller	2.50	7.75	12.50	11.42	9.25	0.50	43.92
La Plata	7.50	3.25	6.75	11.75	7.50	6.75	43.50
Langford Creek	4.78	6.54	1.86	8.38	6.12	11.42	39.10
Lanham	6.75	14.75	0.50	6.25	3.25	7.00	38.50
Laurel	13.25	12.50	21.40	22.50	8.00	24.80	102.45
Leesburg	—	5.25	—	—	—	—	5.25
Leonardtown	3.00	3.75	2.75	12.75	4.75	123.73	150.73
Libertytown	20.75	7.00	17.00	6.75	13.25	13.00	77.75
Lineboro	5.25	6.25	15.00	6.75	28.50	9.25	71.00
Littlestown	7.75	4.50	7.50	8.50	5.48	7.75	41.48
Lonaconing	2.25	6.00	17.67	6.50	9.00	2.05	43.47
Love Point	(0)	(0)	(0)	(0)	29.50	2.25	31.75
Lower Marlboro	112.25	96.48	10.75	40.25	5.71	73.78	339.22
Manchester	4.75	9.00	20.75	16.25	14.50	10.75	76.00
Mardela Springs	5.33	12.90	5.91	9.13	8.15	8.63	50.05
Marion	9.73	3.25	4.75	3.95	5.10	3.72	30.50
Marydel	0.97	—	101.12	—	1.07	—	103.16
Mason-Dixon	7.50	4.50	4.25	6.00	13.50	2.50	38.25
Mathias Point	13.65	3.40	34.00	9.00	35.50	(0)	95.55
McHenry	3.75	5.25	4.00	6.25	2.90	10.00	32.15
Mechanicsville	4.50	16.50	5.25	3.00	5.50	8.23	42.98
Middle River	6.00	4.50	1.16	0.50	5.91	5.16	23.23

Quad	Blocks						Quad Total
	NW	NE	CW	CE	SW	SE	
Millington	6.60	13.31	6.32	40.64	18.69	36.86	122.42
Monie	0.95	8.04	5.98	7.50	3.20	3.21	28.88
Mount Storm	6.75	—	—	—	—	—	6.75
Mount Vernon	(0)	0.63	8.75	10.75	7.00	3.08	30.21
Myersville	12.75	8.50	21.75	11.25	7.00	11.75	73.00
Nanjemoy	6.00	9.17	7.07	2.00	25.15	5.85	55.24
Nanticoke	0.88	2.20	1.75	2.87	(0)	5.30	13.00
New Freedom	8.25	8.75	13.75	19.75	2.50	6.75	59.75
New Windsor	10.75	10.98	8.75	11.00	6.25	9.50	57.23
Newark West	4.78	1.54	3.38	1.00	8.16	4.47	23.33
Ninepin Branch	7.65	10.00	7.83	5.19	6.51	5.81	42.99
Norrisville	0.98	6.75	2.00	21.00	2.25	21.10	54.08
North Beach	37.00	17.00	33.98	30.73	47.00	50.75	216.46
North East	5.00	2.94	4.17	11.16	3.36	3.24	29.87
North of Cove Point	(0)	(0)	(0)	(0)	(0)	(0)	(0)
Oakland	20.00	7.00	7.25	5.25	2.00	5.50	47.00
Ocean City	3.80	—	2.00	—	0.00	—	5.80
Odenton	10.75	14.20	40.25	12.00	28.00	16.50	121.70
Oldtown	36.63	42.73	13.33	26.05	29.25	32.17	180.16
Oxford	10.75	10.75	1.75	5.00	(0)	3.15	31.40
Passapatanzy	—	(0)	—	(0)	—	—	(0)
Patterson Creek	5.48	7.00	7.60	6.53	—	8.00	34.61
Paw Paw	30.75	39.62	27.50	23.20	15.20	(0)	136.27
Perryman	17.67	3.12	24.15	0.00	0.00	0.00	44.94
Phoenix	5.25	2.50	4.25	1.25	3.00	15.20	31.45
Piney Point	32.25	33.96	0.00	24.75	(0)	2.75	93.71
Piscataway	1.30	3.03	4.15	5.15	14.50	2.75	30.88
Pittsville	—	—	5.00	5.25	12.53	7.03	29.81
Pocomoke City	9.79	6.34	8.77	5.22	5.32	6.88	42.32
Point Lookout	3.25	(0)	26.50	(0)	(0)	(0)	29.75
Point No Point	(0)	(0)	103.23	(0)	34.23	(0)	137.46
Point of Rocks	10.50	10.25	12.50	18.50	—	15.75	67.50
Poolesville	22.25	6.50	9.00	4.98	3.23	3.25	49.21
Popes Creek	2.75	2.50	5.25	3.65	8.25	16.98	39.38
Port Tobacco	19.08	5.23	10.75	6.50	4.25	10.62	56.43
Preston	5.53	1.67	2.70	1.95	3.75	2.53	18.13
Price	6.12	4.96	7.39	8.36	6.11	6.91	39.85
Prince Frederick	37.25	46.25	38.75	37.48	35.75	66.50	261.98
Princess Anne	7.32	4.92	3.61	3.37	1.65	4.64	25.51
Public Landing	9.10	9.52	9.30	4.03	4.12	7.65	43.72
Quantico	—	—	—	(0)	—	21.75	21.75
Queenstown	13.48	15.30	9.43	3.20	4.25	27.50	73.16
Reisterstown	8.00	10.75	5.75	5.98	13.25	8.48	52.21
Relay	14.75	10.50	16.50	15.00	11.65	9.25	77.65
Rhodesdale	6.43	3.09	5.78	6.33	8.51	10.57	40.71
Richland Point	(0)	(0)	(0)	(0)	(0)	(0)	(0)
Ridgely	10.13	3.18	11.66	6.73	2.56	3.15	37.41
Rising Sun	2.46	2.59	5.13	6.63	3.64	3.89	24.34
Rock Hall	6.35	4.16	8.50	7.37	8.34	4.98	39.70
Rock Point	4.46	3.00	4.25	4.25	4.00	3.73	23.69
Rockville	5.75	7.25	7.75	6.25	15.50	6.50	49.00
Round Bay	16.75	15.75	15.48	15.73	38.75	17.00	119.46
Saint Clements Island	4.50	8.98	(0)	(0)	(0)	(0)	13.48
Saint George Island	5.75	13.00	(0)	(0)	(0)	(0)	18.75
Saint Marys City	4.75	2.75	2.25	27.50	2.75	68.46	108.46

Quad	Blocks						Quad Total
	NW	NE	CW	CE	SW	SE	
Salisbury	5.75	15.27	5.15	2.02	5.00	3.57	36.76
Sandy Spring	11.20	19.00	4.75	10.50	8.00	5.48	58.93
Sang Run	10.25	6.25	13.75	10.00	29.50	6.00	75.75
Savage	32.25	24.00	15.25	25.93	36.00	28.30	161.73
Saxis	8.66	4.98	(0)	(0)	—	—	13.64
Seaford West	6.40	—	2.05	—	1.13	—	9.58
Selbyville	—	—	4.78	5.55	7.24	5.75	23.32
Seneca	4.48	2.75	16.40	14.25	—	7.50	45.38
Sharptown	5.63	—	7.00	—	8.75	—	21.38
Shepherdstown	14.00	11.73	—	11.25	—	(0)	36.98
Smith Point	(0)	(0)	(0)	(0)	(0)	(0)	(0)
Smithsburg	6.50	6.50	3.75	10.75	7.75	14.75	50.00
Snow Hill	6.99	6.67	9.52	3.83	10.08	5.90	42.99
Solomons Island	63.75	23.00	59.48	117.75	122.50	305.75	692.23
South River	12.73	9.75	3.23	12.50	9.75	28.50	76.46
Sparrows Point	0.33	12.32	0.16	(0)	34.50	(0)	47.31
Spesutie	2.00	(0.15)	2.00	1.00	(0)	3.18	8.33
Sterling	7.71	11.98	1.50	11.98	—	—	33.17
Stratford Hall	(0)	1.50	(0)	(0)	(0)	(0)	1.50
Sudlersville	13.23	3.91	8.55	5.59	6.02	7.48	44.78
Swan Point	(0)	(0)	(0)	1.59	(0)	1.19	2.78
Sykesville	24.48	39.25	16.75	27.00	16.25	24.00	147.73
Table Rock	4.50	4.50	3.25	3.25	6.25	6.25	28.00
Taneytown	11.25	12.75	14.00	10.25	14.00	13.25	75.50
Taylors Island	6.23	10.65	(0)	9.75	(0)	4.08	30.71
Terrapin Sand Point	(0)	(0)	(0)	(0)	(0)	(0)	(0)
Tilghman	1.75	4.00	2.00	(0)	(0)	(0)	7.75
Tingles Island	18.95	7.25	(0)	5.75	(0.50)	3.00	35.45
Towson	8.40	4.03	1.16	11.36	3.61	2.86	31.42
Trappe	6.29	20.71	3.47	4.78	3.57	1.97	40.79
Union Bridge	14.23	6.75	4.25	22.23	5.25	17.75	70.46
Upper Marlboro	2.00	5.00	5.50	5.65	3.00	7.78	28.93
Urbana	9.50	4.50	10.00	5.25	9.75	28.73	67.73
Vienna	—	(1.50)	—	—	—	—	(1.50)
Walkersville	8.25	17.50	16.50	5.75	12.20	5.50	65.70
Wango	13.26	18.62	25.35	7.04	5.93	13.38	83.58
Washington East	12.75	18.75	4.75	6.25	(0)	5.25	47.75
Washington West	1.75	4.00	5.23	—	—	—	10.98
Waterford	—	—	—	1.50	—	7.80	9.30
Westernport	10.50	3.50	7.50	(0)	—	—	21.50
Westminster	6.25	7.48	2.50	5.50	4.25	9.75	35.73
Wetipquin	4.10	5.43	9.20	13.15	7.63	4.53	44.04
Whaleysville	—	—	9.80	5.85	9.29	5.97	30.91
White Marsh	3.12	44.98	23.78	8.70	1.70	4.70	86.98
Whittington Point	0.50	0.00	1.25	—	1.00	—	2.75
Widewater	(0)	17.35	(0)	17.58	(0)	5.50	40.43
Williamsport	27.00	3.00	33.50	6.50	12.75	6.75	89.50
Winfield	11.50	6.73	6.25	9.25	29.00	3.50	66.23
Wingate	8.22	(0.58)	5.70	1.15	2.88	2.12	20.65
Woodbine	12.00	29.75	16.25	15.50	24.00	24.75	122.25
Woodsboro	7.75	15.50	16.25	9.75	12.50	12.50	74.25
Wye Mills	11.11	29.65	7.83	6.42	20.67	5.28	80.96

Appendix C

Species that were confirmed (C) and reported (R) in each county and total number of counties in which each species was documented during the Maryland Amphibian and Reptile Atlas project

Common Name	Allegany	Anne Arundel	Baltimore	Calvert	Caroline	Carroll	Cecil	Charles	Dorchester	Frederick	Garrett	Harford	Howard	Kent	Montgomery	Prince George's	Queen Anne's	Somerset	St. Mary's	Talbot	Washington	Wicomico	Worcester	Total Counties
AMPHIBIANS																								
Jefferson Salamander	C									C	C				C						C			5
Spotted Salamander	C	C	C	C	R	C	C	C	C	C	C	C	C	C	C	C	C		C	C	C			20
Marbled Salamander	C	C	C	C	C	C	C	C	C	C	C		C	C	C	C	C	C	C	C	C	C	C	22
Eastern Tiger Salamander				C										C										2
Hellbender											C													1
Green Salamander											C													1
Northern Dusky Salamander	C	C	C	C		C	C	C		C	C	C	C	C	C	C	C		C		C	C		18
Seal Salamander	C										C													2
Allegheny Mountain Dusky Salamander	C										C													2
Northern Two-lined Salamander	C	C	C	C	C	C	C	C		C	C	C	C	C	C	C	C		C		C			18
Southern Two-lined Salamander																						C		1
Long-tailed Salamander	C		C			C	C			C	C	C	C		C						C			10
Spring Salamander	C									C	C	C									C			5
Four-toed Salamander	C	C	C	C	C	C	C	C		C	C	C	C	C		C	C	C	C	C	C	C	C	21
Eastern Red-backed Salamander	C	C	C	C	C	C	C	C	C	C	C	C	C	C	C	C	C	C	C	C	C	C	C	23
Northern Slimy Salamander	C		C			C	C			C	C	C	C								C			9
Valley and Ridge Salamander	C										C													2
Wehrle's Salamander	C																							1
Mud Salamander		C														C								2
Red Salamander	C	C	C	C			C	C	C		C	C	C	C		C	C		C		C			15
Eastern Newt	C	C	C	C	C	C	C	C	C	C	C	C	C	C	C	C	C		C	C	C			20
American Toad	C	C	C	C		C	C	C	R	C	C	C	C		C	C	R	R	C	R	C	R	C	21
Fowler's Toad	C	C	C	C	C	C	C	C	C	C	C		C	C	C	C	C	C	C	C	C	C	C	22
Eastern Cricket Frog	C	C	C	C	C			C	C	C	C		C	C	C	C	C	C	R	C	C	C	R	20
Cope's Gray Treefrog		C	C	C	C	C	C	C	C	C	C		C	C	C	C	C	C	C	C		C	C	20
Green Treefrog		C	C	C	C	C	C	C	C	C	C		C	C	R	C	C	C	C	C		C	C	20
Barking Treefrog					C											C								2
Gray Treefrog	C	C	C		C	C	C	C		C	C	C	C	C	C	C	C				C			16
Spring Peeper	C	C	C	C	C	C	C	C	C	C	C	C	C	C	C	C	C	C	C	C	C	C	C	23
Upland Chorus Frog	C	C	C	C				C		C		R	R		C	R			C		C			12
New Jersey Chorus Frog				C		C		C							C			C	C	C		C	C	9
Eastern Narrow-mouthed Toad								C											C			C	C	4
American Bullfrog	C	C	C	C	C	C	C	C	C	C	C	C	C	C	C	C	C	C	C	C	C	C	C	23
Green Frog	C	C	C	C	C	C	C	C	C	C	C	C	C	C	C	C	C	C	C	C	C	C	C	23
Pickerel Frog	C	C	C	C	C	C	C	C	C	C	C	C	C	C	C	C	C	C	C	R	C	C	C	23
Southern Leopard Frog		C	C	C	C		C	C	C	C		C	C	C	C	C	C	C	C	C		C	C	19
Wood Frog	C	C	C	C	C	C	C	C		C	C	C	C	C	C	C	C	C	C	C	C	C	C	22

Common Name	Allegany	Anne Arundel	Baltimore	Calvert	Caroline	Carroll	Cecil	Charles	Dorchester	Frederick	Garrett	Harford	Howard	Kent	Montgomery	Prince George's	Queen Anne's	Somerset	St. Mary's	Talbot	Washington	Wicomico	Worcester	Total Counties
Carpenter Frog				C					C								C	C				C	C	6
Eastern Spadefoot		C		C	C		C	C	C			C		C	C	C	C	C	C	C	C	C	C	17
REPTILES																								
Spiny Softshell										C	C													2
Loggerhead Sea Turtle		C		C					R					C			R	R	R	R			C	9
Green Sea Turtle																							C	1
Kemp's Ridley Sea Turtle		C	R	C															C				C	5
Leatherback Sea Turtle																		C	R			R	C	4
Snapping Turtle	C	C	C	C	C	C	C	C	C	C	C	C	C	C	C	C	C	C	C	C	C	C	C	23
Painted Turtle	C	C	C	C	C	C	C	C	C	C	C	C	C	C	C	C	C	C	C	C	C	C	C	23
Spotted Turtle		C	C	C	C	C	C	C	C	C		C	C	C	C	C	C	C	C	C	R	C	C	21
Wood Turtle	C	C	C							C	C	C			C	R					C			9
Bog Turtle			C			C	C					C												4
Northern Map Turtle							C					C												2
Diamond-backed Terrapin		C	C	C				R	C			C		C	C	C	C	C	C			C	C	14
Northern Red-bellied Cooter	C	C	C	C	C	C	C	C	C	C		C	C	C	C	C	C	C	C	C	C	C	C	22
Eastern Box Turtle	C	C	C	C	C	C	C	C	C	C	C	C	C	C	C	C	C	C	C	C	C	C	C	23
Red-eared Slider	C	C	C	C	C	C	C	C		C		C	C	C	C	C	C		C		C	R		18
Eastern Mud Turtle		C	C	C	C		C	C	C			C		C	C	C	C	C	C	C			C	C	17
Eastern Musk Turtle	C	C	C	C	C		C	C	C	C		C	C	C	C	C	C	C	C	R	R	C	C	21
Eastern Fence Lizard	C	C	C	C	C	R	C	C	C	C		C		C	C	C	C	C	C			C	C	19
Coal Skink	C																							1
Common Five-lined Skink	C	C	C	C	C		C	C	C	C		C	C	C	C	C	C	C	C	C	C	C	C	21
Broad-headed Skink		C	R	C			C	C	R	C				C		C	C	C	C	C	C	C	C	16
Little Brown Skink		C		C	C		C	C										C	C	C		C	C	10
Six-lined Racerunner	C	C		C				C											C					5
Scarletsnake														C										1
North American Racer	C	C	C	C	C	C	C	C	C	C	C	C	C	C	C	C	C	C	C	C	C	C	C	23
Eastern Kingsnake		C	R	C	C				C	C			C	C	C	C	C	C	C	C	C		C	C	17
Northern Mole Kingsnake								C										C						2
Eastern Milksnake	C		C			C	C	C	C	C	C	C	C		C				C	C	C		C	15
Rough Greensnake		C	C	C	C		C	C	C	C		C	C	C	C	C	C	C	C	C		C	C	19
Smooth Greensnake	R									R	C													3
Eastern Ratsnake	C	C	C	C	C	C	C	C	C	C	C	C	C	C	C	C	C	C	C	C	C	C	C	23
Red Cornsnake	C		C				R												C		C			5
Common Wormsnake	C	C	C	C	C		C	C	C	C		C	C	C	C	C	C	C	C	C	C	C	C	21
Ring-necked Snake	C	C	C	C	C	C	C	C	C	C	C	C	C	C	C	C	C	C	C	C	C	C	C	23
Rainbow Snake								C																1
Eastern Hog-nosed Snake	C	C	C	C	C	C	C	C	C	C	C		C	R	C	C	C	C	C	C	C	C	C	22
Plain-bellied Watersnake									C													C	C	3
Common Watersnake	C	C	C	C	C	C	C	C	C	C	C	C	C	C	C	C	C	C	C	C	C	C	C	23
Queensnake		C	C	R		C	C	C		C	C	C	C	C	C	C	C				C			15
Dekay's Brownsnake	C	C	C	C	C	C	C	C	C	C	C	C	C	C	C	C	C	C	C	C	C	C	C	23
Red-bellied Snake	C	C	C					C	R	C	C								C	C	C	C	C	12
Eastern Ribbonsnake		C	C	C	C	C	C	C	C	C		C	C	C	C	C	C	C	C	R			C	19
Common Gartersnake	C	C	C	C	C	C	C	C	C	C	C	C	C	C	C	C	C	C	C	C	C	C	C	23
Eastern Smooth Earthsnake	C	C	C	C	C	C	C	C		C		C	C	C	C				C		C	C	C	17
Mountain Earthsnake											C													1
Eastern Copperhead	C	C	C	C			C	C	C	C	C	C	C	C	C	C			C	C	C	C	C	19
Timber Rattlesnake	C									C	C										C			4

Appendix D

Total number of submitted and confirmed records by species for quads, blocks, total sightings, and sightings of adults and juveniles, larvae, and eggs documented during the Maryland Amphibian and Reptile Atlas project

Total sightings may be less than the sum of sightings of adults and juveniles, larvae, and eggs due to the exclusion of redundancies within individual MARA data sheets.

Common Name	Quads	Quads Confirmed	Blocks	Blocks Confirmed	Total Sightings	Total Sightings Confirmed	Sightings Adults and Juveniles	Sightings Larvae	Sightings Eggs
Jefferson Salamander	22	14	36	20	48	24	21	7	27
Spotted Salamander	156	146	417	348	657	413	352	32	311
Marbled Salamander	90	82	203	170	274	190	234	49	3
Eastern Tiger Salamander	2	2	4	4	9	7	3	1	7
Hellbender	1	1	1	1	2	1	2	—	—
Green Salamander	2	2	5	5	7	6	12	—	—
Northern Dusky Salamander	110	104	318	265	436	298	445	8	2
Seal Salamander	18	15	42	27	59	32	59	—	—
Allegheny Mountain Dusky Salamander	19	19	90	88	153	115	153	—	—
Northern Two-lined Salamander	146	136	556	484	836	549	752	100	4
Southern Two-lined Salamander	2	1	4	2	4	2	—	4	—
Long-tailed Salamander	68	65	145	131	179	138	179	1	—
Spring Salamander	22	17	47	32	65	35	34	33	—
Four-toed Salamander	63	57	92	82	107	87	106	—	9
Eastern Red-backed Salamander	201	192	762	674	1,116	778	1,135	—	2
Northern Slimy Salamander	55	52	140	120	197	138	202	—	—
Valley and Ridge Salamander	15	11	31	21	36	23	36	—	—
Wehrle's Salamander	2	2	2	2	2	2	2	—	—
Mud Salamander	4	4	6	6	8	8	8	—	—
Red Salamander	92	89	207	188	237	201	182	65	—
Eastern Newt	89	79	198	147	290	167	295	2	—
SALAMANDER TAXA TOTALS	1,179	1,090	3,306	2,817	4,722	3,214	4,212	302	365
American Toad	169	143	763	621	1,434	723	1,445	9	13
Fowler's Toad	178	161	676	553	1,206	627	1,214	2	—
Eastern Cricket Frog	104	83	278	194	573	222	571	1	1
Cope's Gray Treefrog	162	162	606	606	853	665	853	1	1
Green Treefrog	119	110	375	293	646	329	647	1	—
Barking Treefrog	2	2	5	5	5	5	5	—	—
Gray Treefrog	88	88	317	317	417	340	417	—	—

Common Name	Quads	Quads Confirmed	Blocks	Blocks Confirmed	Total Sightings	Total Sightings Confirmed	Sightings Adults and Juveniles	Sightings Larvae	Sightings Eggs
Unknown Gray Treefrog sp.	11	—	122	—	601	187	598	8	7
Spring Peeper	234	229	1,135	981	2,349	1,077	2,343	9	—
Upland Chorus Frog	39	18	78	24	104	27	104	—	1
New Jersey Chorus Frog	63	62	256	199	461	223	460	1	2
Eastern Narrow-mouthed Toad	16	16	35	35	43	39	43	—	—
American Bullfrog	221	209	913	685	1,609	757	1,614	26	1
Green Frog	231	212	1,064	834	2,082	927	2,109	6	1
Pickerel Frog	187	164	661	476	959	517	969	3	4
Southern Leopard Frog	141	131	566	434	1,130	509	1,127	13	6
Wood Frog	183	158	629	437	946	486	778	51	163
Carpenter Frog	10	8	21	16	32	22	32	—	—
Eastern Spadefoot	94	89	314	279	434	307	437	5	1
FROG & TOAD TAXA TOTALS	2,253	2,045	8,815	6,989	15,884	7,989	15,766	136	201
Spiny Softshell	2	2	2	2	3	2	3	—	—
Loggerhead Sea Turtle	22	9	40	17	51	21	51	—	—
Green Sea Turtle	1	1	2	1	2	1	2	—	—
Kemp's Ridley Sea Turtle	10	3	13	3	16	3	16	—	—
Leatherback Sea Turtle	5	3	7	3	8	3	8	—	—
Snapping Turtle	221	196	757	488	1,154	533	1,178	—	2
Painted Turtle	217	205	771	539	1,377	643	1,442	—	—
Spotted Turtle	104	88	166	122	193	133	198	—	1
Wood Turtle	38	31	65	43	96	59	101	—	—
Bog Turtle	13	13	28	28	49	40	67	—	4
Northern Map Turtle	5	5	7	7	10	10	11	—	—
Diamond-backed Terrapin	72	50	170	92	297	104	301	—	—
Northern Red-bellied Cooter	139	113	315	204	488	224	497	—	—
Eastern Box Turtle	218	209	794	642	1,421	748	1,441	—	1
Red-eared Slider	118	97	237	172	350	198	361	—	—
Eastern Mud Turtle	106	96	242	176	371	197	378	—	—
Eastern Musk Turtle	80	58	124	80	156	90	160	—	—
TURTLE TAXA TOTALS	1,371	1,179	3,740	2,619	6,042	3,009	6,215	—	8
Eastern Fence Lizard	72	61	144	100	228	115	238	—	—
Coal Skink	2	2	2	2	3	3	3	—	—
Common Five-lined Skink	135	110	357	228	572	270	593	—	1
Broad-headed Skink	52	38	86	59	121	71	124	—	—
Little Brown Skink	39	30	60	37	66	38	66	—	—
Six-lined Racerunner	11	6	19	8	37	8	38	—	—
LIZARD TAXA TOTALS	311	247	668	434	1,027	505	1,062	—	1
Scarletsnake	1	1	1	1	1	1	1	—	—
North American Racer	181	155	449	304	647	334	647	—	5
Eastern Kingsnake	71	58	124	90	160	98	160	—	—

Common Name	Quads	Quads Confirmed	Blocks	Blocks Confirmed	Total Sightings	Total Sightings Confirmed	Sightings Adults and Juveniles	Sightings Larvae	Sightings Eggs
Northern Mole Kingsnake	2	2	3	3	4	4	4	—	—
Eastern Milksnake	84	76	165	130	204	140	208	—	—
Rough Greensnake	115	102	248	183	331	199	331	—	—
Smooth Greensnake	11	4	16	6	21	7	21	—	—
Eastern Ratsnake	225	206	827	631	1,332	718	1,350	—	—
Red Cornsnake	11	9	17	13	20	15	20	—	—
Common Wormsnake	130	117	290	216	409	231	412	—	—
Ring-necked Snake	169	149	426	327	521	343	525	—	2
Rainbow Snake	2	2	3	3	4	4	4	—	—
Eastern Hog-nosed Snake	93	73	162	116	199	125	201	—	—
Plain-bellied Watersnake	11	9	28	20	38	25	38	—	—
Common Watersnake	219	192	683	463	1,008	494	1,028	—	—
Queensnake	56	51	81	70	97	74	98	—	—
Dekay's Brownsnake	135	127	297	261	367	288	368	—	—
Red-bellied Snake	35	30	59	40	68	43	68	—	—
Eastern Ribbonsnake	56	52	75	65	96	77	96	—	—
Common Gartersnake	207	188	709	532	1,014	580	1,018	—	—
Smooth Earthsnake	60	56	95	87	125	100	128	—	—
Eastern Copperhead	106	95	229	174	345	193	349	—	—
Timber Rattlesnake	30	26	59	45	131	67	138	—	—
SNAKE TAXA TOTALS	2,010	1,780	5,046	3,780	7,142	4,160	7,213	—	7
Exotic Taxa	52	52	71	71	84	84	84	—	—

References

Adamovicz, L., E. Bronson, K. Barrett, and S. L. Deem. 2015. "Health assessment of free-living eastern box turtles (*Terrapene carolina carolina*) in and around the Baltimore Zoo 1996-2011." *Journal of Zoo and Wildlife Medicine* 46:39-51.

Akin, J. A. 1998. "Intra- and inter-sexual aggression in the ground skink (*Scincella lateralis*)." *Canadian Journal of Zoology* 76:87-93.

Albaugh, S. J. 2008. "Habitat comparison of *Pseudacris f. feriarum* and *Pseudacris c. crucifer* with emphasis on associated plant communities and distribution of *Clemmys guttata* and *Pseudacris f. feriarum* in West Virginia." MS thesis. Marshall University, Huntington, WV.

Albers, P. H., and R. M. Prouty. 1987. "Survival of spotted salamander eggs in temporary woodland ponds of coastal Maryland." *Environmental Pollution* 46:45-61.

Albers, P. H., L. Sileo, and B. M. Mulhern. 1986. "Effects of environmental contaminants on snapping turtles of a tidal wetland." *Archives of Environmental Contamination and Toxicology* 15:39-49.

Alderton, D. 1988. *Turtles and Tortoises of the World*. New York: Facts on File Publications.

Alexander, D. G. 1965. "An ecological study of the swamp cricket frog (*Pseudacris nigrita feriarum* (Baird)) with comparative notes on two other hylids in the Chapel Hill, North Carolina region." PhD dissertation. University of North Carolina, Chapel Hill.

Alexander, W. P. 1927. "The Allegheny hellbender and its habitat." *Buffalo Society of Natural Science* 7(10):13-18.

Allender, M. C., D. B. Raudabaugh, F. H. Gleason, and A. N. Miller. 2015. "The natural history, ecology, and epidemiology of *Ophidiomyces ophiodiicola* and its potential impact on free-ranging snake populations." *Fungal Ecology* 17:187-196.

Alroy, J. 2015. "Current extinction rates of reptiles and amphibians." *Proceedings of the National Academy of Sciences of the United States of America* 112:13,003-13,008.

Anderson, J. D., D. D. Hassinger, and G. H. Dalrymple. 1971. "Natural mortality of eggs and larvae of *Ambystoma t. tigrinum*." *Ecology* 52:1107-1112.

Anderson, K. P. 2014. "Impacts of human recreation and hydroelectric flow regime on basking behavior of northern map turtles, *Graptemys geographica*." MS thesis. Towson University, Towson, MD.

Anderson, K. P., N. W. Byer, R. J. McGehee, and T. Richards-Dimitrie. 2015. "A new system for marking hatchling turtles using visible implant elastomer." *Herpetological Review* 46:25-27.

Anderson, K., and H. G. Dowling. 1982. "Geographic distribution: *Hyla gratiosa* (barking treefrog)." *Herpetological Review* 13:130.

Anderson, P. K. 1954. "Studies in the ecology of the narrow-mouthed toad, *Microhyla carolinensis carolinensis*." *Tulane Studies in Zoology* 2:15-46.

Andrews, J. S. 2013. *The Vermont Reptile and Amphibian Atlas: 2013 Update*. Middlebury, VT: The Vermont Reptile and Amphibian Atlas Project.

Angle, J. P. 1969. "The reproductive cycle of the northern ravine salamander, *Plethodon richmondi richmondi*, in the Valley and Ridge Province of Pennsylvania and Maryland." *Journal of the Washington Academy of Sciences* 59:192-202.

Applegate, R. D. 1990. "Natural history notes: *Rana catesbeiana*, *Rana palustris* (bullfrog, pickerel frog): predation." *Herpetological Review* 21:90-91.

Arber, E. 1884. *Captain John Smith, of Willoughby by Alford, Lincolnshire; president of Virginia, and Admiral of New England. Works, 1608-1631*. The English Scholar's Library Edition, no. 16. Chillworth & London: Unwin Bros, Gresham Press.

Arndt, R. G. 1975. "The occurrence of barnacles and algae on the red-bellied turtle, *Chrysemys r. rubriventris* (Le Conte)." *Journal of Herpetology* 9:357-359.

Arndt, R. G. 1977. "Notes on the natural history of the bog turtle, *Clemmys muhlenbergii* (Schoepf), in Delaware." *Chesapeake Science* 16:67-76.

Arndt, R. G. 1989. "Notes on the natural history and status of the tiger salamander, *Ambystoma tigrinum*, in Delaware." *Bulletin of the Maryland Herpetological Society* 25:1-18.

Arndt, R. G., and J. F. White. 1988. "Geographic distribution: *Hyla gratiosa* (barking treefrog)." *Herpetological Review* 19:16.

Arnold, S. J. 1976. "Sexual behavior, sexual interference, and sexual defense in the salamanders *Ambystoma maculatum*, *Ambystoma tigrinum*, and *Plethodon jordani*." *Zeitschrift für Tierpsychologie* 42: 247-300.

Arnold, S. J. 1977. "The evolution of courtship behavior in New World salamanders with some comments on Old World salamandrids." In *The Reproductive Biology of Amphibians*, edited by D. H. Taylor and S. I. Guttman, 141-183. New York: Plenum Press.

Arnold, S. J., and R. J. Wassersug. 1978. "Differential predation on metamorphic anurans by garter snakes (*Thamnophis*): social behavior as a possible defense." *Ecology* 59:1014-1022.

Ashton, R., R. Chance, R. Franz, J. D. Groves, J. D. Hardy, Jr., H. S. Harris, Jr., W. Hays, W. Hildebrand, D. S. Lee, W. S. Sipple, R. Standley, and R. G. Tuck, Jr. 2007. "Delusions of science: concerns regarding the unwarranted introduction of pine snakes to the Delmarva Peninsula of Maryland." *Bulletin of the Maryland Herpetological Society* 43:147-158.

Babbitt, L. H. 1937. "The amphibia of Connecticut." *Connecticut State Geological and Natural History Survey Bulletin*, no. 57.

Bachmann, M. D., R. G. Carlton, J. M. Burkholder, and R. G. Wetzel. 1986. "Symbiosis between salamander eggs and green algae: microelectrode measurements inside eggs demonstrate effect of photosynthesis on oxygen concentration." *Canadian Journal of Zoology* 64:1586-1588.

Bahret, R. 1996. "Ecology of lake dwelling *Eurycea bislineata* in the Shawangunk Mountains, New York." *Journal of Herpetology* 30:399-401.

Bailey, J. E. 1992. "An ecological study of the Cumberland Plateau salamander, *Plethodon kentucki* Mittleman, in West Virginia." MS thesis. Marshall University, Huntington, WV.

Baird, S. F. 1850. "Registry of periodical phenomena." Published by the Smithsonian Institution. Text of announcement reprinted in G. B. Goode, "Biographies of American naturalists I: the published writings of Spencer Fullerton Baird: 1843-1882," *Bulletin of the United States National Museum*, vol. 23, no. 20, 1883.

Baird, S. F., and C. Girard. 1853. *Catalogue of North American Reptiles in the Museum of the Smithsonian Institution. Part I. Serpents.* Smithsonian miscellaneous collections 2. Washington, DC: Smithsonian Institution.

Bank, M. S., J. B. Crocker, S. Davis, D. K. Brotherton, R. Cook, J. Behler, and B. Connery. 2006. "Population decline of northern dusky salamanders at Acadia National Park, Maine, USA." *Biological Conservation* 130:230-238.

Barbour, R. W., M. J. Harvey, and J. W. Hardin. 1969. "Home range, movements, and activity of the eastern worm snake, *Carphophis amoenus amoenus.*" *Ecology* 50:470-476.

Barrow, L. N., H. F. Ralicki, S. A. Emme, and E. Moriarty Lemmon. 2014. "Species tree estimation of North American chorus frogs (*Hylidae: Pseudacris*) with parallel tagged amplicon sequencing." *Molecular Phylogenetics and Evolution* 75:78-90.

Barthalmus, G. T., and E. D. Bellis. 1969. "Homing in the northern dusky salamander, *Desmognathus fuscus fuscus* (Rafinesque)." *Copeia* 1969:148-153.

Barthalmus, G. T., and E. D. Bellis. 1972. "Home range, homing and the homing mechanism of the salamander, *Desmognathus fuscus.*" *Copeia* 1972:632-642.

Barthalmus, G. T., and I. R. Savidge. 1974. "Time: an index of distance as a barrier to salamander homing." *Journal of Herpetology* 8:251-254.

Bartram, J. 1734 [1740]. "XIX. A Letter from John Bartram, M. D. to Peter Collinson, F. R. S. concerning a cluster of small teeth observed by him at the root of each fang or great tooth in the head of a rattle-snake, upon dissecting it." Philosophical Transactions of the Royal Society of London 41:358-359.

Bascietto, J. J., and L. W. Adams. 1983. "Frogs and toads of stormwater management basins in Columbia, Maryland." *Bulletin of the Maryland Herpetological Society* 19:58-60.

Batts, B. S. 1961. "Intertidal fishes as food of the common garter snake." *Copeia* 1961:350-351.

Beamer, D. A., and M. J. Lannoo. 2005a. "*Plethodon glutinosus* (Green, 1818), northern slimy salamander." In *Amphibian Declines: The Conservation Status of United States Species*, edited by M. J. Lannoo, 808-811. Berkeley: University of California Press.

Beamer, D. A., and M. J. Lannoo. 2005b. "*Plethodon virginia* Highton, 1999, Shenandoah salamander." In *Amphibian Declines: The Conservation Status of United States Species*, edited by M. J. Lannoo, 850-852. Berkeley: University of California Press.

Beane, J. C. 1990. "Life history notes: *Rana palustris* (pickerel frog): predation." *Herpetological Review* 21:59.

Beane, J. C., A. L. Braswell, J. C. Mitchell, W. M. Palmer, and J. R. Harrison, III. 2010. *Amphibians and Reptiles of the Carolinas and Virginia*, 2nd ed. Chapel Hill: University of North Carolina Press.

Becker, B. M., and M. A. Paulissen. 2012. "Sexual dimorphism in head size in the little brown skink (*Scincella lateralis*)." *Herpetological Conservation and Biology* 7:109-114.

Behle, W. H., and R. J. Erwin. 1962. "The green frog (*Rana clamitans*) established at West Ogden, Weber County, Utah." *Utah Academy of Sciences, Arts, and Letters Proceedings* 39:74-76.

Behler J. L., and F. W. King. 1979. *National Audubon Society Field Guide to North American Reptiles and Amphibians*. New York: Knopf Publishers.

Bell, S. L. M., T. B. Herman, and R. J. Wassersug. 2007. "Ecology of *Thamnophis sauritus* (eastern ribbon snake) at the northern limit of its range." *Northeastern Naturalist* 14:279-292.

Beltz, E. 2006. "Scientific and common names of the reptiles and amphibians of North America explained." Online at ebeltz.net/herps/etymain.html.

Bennett, D. H. 1972. "Notes on the terrestrial wintering of mud turtles (*Kinosternon subrubrum*)." *Herpetologica* 28:245-246.

Bennett, D. H., J. W. Gibbons, and J. C. Franson. 1970. "Terrestrial activity in aquatic turtles." *Ecology* 51:738-740.

Bennett, M. 2002. *Use of Radio Telemetry to Measure Home Range in Red-bellied Turtle*. Technical Report of the Jug Bay Wetlands Sanctuary. Lothian, MD: Jug Bay Wetlands Sanctuary.

Berven, K. A. 1982. "The genetic basis of altitudinal variation in the wood frog Rana sylvatica I: an experimental analysis of larval development." Evolution 36:962-983.

Birdsall C. W., C. E. Grue, and A. Anderson. 1986. "Lead concentrations in bullfrog, *Rana catesbeiana* and green frog, *R. clamitans*, tadpoles inhabiting highway drainages." *Environmental Pollution (Series A)* 40:233-247.

Bishop, S. C. 1928. "Notes on some amphibians and reptiles from the southeastern states with the description of a new salamander from North Carolina." *Journal of the Elisha Mitchell Scientific Society* 43:153-170.

Bishop, S. C. 1941. *Salamanders of New York*. Bulletin no. 324. Albany: New York State Museum.

Bishop, S. C. 1943. *Handbook of Salamanders: The Salamanders of the United States, of Canada, and of Lower California*. Ithaca, NY: Comstock Publishing Company.

Black, I. H., and K. L. Gosner. 1958. "The barking tree frog, *Hyla gratiosa* in New Jersey." *Herpetologica* 13:254-255.

Blanchard, F. N. 1921. *A Revision of the Kingsnakes: Genus Lampropeltis*. United States National Museum Bulletin no. 114. Washington, DC: Government Printing Office.

Blanchard, F. N. 1937. "Data on the natural history of the redbellied snake, *Storeria occipitomaculata* (Storer), in northern Michigan." *Copeia* 1937:151-162.

Blanchard, F. N., M. R. Gilreath, and F. C. Blanchard. 1979. "The eastern ring-neck snake (*Diadophis punctatus edwardsii*) in northern Michigan (Reptilia, Serpentes, Colubridae)." *Journal of Herpetology* 13:377-402.

Bloomer, T. J. 1978. "Hibernacula congregating in the *Clemmys* genus." *Journal of the Northern Ohio Association of Herpetologists* 4: 37-42.

Blouin-Demers, G., K. A. Prior, and P. J. Weatherhead. 2002. "Comparative demography of black rat snakes (*Elaphe obsoleta*) in Ontario and Maryland." *Journal of Zoology* 256:1-10.

Bogart, J. P. 1980. "Evolutionary implications of polyploidy in amphibians and reptiles." In *Polyploidy: Biological Relevance*, edited by W. H. Lewis, 341-378. New York: Plenum Press.

Bonin, J., J. L. DesGranges, J. Rodrigue, and M. Ouellet. 1997. "Anuran species richness in agricultural landscapes of Quebec: foreseeing long-term results of road call surveys." *Amphibians in Decline: Canadian Studies of a Global Problem* 1:246-257.

Boyer, D. R. 1965. "Ecology of the basking habit in turtles." *Ecology* 46:99-118.

Brady, M. K. 1924. "Muhlenberg's turtle near Washington." *Copeia* 135:92.

Brady, M. K. 1937. "Natural history of Plummer's Island, Maryland. V. Reptiles and amphibians." *Proceedings of the Biological Society of Washington* 50:137-140.

Branch, M. P., editor. 2004. *Reading the Roots: American Nature Writing before Walden*. Athens: University of Georgia Press.

Brand, A. B., and J. W. Snodgrass. 2010. "Value of artificial habitats for amphibian reproduction in altered landscapes." *Conservation Biology* 24:295-301.

Brand, A. B., J. W. Snodgrass, M. T. Gallagher, R. E. Casey, and R. Van Meter. 2010. "Lethal and sublethal effects of embryonic and larval exposure of *Hyla versicolor* to stormwater pond sediments." *Archives of Environmental Contamination and Toxicology* 58:325-331.

Brandon, R. A., and J. E. Huheey. 1975. "Diurnal activity, avian predation, and the question of warning coloration and cryptic coloration in salamanders." *Herpetologica* 31:252-255.

Brandon, R. A., and J. E. Huheey. 1981. "Toxicity in the plethodontid salamanders *Pseudotriton ruber* and *Pseudotriton montanus* (Amphibia, Caudata)." *Toxicon* 19:25-31.

Brandon, R. A., G. M. Labanick, and J. E. Huheey. 1979. "Relative palatability, defensive behavior, and mimetic relationships of red salamanders (*Pseudotriton ruber*), mud salamanders (*Pseudotriton montanus*), and red efts (*Notophthalmus viridescens*)." *Herpetologica* 35:289-303.

Branson, B. A., and E. C. Baker. 1974. "An ecological study of the queen snake, *Regina septemvittata* (Say) in Kentucky." *Tulane Studies in Zoology and Botany* 4:153-171.

Brennan, L. A., and W. P. Kuvlesky, Jr. 2005. "North American grassland birds: an unfolding conservation crisis?" *Journal of Wildlife Management* 69:1-13.

Bridges, V. E., C. A. Kopral, and R. A. Johnson. 2001. *The Reptile and Amphibian Communities*. Fort Collins, CO: US Department of Agriculture, APHIS, Center for Emerging Issues Centers for Epidemiology and Animal Health.

Brimley, C. S. 1941. "The amphibians and reptiles of North Carolina." *Carolina Tips* 4:23.

Brodie, E. D., Jr. 1968. "Investigations on the skin toxin of the red-spotted newt, Notophthalmus viridescens viridescens." American Midland Naturalist 80:276-280.

Brodie, E. D., Jr., R. T. Norwak, and W. R. Harvey. 1979. "The effectiveness of antipredator secretions and behavior of selected salamanders against shrews." *Copeia* 1979:270-274.

Brodman, R. 1995. "Annual variation in breeding success of two syntopic species of *Ambystoma* salamanders." *Journal of Herpetology* 29:111-113.

Brooks, G. R., Jr. 1964. "An analysis of the food habits of the bullfrog, *Rana catesbeiana*, by body size, sex, month and habitat." *Virginia Journal of Science* 20:173-186.

Brooks, G. R., Jr. 1967. "Population ecology of the ground skink, *Lygosoma laterale* [Say]." *Ecological Monographs* 37:71-87.

Brooks, M. 1945. "Notes on amphibians from Bickle's Knob, West Virginia." *Copeia* 1945:231.

Brown, A. 2010. "Maryland's everglades pythons." Unpublished account in the *Maryland Amphibian & Reptile Atlas Newsletter*, December, p. 4.

Brown, A. E. 1901. "A review of the genera and species of American snakes, North of Mexico." *Proceedings of the Academy of Natural Sciences of Philadelphia* 53:10-110.

Brown, E. E. 1956. "Nests and young of the six-lined racerunner *Cnemidophorus sexlineatus* Linnaeus." *Journal of the Elisha Mitchell Scientific Society* 72:30-40.

Brown, E. E. 1979. "Some snake food records from the Carolinas." *Brimleyana* 1:113-124.

Brown, G. P., and R. J. Brooks. 1993. "Sexual and seasonal differences in activity in a northern population of snapping turtles, *Chelydra serpentina*." *Herpetologica* 49:311-318.

Bruce, R. C. 1968. "The role of the Blue Ridge Embayment in the zoogeography of green salamander, *Aneides aeneus*." *Herpetologica* 24:185-194.

Bruce, R. C. 1972a. "The larval life of the red salamander, *Pseudotriton ruber*." *Journal of Herpetology* 6:43-51.

Bruce, R. C. 1972b. "Variation in the life cycle of the salamander *Gyrinophilus porphyriticus*." *Herpetologica* 28:230-245.

Bruce, R. C. 1975. "Reproductive biology of the mud salamander, *Pseudotriton montanus*, in western South Carolina." *Copeia* 1975: 130-137.

Bruce, R. C. 1978. "Reproductive biology of the salamander *Pseudotriton ruber* in the southern Blue Ridge Mountains." *Copeia* 1978: 417-423.

Bruce, R. C. 1980. "A model of the larval period of the spring salamander, *Gyrinophilus porphyriticus*, based on size-frequency distributions." *Herpetologica* 36:78-86.

Brunner, J. L., K. E. Barnett, C. J. Gosier, S. A. McNulty, M. J. Rubbo, and M. B. Kolozsvary. 2011. "Ranavirus infection in dieoffs of vernal pool amphibians in New York, USA." *Herpetological Review* 42:76-79.

Buhlmann, K. A., J. C. Mitchell, and M. G. Rollins. 1997. "New approaches for the conservation of bog turtles, *Clemmys muhlenbergii*, in Virginia." In *Proceedings: Conservation, Restoration, and Management of Tortoises and Turtles—An International Conference*, edited by J. Van Abbema, 359-363. New York: New York Turtle and Tortoise Society/Wildlife Conservation Society Turtle Recovery Program.

Buhlmann, K., T. Tuberville, and W. Gibbons. 2008. *Turtles of the Southeast*. Athens: University of Georgia Press.

Bulte, G. 2000. *Population Ecology of Painted Turtles* (Chrysemys picta picta) *in a Beaver Pond*. Technical Report of the Jug Bay Wetlands Sanctuary. Lothian, MD: Jug Bay Wetlands Sanctuary.

Burbrink, F. T. 2001. "Systematics of the eastern ratsnake complex (*Elaphe obsoleta*)." *Herpetological Monographs* 15:1-53.

Burbrink, F. T., R. Lawson, and J. B. Slowinski. 2000. "Mitochondrial DNA phylogeography of the polytypic North American rat snake (*Elaphe obsoleta*): a critique of the subspecies concept." *Evolution* 54:2107-2118.

Burbrink, F. T., C. A. Phillips, and E. J. Heske. 1998. "A riparian zone in southern Illinois as a potential dispersal corridor for reptiles and amphibians." *Biological Conservation* 86:107-115.

Burke, V. J., J. W. Gibbons, and J. L. Greene. 1994. "Prolonged nesting forays by common mud turtles (*Kinosternon subrubrum*)." *American Midland Naturalist* 131:190-195.

Burke, V. J., S. J. Morreale, and A. G. Rhodin. 1993. "Testudines, *Lepidochelys kempii* Kemp's ridley sea turtle and *Caretta caretta* loggerhead sea turtle diet." *Herpetological Review* 24:31-32.

Burkett, R. D. 1984. "An ecological study of the cricket frog *Acris crepitans*." In *Vertebrate Ecology and Systematics: A Tribute to Henry S. Fitch*, edited by R. A. Seigel, L. E. Hunt, J. L. Knight, L. Malaret, and N. L. Zuschlag, 89-103. Lawrence: Museum of Natural History, University of Kansas.

Burton, T. M. 1976. "An analysis of the feeding ecology of the salamanders (Amphibia: Urodela) of the Hubbard Brook Experimental Forest, New Hampshire." *Journal of Herpetology* 10: 187-204.

Burton, T. M., and G. E. Likens. 1975a. "Energy flow and nutrient cycling in salamander populations in the Hubbard Brook Experimental Forest, New Hampshire." *Ecology* 56:1068-1080.

Burton, T. M., and G. E. Likens. 1975b. "Salamander populations and biomass in the Hubbard Brook Experimental Forest, New Hampshire." *Copeia* 1975:541-546.

Bury, R. B. 1979. *Review of the Ecology and Conservation of the Bog Turtle,* Clemmys muhlenbergii. US Fish and Wildlife Service, Special Scientific Report—Wildlife no. 219. Washington, DC: Government Printing Office.

Bury, R. B., and J. A. Whelan. 1984. *Ecology and Management of the Bullfrog*. US Fish and Wildlife Service, Resource Publication no. 155. Washington, DC: Government Printing Office.

Bushmann, P. 2000. *Abundance and Population Structure of Fishes of Cove Point Marsh*. Prince Frederick, MD: Cove Point Natural Heritage Trust.

Butterfield, B. P., M. J. Lannoo, and P. Nanjappa. 2005. "*Rana sphenocephala* Cope, 1886, southern leopard frog." In *Amphibian Declines: The Conservation Status of United States Species,* edited by M. J. Lannoo, 586-587. Berkeley: University of California Press.

Buxbaum, R. 1942. "The herpetofauna of the Gwynns Falls area." *Natural History Society of Maryland, Junior Division Bulletin* 6:1-5.

Byer, N. W. 2015. "Movement patterns, nesting ecology, and nest-site selection of the federally-listed bog turtle in Maryland." MS thesis. Towson University, Towson, MD.

Cagle, F. R. 1937. "Egg laying habits of the slider turtle (*Pseudemys troostii*), the painted turtle (*Chrysemys picta*), and the musk turtle (*Sternotherus odoratus*)." *Journal of the Tennessee Academy of Sciences* 12:87-95.

Campbell, H. W. 1960. "The bog turtle in Maryland." *Maryland Naturalist* 30:15-16.

Canterbury, R. A., and T. K. Pauley. 1994. "Time of mating and egg deposition of West Virginia populations of the salamander *Aneides aeneus*." *Journal of Herpetology* 28:431-434.

Capps, J. 2002. *Habitat Use and Movements of Spotted Turtles Based on Radio Telemetry*. Technical Report of the Jug Bay Wetlands Sanctuary. Lothian, MD: Jug Bay Wetlands Sanctuary.

Capps, K. 2001. *An Experimental study of Nest Density and Nest Predation Using artificial Painted Turtle Nests and Eggs*. Technical Report of the Jug Bay Wetlands Sanctuary. Lothian, MD: Jug Bay Wetlands Sanctuary.

Carpenter, C. C. 1952. "Comparative ecology of the common garter snake (*Thamnophis s. sirtalis*), the ribbon snake (*Thamnophis s. sauritis*), and Butler's garter snake (*Thamnophis butleri*) in mixed populations." *Ecological Monographs* 22:235-258.

Carpenter, C. C. 1953. "A study of hibernacula and hibernating associations of snakes and amphibians in Michigan." *Ecology* 34: 74-80.

Carr, A. F., Jr. 1940. "A contribution to the herpetology of Florida." *Biological Sciences Series* (University of Florida) 3:1-118.

Carr, A. 1952. *Handbook of Turtles*. Ithaca, NY: Cornell University Press.

Carr, A. F., Jr. 1987. "New perspectives on the pelagic stage of sea turtle development." *Conservation Biology* 1:103-121.

Carr, A. F., and A. B. Meylan. 1980. "Evidence of passive migrations of green turtle hatchlings in *Sargassum*." *Copeia* 1980: 366-368.

Carriere, M., G. Bulte, and G. Blouin-Demers. 2009. "Spatial ecology of northern map turtles (*Graptemys geographica*) in a lotic and a lentic habitat." *Journal of Herpetology* 43:597-604.

Caruso, N. M., and K. R. Lips. 2013. "Truly enigmatic declines in terrestrial salamander populations in Great Smoky Mountains National Park." *Diversity and Distributions* 19:38-48.

Carver, D. 1970. "A new state record! . . . No." *Bulletin of the Maryland Herpetological Society* 6:16.

Casey, R. E., A. N. Shaw, L. R. Massal, and J. W. Snodgrass. 2005. "Multimedia evaluation of trace metal distribution within stormwater retention ponds in suburban Maryland, USA." *Bulletin of Environmental Contamination and Toxicology* 74:273-280.

Casper, G. S. 1996. *Geographic Distributions of the Amphibians and Reptiles of Wisconsin*. Milwaukee, WI: Milwaukee Public Museum.

CBP (Chesapeake Bay Program). 2004. *The State of the Chesapeake Bay and Its Watershed: A Report to the Citizens of the Bay Region*. Washington, DC: US Environmental Protection Agency.

Chalmers, R. J. 2006. *Eastern Mud Salamander* (Pseudotriton montanus) *Status in Maryland*. Annapolis: Maryland Department of Natural Resources.

Chase, J. D., K. R. Dixon, J. E. Gates, D. Jacobs, and G. J. Taylor. 1989. "Habitat characteristics, population size, and home range of the bog turtle, *Clemmys muhlenbergii*, in Maryland." *Journal of Herpetology* 23:356-362.

Chipley, R. M., G. H. Fenwick, and D. D. Boone. 1984. "The Maryland Natural Heritage Program." In *Threatened and Endangered Plants and Animals of Maryland*, edited by A. W. Norden, D. C. Forester, and G. H. Fenwick, 33–42. Natural Heritage Program Special Publication 84-1. Annapolis: Maryland Department of Natural Resources.

Churchill, T. A., and K. B. Storey. 1996. "Organ metabolism and cryoprotectant synthesis during freezing in spring peepers *Pseudacris crucifer*." *Copeia* 1996:517–525.

CITES (Convention on International Trade in Endangered Species of Wild Fauna and Flora). 2017. "Appendices I, II, and III." Online at https://www.cites.org/eng/app/appendices.php.

Clark, A. H. 1930. "Records of the wood tortoise (*Clemmys insculpta*) in the vicinity of the District of Columbia." *Proceedings of the Biological Society of Washington* 43:13–15.

Clark, D. R., Jr. 1976. "Ecological observations on a Texas population of six-lined racerunners, *Cnemidophorus sexlineatus* (Reptilia, Lacertilia, Teiidae)." *Journal of Herpetology* 10:133–138.

Clark, D. R., Jr. 2005. "The strategy of tail-autotomy in the ground skink, *Lygosoma laterale*." *Journal of Experimental Zoology* 176: 295–302.

Clark, D. R., Jr., C. M. Bunck, and R. J. Hall. 1997. "Female reproductive dynamics in a Maryland population of ringneck snakes (*Diadophis punctatus*)." *Journal of Herpetology* 31:476–483.

Clark, J. R. 1997. "Final rule to list the northern population of the bog turtle as threatened and the southern population as threatened due to similarity of appearance." *Federal Register* 62: 59605–59622.

Clark, R. D. 1974. "Activity and movement patterns in a population of Fowler's toad, *Bufo woodhousei fowleri*." *American Midland Naturalist* 92:257–274.

Clark, R. W., W. S. Brown, R. Stechert, and K. R. Zamudio. 2008. "Integrating individual behavior and landscape genetics: the population structure of timber rattlesnake hibernacula." *Molecular Ecology* 17:719–730.

Clark, R. W., W. S. Brown, R. Stechert, and K. R. Zamudio. 2010. "Don't tread on them: roads, interrupted dispersal, and genetic diversity in timber rattlesnakes (*Crotalus horridus*)." *Conservation Biology* 24:1059–1069.

Clark, R. W., M. N. Marchand, B. J. Clifford, R. Stechert, and S. Stephens. 2011. "Decline of an isolated timber rattlesnake (*Crotalus horridus*) population: interactions between climate change, disease, and loss of genetic diversity." *Biological Conservation* 144: 886–891.

Clausen, H. J. 1936. "The effects of aggregation on the respiratory metabolism of the brown snake, *Storeria dekayi*." *Journal of Cellular Physiology* 8:367–386.

Clawson, M. E., and T. S. Baskett. 1982. "Herpetofauna of the Ashland Wildlife Area, Boone County, Missouri." *Transactions of the Missouri Academy of Science* 16:5–16.

Clifford, M. J. 1976. "Relative abundance and seasonal activity of snakes in Amelia County." *Virginia Herpetological Society Bulletin* 79:4–6.

Coale, W. E., J. H. Quimby, and H. R. Hazelhurst. 1838. "Directions for collecting and preserving objects of natural history by the Maryland Academy of Science and Literature." Maryland Historical Society files, Baltimore.

Cochran, J. D. 1978. "A note on the behavior of the diamondback terrapin, *Malaclemys t. terrapin* (Schoepff) in Maryland." *Bulletin of the Maryland Herpetological Society* 14:100.

Collins, J. P., and A. Storfer. 2003. "Global amphibian declines: sorting the hypotheses." *Diversity and Distributions* 9:89–98.

Collins, J. T. 1965. "A population study of *Ambystoma jeffersonianum*." *Journal of the Ohio Herpetological Society* 5:61.

Collins, J. T. 1974. "Observations on reproduction in the southern coal skink (*Eumeces anthracinus pluvialis* Cope)." *Transactions of the Kansas Academy of Science* 77:126–127.

Colton, H. S. 1909. "Peale's Museum." *Popular Science Monthly* 75: 221–238.

Conant, R. 1943a. "The milk snakes of the Atlantic Coastal Plain." *Proceedings of the New England Zoological Club* 22:3–24.

Conant, R. 1943b. "*Natrix erythrogaster erythrogaster* in the northeastern part of its range." *Herpetologica* 2:83–86.

Conant, R. 1945. *An Annotated Check List of the Amphibians and Reptiles of the Del-Mar-Va Peninsula*. Wilmington: Society of Natural History of Delaware.

Conant, R. 1947a. "The carpenter frog in Maryland." *Maryland: A Journal of Natural History* 17:72–73.

Conant, R. 1947b. *Reptiles and Amphibians of the Northeastern States*. Philadelphia: Zoological Society of Philadelphia.

Conant, R. 1957. "The eastern mud salamander, *Pseudotriton montanus montanus*: a new state record for New Jersey." *Copeia* 1957: 52–153.

Conant, R. 1958a. *A Field Guide to the Reptiles and Amphibians of the United States and Canada East of the 100th Meridian*. Boston: Houghton Mifflin.

Conant, R. 1958b. "Notes on the herpetology of the Delmarva Peninsula." *Copeia* 1958:50–52.

Conant, R. 1975. *A Field Guide to the Reptiles and Amphibians of Eastern and Central North America*, 2nd ed. Boston: Houghton Mifflin.

Conant, R. 1993. "The Delmarva Peninsula." *Maryland Naturalist* 37: 7–21.

Conant, R., and J. T. Collins. 1998. *A Field Guide to Reptiles and Amphibians: Eastern and Central North America*, 3rd ed., expanded. Boston and New York: Houghton Mifflin.

Conaway, C. H., and D. E. Metter. 1967. "Skin glands associated with breeding in *Microhyla carolinensis*." *Copeia* 1967:672–673.

Cooper, J. E. 1949. "Additional records for *Clemmys muhlenbergii* from Maryland." *Herpetologica* 5:75–76.

Cooper, J. E. 1953. "Notes on the amphibians and reptiles of southern Maryland." *Maryland Naturalist* 23:90–100.

Cooper, J. E. 1956. "An annotated list of the amphibians and reptiles of Anne Arundel County, Maryland." *Maryland Naturalist* 26:6–8.

Cooper, J. E. 1958. "The snake *Haldea valeriae pulchra* in Maryland." *Herpetologica* 14:121–122.

Cooper, J. E. 1959. "The turtle *Pseudemys scripta* feral in Maryland." *Herpetologica* 15:44.

Cooper, J. E. 1960a. "Another rainbow snake, *Abastor erythrogrammus*, from Maryland." *Chesapeake Science* 1:203–204.

Cooper, J. E. 1960b. "Distributional survey V: Maryland and the District of Columbia." *Bulletin of the Philadelphia Herpetological Society* 8(3):18–24.

Cooper, J. E. 1960c. "Notes on cave-associated vertebrates." *Baltimore Grotto News* 3:10.

Cooper, J. E. 1965a. "Cave-associated herpetozoa. I. An annotated dichotomous key to the adult cave-associated salamanders of Maryland, Pennsylvania, Virginia, and West Virginia." *Baltimore Grotto News* 8:150-163.

Cooper, J. E. 1965b. "Distributional survey: Maryland and the District of Columbia." *Bulletin of the Maryland Herpetological Society* 1(1):3-14. Reprinted from *PHS: Bulletin of the Philadelphia Herpetological Society* 8(3):18-24, 1960, with additions by H. S. Harris, Jr.

Cooper, J. E. 1966. "Errata: 'Snakes of Maryland' by Howard A. Kelly, Audrey W. Davis and H. C. Robinson." Natural History Society of Maryland, Baltimore, 1936. *Bulletin of the Maryland Herpetological Society* 2(3):1-4.

Cooper, J. E. 1969. "A red-bellied water snake from Maryland's western coastal plain." *Journal of Herpetology* 3:185-186.

Cooper, J. E., and F. Groves. 1959. "The rattlesnake, *Crotalus horridus*, in the Maryland piedmont." *Herpetologica* 15:33-34.

Cooper, J. E., and T. Hunt. 1970. "*Hyla femoralis* in Maryland, revisited." *Bulletin of the Maryland Herpetological Society* 6:14-15.

Cooper, W. E., and L. J. Vitt. 1997. "Maximizing male reproductive success in the broad-headed skink (*Eumeces laticeps*): preliminary evidence for mate guarding, size-assortative pairing and opportunistic extra-pair mating." *Amphibia-Reptilia* 18:59-73.

Cope, E. D. 1873. "Sketch of the zoology of Maryland." In *Walling and Gray's New Topographical Atlas of Maryland*, 16-18. Philadelphia: O. W. Gray.

Cope, E. D. 1875. *Check-list of North American Batrachia and Reptilia; with a Systematic List of the Higher Groups, and an Essay on Geographical Distribution. Based on the Specimens Contained in the United States National Museum.* Smithsonian Miscellaneous Collections, vol. 1. Washington, DC: Smithsonian Institution.

Cope, E. D. 1889. "The batrachia of North America." *U.S. National Bulletin*, no. 34. Washington, DC: Smithsonian Institution.

Cope, E. D. 1900. "The crocodilans, lizards, and snakes of North America." In *Report of the U.S. National Museum for 1898*, 153-1270. Washington, DC: Smithsonian Institution.

Cordero, G. A. 2009. "*Clemmys guttata* (spotted turtle) displacement." *Herpetological Review* 40:74.

Cordero, G. A., M. Quinlan, S. Blackstone, and C. W. Swarth. 2012a. "*Kinosternon subrubrum* (eastern mud turtle) overwintering." *Herpetological Review* 43:327.

Cordero, G. A., R. Reeves, and C. W. Swarth. 2012b. "Long distance aquatic movement and home-range size of an eastern mud turtle, *Kinosternon subrubrum*, population in the mid-Atlantic region of the United States." *Chelonian Conservation and Biology* 11:121-124.

Cordero, G. A., and C. W. Swarth. 2010. "Notes on the movement and aquatic behavior of some Kinosternid turtles." *Acta Zoológica Mexicana* 26:233-235.

Costanzo, J. P. 1989. "Effects of humidity, temperature and submergence behavior on survivorship and energy use in hibernating garter snakes, *Thamnophis sirtalis*." *Canadian Journal of Zoology* 67:2486-2492.

Costanzo, J. P., E. L. Richard, and G. R. Ultsch. 2008. "Physiological ecology of overwintering in hatchling turtles." *Journal of Experimental Zoology* 309:297-379.

CREARM (Committee on Rare and Endangered Amphibians and Reptiles of Maryland). 1973. "Endangered amphibians and reptiles of Maryland: a special report." *Bulletin of the Maryland Herpetological Society* 9:42-99.

Croshaw, D. A. 2008. "Eastern worm snake *Carphophis amoenus*." In *Amphibians and Reptiles of Georgia*, edited by J. B. Jensen, C. D. Camp, W. Gibbons, and M. J. Elliott, 328-329. Athens: University of Georgia Press.

Crother, B. I., editor. 2008. *Scientific and Standard English Names of Amphibians and Reptiles of North America North of Mexico*, 6th ed. SSAR Herpetological Circular no. 37. Salt Lake City, UT: Society for the Study of Amphibians and Reptiles.

Crother, B. I., editor. 2017. *Scientific and Standard English Names of Amphibians and Reptiles of North America North of Mexico, with Comments Regarding Confidence in Our Understanding*, 8th ed. SSAR Herpetological Circular no. 43. Salt Lake City, UT: Society for the Study of Amphibians and Reptiles.

Crother, B. I., J. Boundy, F. T. Burbrink, J. A. Campbell, and R. A. Pyron. 2012. "Squamata—snakes." In *Scientific and Standard English Names of Amphibians and Reptiles of North America North of Mexico, with Comments Regarding Confidence in Our Understanding*, 7th ed., edited by B. I. Crother, 52-72. SSAR Herpetological Circular no. 39. Salt Lake City, UT: Society for the Study of Amphibians and Reptiles.

Crother, B. I., J. Boundy, F. T. Burbrink, and S. Ruane. 2017. "Squamata (in part)—snakes." In *Scientific and Standard English Names of Amphibians and Reptiles of North America North of Mexico, with Comments Regarding Confidence in Our Understanding*, 8th ed., edited by B. I. Crother, 59-80. SSAR Herpetological Circular no. 43. Salt Lake City, UT: Society for the Study of Amphibians and Reptiles.

Crumrine, P., and S. Bartimo. 1997a. *A summary of Turtle Trapping: Mark's Pond, Jug Bay Wetlands Sanctuary.* Technical Report of the Jug Bay Wetlands Sanctuary. Lothian, MD: Jug Bay Wetlands Sanctuary.

Crumrine, P., and S. Bartimo. 1997b. "Two dozen turtles in one little pond." *Marsh Notes: Newsletter of the Jug Bay Wetlands Sanctuary* 12:3.

Cunningham, H. R., L. J. Rissler, and J. J. Apodaca. 2009. "Competition at the range boundary in the slimy salamander: using reciprocal transplants for studies on the role of biotic interactions in spatial distributions." *Journal of Animal Ecology* 78: 52-62.

Cunningham, H. R., and W. H. Smith. 2010. "*Plethodon glutinosus* reproduction and behavior." *Herpetological Review* 41:190-191.

Cupp, P. V., Jr. 1991. "Aspects of the life history and ecology of the green salamander, *Aneides aeneus*." *Journal of the Tennessee Academy of Sciences* 66:171-174.

Cupp, P. V., Jr. 1994. "Salamanders avoid chemical cues from predators." *Animal Behaviour* 48:232-235.

Curless, S. 2001. *A Red-bellied Turtle* (Pseudemys rubriventris) *nesting behavior study.* Technical Report of the Jug Bay Wetlands Sanctuary. Lothian, MD: Jug Bay Wetlands Sanctuary.

Currylow, A. F., A. J. Johnson, and R. N. Williams. 2014. "Evidence of ranavirus infections among sympatric larval amphibians and box turtles." *Journal of Herpetology* 48:117-121.

Czarnowsky, R. 1975. "A new county record for *Gastrophryne car-*

olinensis in Maryland." *Bulletin of the Maryland Herpetological Society* 11:185-186.

Danstedt, R. T., Jr. 1975. "Local geographic variation in demographic parameters and body size of *Desmognathus fuscus* (Amphibia: Plethodontidae)." *Ecology* 56:1054-1067.

Danstedt, R. T., Jr. 1979. "A demographic comparison of two populations of the dusky salamander (*Desmognathus fuscus*) in the same physiographic province." *Herpetologica* 35:164-168.

Davis-Chase, J. 1983. "Habitat characteristics, population size, and home range of the bog turtle, *Clemmys muhlenbergii*, in Maryland." MS thesis. Frostburg State University, Frostburg, MD.

Davis Rabosky, A. R., C. L. Cox, D. L. Rabosky, P. O. Title, I. A. Holmes, A. Feldman, and J. A. McGuire. 2016. "Coral snakes predict the evolution of mimicry across New World snakes." *Nature Communications* 7:11484.

Dawson, S. A. 1984. "The status of the bog turtle (*Clemmys muhlenbergi*) in Maryland." In *Threatened and Endangered Plants and Animals of Maryland*, edited by A. W. Norden, D. C. Forester, and G. H. Fenwick, 360-362. Natural Heritage Program Special Publication 84-1. Annapolis: Maryland Department of Natural Resources.

DeGregorio, B. 2002. *Basking Behavior of Queen Snakes and Northern Water Snakes in a Tidal Environment*. Technical Report of the Jug Bay Wetlands Sanctuary. Lothian, MD: Jug Bay Wetlands Sanctuary.

Delzell, D. E. 1958. "Spatial movement and growth of *Hyla crucifer*." PhD dissertation. University of Michigan, Ann Arbor.

deMaynadier, P. G., and M. L. Hunter, Jr. 1998. "Effects of silvicultural edges on the distribution and abundance of amphibians in Maine." *Conservation Biology* 12:340-352.

Dennis, D. M. 1962. "Notes on the nesting habits of *Desmognathus fuscus* (Raf.) in Licking Co., Ohio." *Journal of the Ohio Herpetological Society* 3:28-35.

Denton Journal. 1878. "Scraps." *Denton Journal* 30(5):3.

de Quieroz, K., T. W. Reeder, and A. D. Leaché. 2017. "Squamata (in part)—lizards." In *Scientific and Standard English Names of Amphibians and Reptiles of North America North of Mexico, with Comments Regarding Confidence in Our Understanding*, 8th ed., edited by B. I. Crother, 35-55. SSAR Herpetological Circular no. 43. Salt Lake City, UT: Society for the Study of Amphibians and Reptiles.

Dickerson, M. C. 1906. *The Frog Book: North American Toads and Frogs, with a Study of the Habits and Life Histories of Those in the Northeastern States*. New York: Doubleday, Page & Company.

Dickey, R., S. Matthews, and C. W. Swarth. 2009. "*Pseudemys rubriventris* (northern red-bellied cooter), hatchling behavior." *Herpetological Review* 40:337-338.

Dickson, J. D., III. 1948. "Observations on the feeding habits of the scarlet snake." *Copeia* 1948:216-217.

Dinkelacker, S. A. 2000. "Microhabitat selection and spatial ecology of syntopic *Clemmys muhlenbergii*, *Clemmys guttata* and *Chelydra serpentina* in Maryland." MS thesis. Frostburg State University, Frostburg, MD.

Ditmars, R. L. 1907. *The Reptile Book: A Comprehensive Popularized Work on the Structure and Habits of the Turtles, Tortoises, Crocodilians, Lizards and Snakes Which Inhabit the United States and Northern Mexico*. New York: Doubleday, Page & Company.

Ditmars, R. L. 1939. *A Field Book of North American Snakes*. New York: Doubleday & Company.

Ditmars, R. L. 1944. *Serpents of the Northeastern States: A Guide to the Venomous and Non-venomous Species of the North Atlantic and New England Areas*. New York: New York Zoological Society.

Dodd, C. K., Jr. 1988. "Synopsis of the biological data on the loggerhead sea turtle *Caretta caretta* (Linnaeus 1758)." *U.S. Fish and Wildlife Service Biological Report* 88(14).

Dodd, C. K., Jr. 2001. *North American Box Turtles: A Natural History*. Norman: University of Oklahoma Press.

Dodd, C. K., Jr. 2003. *Monitoring Amphibians in Great Smokey Mountains National Park*. US Geological Survey Circular no. 1258. Tallahassee, FL: USGS.

Dodd, C. K., Jr. 2004. *The Amphibians of Great Smoky Mountains National Park*. Knoxville: University of Tennessee Press.

Dodd, C. K., Jr. 2013a. *Frogs of the United States and Canada*, vol. 1. Baltimore: Johns Hopkins University Press.

Dodd, C. K., Jr. 2013b. *Frogs of the United States and Canada*, vol. 2. Baltimore: Johns Hopkins University Press.

Dodd, C. K., Jr., M. L. Griffey, and J. D. Croser. 2001. "The cave associated amphibians of the Great Smoky Mountains National Park: review and monitoring." *Journal of Elisha Mitchell Scientific Society* 117:139-149.

Dorcas, M., and W. Gibbons. 2008. *Frogs and Toads of the Southeast*. Athens: University of Georgia Press.

Douglas, M. E. 1979. "Migration and sexual selection in *Ambystoma jeffersonianum*." *Canadian Journal of Zoology* 57:2303-2310.

Downs, F. L. 1989. "Family Ambystomatidae." In *Salamanders of Ohio*, edited by R. A. Pfingsten and F. L. Downs, 87-172. Ohio Biological Survey Bulletin, New Series 7, no. 2. Columbus: College of Biological Sciences, Ohio State University.

Dreher, S. 1999. "The basking behavior of the red-bellied turtle (*Pseudemys rubriventris*) and the painted turtle (*Chrysemys picta*) on the Patuxent River in Maryland." Senior thesis. College of Wooster, Wooster, MA.

Ducatel, J. T. 1837. "Outline of the physical geography of Maryland, embracing its prominent geological features." *Transactions of the Maryland Academy of Science and Literature* 2:24-54.

Dukehart, J. P. 1884. "Transfer of soft-shell terrapin from Ohio to the Potomac River." *Bulletin of the U.S. Fishery Commission* 4:143.

Dundee, H. A. 2001. "The etymological riddle of the ridley sea turtle." *Marine Turtle Newsletter* 58:10-12.

Dundee, H. A., and D. S. Dundee. 1965. "Observations on the systematics and ecology of *Cryptobranchus* from the Ozark Plateaus of Missouri and Arkansas." *Copeia* 1965:369-70.

Dunn, A. M., and M. A. Weston. 2008. "A review of bird atlases of the world and their application." *Emu* 108:42-67.

Dunn, E. R. 1916. "Two new salamanders of the genus *Desmognathus*." *Proceedings of the Biological Society of Washington* 29:73-76.

Dunn, E. R. 1937. "The status of *Hyla evittata* Miller." *Proceedings of the Biological Society of Washington* 50:9-10.

Dunson, W. A. 1986. "Estuarine populations of the snapping turtle (*Chelydra*) as a model for the evolution of marine adaptations in reptiles." *Copeia* 1986:741-756.

Durner, G. M. 1991. "Home range and habitat use of black rat snakes on Remington Farms, Maryland." MS thesis. Frostburg State University, Frostburg, MD.

Durner, G. M., and J. E. Gates. 1991. "Home range and habitat use of black rat snakes on Remington Farms, MD." Paper presented at 1991 Joint Annual Meeting, 34th Annual Meeting of the Society for the Study of Amphibians and Reptiles (SSAR) and 39th Annual Meeting of the Herpetologists' League, University Park, PA.

Durner, G. M., and J. E. Gates. 1993. "Spatial ecology of black rat snakes on Remington Farms, Maryland." *Journal of Wildlife Management* 57:812-826.

Dyrkacz, S. 1981. "Recent instances of albinism in North American amphibians and reptiles." *SSAR Herpetological Circular*, no. 11.

Earll, R. E. 1887. "Part X. Maryland and its fisheries." In *The Fisheries and Fishery Industries of the United States. Section II: A Geographical Review of the Fisheries Industries and Fishing Communities for the Year 1880*, edited by G. B. Goode and staff, 421-448. Washington, DC: Government Printing Office.

Edgren, R. A., Jr. 1955. "The natural history of the hog-nosed snakes, genus *Heterodon*: a review." *Herpetologica* 11:105-117.

Edwards, J., Jr. 1981. *A Brief Description of the Generalized Geology of Maryland*. Baltimore: Maryland Geological Survey.

Eigenbrod, F., S. J. Hecnar, and L. Fahrig. 2008. "The relative effects of road traffic and forest cover on anuran populations." *Biological Conservation* 141:35-46.

Elick, G. E., and J. A. Sealander. 1972. "Comparative water loss in relation to habitat selection in small colubrid snakes." *American Midland Naturalist* 88:429-439.

Elliott, L., C. Gerhardt, and C. Davidson. 2009. *The Frogs and Toads of North America: A Comprehensive Guide to Their Identification, Behavior, and Calls*. New York: Houghton Mifflin Harcourt.

Ellison, W. G., editor. 2010. *Second Atlas of the Breeding Birds of Maryland and the District of Columbia*. Baltimore: Johns Hopkins University Press.

Enge, K. M., and J. D. Sullivan. 2000. "Seasonal activity of the scarlet snake, *Cemophora coccinea*, in Florida." *Herpetological Review* 31:82-83.

EPA (Environmental Protection Agency). 2016. *What Climate Change Means for Maryland*. Environmental Protection Agency 430-F-16-022. Washington, DC: EPA.

Ernst, C. H. 1977. "Biological notes on the bog turtle, *Clemmys muhlenbergii*." *Herpetologica* 33:241-246.

Ernst, C. H. 2001. Some ecological parameters of the wood turtle, Clemmys insculpta, in southeastern Pennsylvania. Chelonian Conservation and Biology 4:94-99.

Ernst, C. H., T. S. B. Akre, J. C. Wilgenbusch, T. P. Wilson, and K. Mills. 1999. "Shell disease in turtles in the Rappahannock River, Virginia." *Herpetological Review* 30:214-215.

Ernst, C. H., and R. W. Barbour. 1972. *Turtles of the United States*. Lexington: University Press of Kentucky.

Ernst, C. H., and R. W. Barbour. 1989. *Snakes of Eastern North America*. Fairfax, VA: George Mason University Press.

Ernst, C. H., S. C. Belfit, S. W. Sekscienski, and A. F. Laemmerzahl. 1997. "The amphibians and reptiles of Ft. Belvoir and northern Virginia." *Bulletin of the Maryland Herpetological Society* 33:1-62.

Ernst, C. H., and E. M. Ernst. 2003. *Snakes of the United States and Canada*. Washington, DC, and London: Smithsonian Books.

Ernst, C. H., and M. J. Gilroy. 1979. "Are leatherback turtles, *Der-mochelys coriacea*, common along the middle Atlantic coast?" *Bulletin of the Maryland Herpetological Society* 15:16-19.

Ernst, C. H., and J. E. Lovich. 2009. *Turtles of the United States and Canada*. Baltimore: Johns Hopkins University Press.

Ernst, C. H., J. E. Lovich, and R. W. Barbour. 1994. *Turtles of the United States and Canada*. Washington, DC: Smithsonian Institution Press.

Ernst, C. H., R. T. Zappalorti, and J. E. Lovich. 1989. "Overwintering sites and thermal relations of hibernating bog turtles, *Clemmys muhlenbergii*." *Copeia* 1989:761-764.

Etheridge, K. L., L. C. Wit, and J. C. Sellers. 1983. "Hibernation in the lizard *Cnemidophorus sexlineatus*." *Copeia* 1983:206-214.

Etheridge, K. L., L. C. Wit, J. C. Sellers, and S. E. Trauth. 1986. "Seasonal changes in reproductive condition and energy stores in *Cnemidophorus sexlineatus*." *Journal of Herpetology* 20:554-559.

Evans, J., A. Norden, F. Cresswell, K. Insley, and S. Knowles. 1997. "Sea turtle strandings in Maryland, 1991 through 1995." *Maryland Naturalist* 41:23-34.

Evans, P. D. 1942. "A method of fishing used by water snakes." *Chicago Naturalist* 5:53-55.

Faccio, S. D. 2003. "Postbreeding emigration and habitat use by Jefferson and spotted salamanders in Vermont." *Journal of Herpetology* 37:479-489.

Fairchild, L. 1984. "Male reproductive tactics in an explosive breeding toad population." *American Zoology* 24:407-418.

Farnsworth, S. D., and R. A. Seigel. 2013. "Responses, movements, and survival of relocated box turtles during the construction of the Inter-County Connector Highway in Maryland." *Journal of the Transportation Research Board* 2362:1-8.

Farrell, J. L., and T. E. Graham. 1991. "Ecological notes on the turtle *Clemmys insculpta* in northwestern New Jersey." *Journal of Herpetology* 25:1-9.

Feinberg, J. A., and R. L. Burke. 2003. "Nesting ecology and predation of diamondback terrapins, *Malaclemys terrapin*, at Gateway National Recreation Area, New York." *Journal of Herpetology* 37: 517-526.

Feinberg, J. A., C. E. Newman, G. J. Watkins-Colwell, M. D. Schlesinger, B. Zarate, B. R. Curry, H. B. Schaffer, and J. Burger. 2014. "Cryptic diversity in metropolis: confirmation of a new leopard frog species (Anura: Ranidae) from New York City and surrounding Atlantic coast regions." *PLoS ONE* 9(10): e108213. doi:10.1371/journal.pone.0108213.

Fellers, G. M. 1979. "Mate selection in the gray treefrog, *Hyla versicolor*." *Copeia* 1979:286-290.

Ferebee, K. B., and P. F. P. Henry. 2008. "Movements and distribution of *Terrapene carolina* in a large urban area, Rock Creek National Park, Washington, D.C." In *Urban Herpetology*, edited by J. C. Mitchell, R. E. Jung, and B. Bartholomew, 365-372. Salt Lake City, UT: Society for the Study of Amphibians and Reptiles.

Fisher, W. H. 1932. "Maryland snakes checklist." Unpublished manuscript. Dr. Howard Kelly file, Natural History Society of Maryland Archives, Baltimore.

Fitch, H. S. 1958. "A natural history of the six-lined racerunner (*Cnemidophorus sexlineatus*)." *University of Kansas Publications, Museum of Natural History* 11:11-62.

Fitch, H. S. 1963. "Natural history of the racer *Coluber constrictor*."

University of Kansas Publications, Museum of Natural History 15: 351-468.

Fitch, H. S., and A. V. Fitch. 1967. "Physical tolerances of eggs of lizards and snakes." *Ecological Society of America* 48:160-165.

Fitch, H. S., and P. L. von Achen. 1977. "Spatial relationships and seasonality in the skinks *Eumeces fasciatus* and *Scincella laterale.*" *Herpetologica* 33:303-313.

Flageole, S., and R. Leclair, Jr. 1992. "Demographic study of a yellow-spotted salamander (*Ambystoma maculatum*) population by means of the skeleton-chronological method." *Canadian Journal of Zoology* 70:740-747.

Fogel, M. L., J. Sage, and C. W. Swarth. 2002. "Complex dietary analysis of red-bellied turtles from hatchling to adult based on stable isotope analysis." Abstract. Isotopes in Ecology Conference, Flagstaff, AZ.

Foley, D. H., III, and S. A. Smith. 1999. *Comparison of Two Herpetofaunal Inventory Methods and an Evaluation of Their Use in a Volunteer-based Statewide Reptile and Amphibian Atlas Project.* Final report. Wye Mills: Maryland Department of Natural Resources.

Fontanella, F. M., C. R. Feldman, M. E. Siddall, and F. T. Burbrink. 2008. "Phylogeography of *Diadophis punctatus*: extensive lineage diversity and repeated patterns of historical demography in a trans-continental snake." *Molecular Phylogenetics and Evolution* 46:1049-1070.

Fontenot, C. L., Jr. 2008. "Variation in pupil diameter in North American gartersnakes (*Thamnophis*) is regulated by immersion in water, not by light intensity." *Vision Research* 48:1663-1669.

Force, E. R. 1925. "Notes on reptiles and amphibians of Okmulgee County, Oklahoma." *Copeia* 141:25-27.

Forester, D. C. 1992. "The importance of saturated forested wetlands to the perpetuation of amphibian populations." In *Proceedings of a Workshop on Saturated Forested Wetlands in the Mid-Atlantic Region: The State of the Science*, 16-23. Annapolis, MD: US Fish and Wildlife Service.

Forester, D. C. 1999. "C & O Canal National Historic Site amphibian survey: Georgetown (mile post 0) to the Appalachian Trail (mile post 60)." Report to the C & O Canal National Historic Site, Hagerstown, MD.

Forester, D. C. 2000. "Amphibian inventory Chesapeake and Ohio Canal and National Historic Park." Report to the National Park Service, Washington, DC.

Forester, D. C., M. Cameron, and J. D. Forester. 2008. "Nest and egg recognition by salamanders in the genus *Desmognathus*: a comprehensive re-examination." *Ethology* 114:965-971.

Forester, D. C., S. Knoedler, and R. Sanders. 2003. "Life history and status of the mountain chorus frog (*Pseudacris brachyphona*) in Maryland." *Maryland Naturalist* 46:1-15.

Forester, D. C., and D. La Pasha. 1982. "Failure of orientation to frog calls by migrating spotted salamanders." *Bulletin of the Maryland Herpetological Society* 18:143-151.

Forester, D. C., and D. V. Lykens. 1991. "Age structure in a population of red-spotted newts from the Allegheny Plateau of Maryland." *Journal of Herpetology* 25:373-376.

Forester, D. C., and R. W. Miller. 1992. "Conducting a survey and documenting the occurrence of rare herptiles at the Naval Surface Warfare Center, Indian Head Division, Indian Head, Mary-

land." Final report to the Maryland Department of Natural Resources, Natural Heritage Program, Annapolis, MD.

Forester, D. C., J. W. Snodgrass, K. Marsalek, and Z. Lanham. 2006. "Post-breeding dispersal and summer home range of female American toads (*Bufo americanus*)." *Northeastern Naturalist* 13: 59-72.

Forester, D. C., and K. Thompson. 1998. "Gauntlet behavior as a male sexual strategy in *Bufo americanus* (Amphibia: Bufonidae)." *Behaviour* 135:99-119.

Fowler, H. W. 1915. "Some amphibians and reptiles of Cecil County, Maryland." *Copeia* 23:37-40.

Fowler, H. W. 1925. "Records of amphibians and reptiles for Delaware, Maryland and Virginia. II. Maryland." *Copeia* 145:61-64.

Fowler, J. A. 1941. "The occurrence of *Pseudotriton montanus montanus* in Maryland." *Copeia* 1941:181.

Fowler, J. A. 1943a. "Another false map turtle from the District of Columbia vicinity." *Proceedings of the Biological Society of Washington* 56:168.

Fowler, J. A. 1943b. "A distributional study of the amphibians and reptiles of the District of Columbia and vicinity with emphasis on physiography." MA thesis. George Washington University, Washington, DC.

Fowler, J. A. 1944. "A collection of salamanders from Allegany County, Maryland." *Maryland: A Journal of Natural History* 14: 90-92.

Fowler, J. A. 1945. *The Amphibians and Reptiles of the National Capital Parks and the District of Columbia Region*. Washington, DC: US Department of Interior National Park Service.

Fowler, J. A. 1946. "The eggs of *Pseudotriton montanus montanus.*" *Copeia* 1946:105.

Fowler, J. A. 1947a. "The hellbender (*Cryptobranchus alleganiensis*) in Maryland." *Maryland: A Journal of Natural History* 17:14-17.

Fowler, J. A. 1947b. "Record for *Aneides aeneus* in Virginia." *Copeia* 1947:144.

Fowler, J. A. 1947c. "The upland chorus frog in Maryland." *Maryland: A Journal of Natural History* 17:73, 76-77.

Fowler, J. A. 1952. "The eggs of *Plethodon dixi.*" *National Speleological Society Bulletin* 14:61.

Fowler, J. A. 1969. "A note concerning the presumed occurrence of *Hyla femoralis* in Maryland." *Bulletin of the Maryland Herpetological Society* 5:80-81.

Fowler, J. A., and G. Orton. 1947. "The occurrence of *Hyla femoralis* in Maryland." *Maryland: A Journal of Natural History* 17:6-7.

Fowler, J. A., and C. J. Stine. 1953. "A new county record for *Microhyla carolinensis carolinensis* in Maryland." *Herpetologica* 9: 167-168.

Franz, R. 1964. "Eggs of the long-tailed salamander from a Maryland cave." *Herpetologica* 20:216.

Franz, R. 1967a. "Notes on the long-tailed salamander, *Eurycea longicauda* (Green), in Maryland caves." *Bulletin of the Maryland Herpetological Society* 3:1-6.

Franz, R. 1972. "Dusky salamanders and the Alleghany Front of Maryland." *Bulletin of the Maryland Herpetological Society* 8:61-64.

Franz, R., and H. Harris. 1965. "Mass transformation and movement of larval long-tailed salamanders, *Eurycea longicauda longicauda* (Green)." *Journal of the Ohio Herpetological Society* 5:32.

Fraser, D. F. 1976a. "Coexistence of salamanders of the genus

Plethodon: a variation on the Santa Rosalia theme." *Ecology* 57: 238-251.

Fraser, D. F. 1976b. "Empirical evaluation of the hypothesis of food competition in salamanders of the genus *Plethodon*." *Ecology* 57:459-471.

Frazer, N. B., J. W. Gibbons, and J. L. Greene. 1991. "Life history and demography of the common mud turtle *Kinosternon subrubrum* in South Carolina, USA." *Ecology* 72:2218-2231.

Frese, D. P. 1996. *Maryland Manual: A Guide to Maryland Government*. Annapolis: Maryland State Archives.

Frick, G. F., J. L. Reveal, C. R. Broom, and M. L. Brown. 1987. "The practice of Dr. Andrew Scott of Maryland and North Carolina." *Maryland Historical Magazine* 82:123-141.

Gage, S. H. 1891. "The life history of the vermilion-spotted newt (*Diemyctylus viridescens* Raf.)." *American Naturalist* 25:1084-1103.

Galbraith, D. A., and R. J. Brooks. 1989. "Age estimates for snapping turtles." *Journal of Wildlife Management* 53:502-508.

Galbraith, D. A., B. N. White, R. J. Brooks, and P. T. Boag. 1993. "Multiple paternity in clutches of snapping turtles (*Chelydra serpentina*) detected using DNA fingerprints." *Canadian Journal of Zoology* 71:318-324.

Gallagher, M. T., J. W. Snodgrass, A. B. Brand, R. E. Casey, S. M. Lev, and R. J. Van Meter. 2014. "The role of pollutant accumulation in determining the use of stormwater ponds by amphibians." *Wetlands Ecology and Management* 5:551-564.

Garber, S. D., and J. Burger. 1995. "A 20-year study documenting the relationship between turtle decline and human recreation." *Ecological Applications* 5:1151-1162.

Garman, H. 1892. "A synopsis of the reptiles and amphibians of Illinois." *Illinois State Laboratory of Natural History Bulletin* 3:215-388.

Garman, S. 1883. *The Reptiles and Batrachians of North America: Kentucky Geological Survey Stereotyped for the Survey by Major, Johnston & Barrett.* Frankfort, KY: Yeoman Press.

Garton, J. S., and R. A. Brandon. 1975. "Reproductive ecology of the green treefrog, *Hyla cinerea*, in southern Illinois (Anura: Hylidae)." *Herpetologica* 31:150-161.

Gates, J. E., C. H. Hocutt, and J. R. Stauffer, Jr. 1984. "The status of the hellbender (*Cryptobranchus alleganiensis*) in Maryland." In *Threatened and Endangered Plants and Animals of Maryland*, edited by A. W. Norden, D. C. Forester, and G. H. Fenwick, 329-335. Natural Heritage Program Special Publication 84-1. Annapolis: Maryland Department of Natural Resources.

Gates, J. E., C. H. Hocutt, J. R. Stauffer, Jr., and G. J. Taylor. 1985a. "The distribution and status of *Cryptobranchus alleganiensis* in Maryland." *Herpetological Review* 16:17-18.

Gates, J. E., R. H. Stouffer, Jr., J. R. Stauffer, Jr., and C. H. Hocutt. 1985b. "Dispersal patterns of translocated *Cryptobranchus alleganiensis* in a Maryland stream." *Journal of Herpetology* 19:436-438.

Gates, J. E., and E. L. Thompson. 1981. "Breeding habitat association of spotted salamanders (*Ambystoma maculata*) in western Maryland." *Journal of the Elisha Mitchell Scientific Society* 97:209-216.

Gauza, R., and D. Smith. 2010. *Maryland Amphibian and Reptile Atlas (MARA) Training Handbook*. Annapolis: Maryland Department of Natural Resources, and Baltimore: Natural History Society of Maryland.

Gayou, D. C. 1984. "Effects of temperature on the mating call of *Hyla versicolor*." *Copeia* 1984:733-738.

Geddes, T. H. 1981. "An annotated checklist to the snakes of the Moyaone Reserve, Prince George's County, and Charles County, Maryland." *Bulletin of the Maryland Herpetological Society* 17:93-101.

Gehlbach, F. R. 1970. "Death-feigning and erratic behavior in Leptotyphlopid, Colubrid, and Elapid snakes." *Herpetologica* 26: 24-34.

Gerhardt, H. C. 1982. "Sound pattern recognition in some North American treefrogs (Anura: Hylidae): implications for mate choice." *American Zoologist* 22:581-595.

Gerhardt, H. C. 1994. "Reproductive character displacement of female mate choice in the gray treefrog, *Hyla chrysoscelis*." *Animal Behaviour* 47:959-969.

Gerhardt, H. C. 2005. "Advertisement-call preferences in diploid-tetraploid treefrogs (*Hyla chrysoscelis* and *Hyla versicolor*): implications for mate choice and the evolution of communication systems." *Evolution* 59:395-408.

Giannini, T. 2010. *Analysis of Eastern Mud Turtle* (Kinosternon subrubrum subrubrum) *and Common Musk Turtle* (Sternotherus odoratus) *Diet in a Beaver Pond and Tidal Marsh Using Stable Isotope Techniques.* Technical Report of the Jug Bay Wetlands Sanctuary. Lothian, MD: Jug Bay Wetlands Sanctuary.

Gibbons, J. W., and M. E. Dorcas. 2004. *North American Watersnakes: A Natural History*. Norman: University of Oklahoma Press.

Gibbons, J. W., V. J. Burke, J. E. Lovich, R. D. Semlitsch, T. D. Tuberville, J. R. Bodie, J. L. Greene, P. H. Niewiarowski, H. H. Whiteman, D. E. Scott, J. H. K. Pechmann, C. R. Harrison, S. H. Bennett, J. D. Krenz, M. S. Mills, K. A. Buhlmann, J. R. Lee, R. A. Seigel, A. D. Tucker, T. M. Mills, T. Lamb, M. E. Dorcas, J. D. Congdon, M. H. Smith, D. H. Nelson, M. B. Dietsch, H. G. Hanlin, J. A. Ott, and D. J. Karapatakis. 1997. "Perceptions of species abundance, distribution, and diversity: lessons from four decades of sampling on a government-managed reserve." *Environmental Management* 21:259-268.

Gibbons, J. W., J. W. Coker, and T. M. Murphy, Jr. 1977. "Selected aspects of the life history of the rainbow snake (*Farancia erytrogramma*)." *Herpetologica* 33:276-281.

Gibbons, J. W., J. Greene, and T. Mills. 2009. *Lizards and Crocodilians of the Southeast*. Athens: University of Georgia Press.

Gibbons, J. W., D. E. Scott, T. J. Ryan, K. A. Buhlmann, T. D. Tuberville, B. S. Metts, J. L. Greene, T. Mills, Y. Leiden, S. Poppy, and C. T. Winne. 2000. "The global decline of reptiles, déjà vu amphibians." *Bioscience* 50:653-666.

Gibbons, J. W., and R. D. Semlitsch. 1991. *Guide to the Reptiles and Amphibians of the Savannah River Site*. Athens: University of Georgia Press.

Gibbons, W., and M. Dorcas. 2005. *Snakes of the Southeast*. Athens: University of Georgia Press.

Gibbs, J. P. 1998a. "Amphibian movements in response to forest edges, roads, and streambeds in southern New England." *Journal of Wildlife Management* 62:584-589.

Gibbs, J. P. 1998b. "Distribution of woodland amphibians along a forest fragmentation gradient." *Landscape Ecology* 13:263-268.

Gibbs, J. P., A. R. Breisch, P. K. Ducey, G. Johnson, J. L. Behler, and

R. Bothner. 2007. *The Amphibians and Reptiles of New York State: Identification, Natural History, and Conservation*. New York: Oxford University Press.

Gibbs, J. P., and D. A. Steen. 2005. "Trends in sex ratios of turtles in the United States: implications of road mortality." *Conservation Biology* 19:552-556.

Gibson, J. D., and P. Sattler. 2008. "Amphibians and reptiles." In A. V. Evans, "The 2006 Potomac Gorge Bioblitz." *Banisteria* 32: 63-68.

Giery, S. T., and R. S. Ostfeld. 2007. "The role of lizards in the ecology of Lyme disease in two endemic zones of the northeastern United States." *Journal of Parasitology* 93:511-517.

Gill, D. E. 1979. "Density dependence and homing behavior in adult red-spotted newts *Notophthalmus viridescens* (Rafinesque)." *Ecology* 60:800-813.

Gill, D. E. 1985. "Interpreting breeding patterns from census data: solution to the Husting dilemma." *Ecology* 66:344-354.

Given, M. F. 1987. "Vocalizations and acoustic interactions of the carpenter frog, *Rana virgatipes*." *Herpetologica* 43:467-481.

Given, M. F. 1988. "Territoriality and aggressive interactions of male carpenter frogs, *Rana virgatipes*." *Copeia* 1988:411-421.

Given, M. F. 1990. "Spatial distribution and vocal interaction in *Rana clamitans* and *R. virgatipes*." *Journal of Herpetology* 24:377-382.

Given, M. F. 1999. "Distribution records of *Rana virgatipes* and associated anuran species along Maryland's eastern shore." *Herpetological Review* 30:144-146.

Godfrey, A. E., and E. T. Cleaves. 1991. "Landscape analysis: theoretical considerations and practical needs." *Environmental Geology and Water Sciences* 17:141-155.

Godfrey, M. H., and N. Mrosovsky. 1997. "Estimating the time between hatching of sea turtles and their emergence from the nest." *Chelonian Conservation Biology* 2:581-585.

Godley, J. S., R. W. McDiarmid, and N. N. Rojas. 1984. "Estimating prey size and number in crayfish-eating snakes, genus *Regina*." *Herpetologica* 40:82-88.

Goin, C. J., and O. B. Goin. 1962. *Introduction to Herpetology*. San Franciso: W. H. Freeman and Company.

Goodwin, O. K., and J. T. Wood. 1953. "Note on egg-laying of the four-toed salamander, *Hemidactylium scutatum* (Schlegel), in eastern Virginia." *Virginia Journal of Science* 4:65-66.

Gore, J. A. 1983. "The distribution of desmognathine larvae (Amphibia: Plethodontidae) in coal surface impacted streams of the Cumberland Plateau, USA." *Journal of Freshwater Ecology* 2: 13-23.

Gordon, R. E. 1952. "A contribution to the life history and ecology of the plethodontids salamander *Aneides aeneus* (Cope and Packard)." *American Midland Naturalist* 47:666-701.

Gordon, R. E., and R. L. Smith. 1949. "Notes on the life history of the salamander *Aneides aeneus*." *Copeia* 1949:173-175.

Gosner, K. L., and I. H. Black. 1957. "The effects of acidity on the development and hatching of New Jersey frogs." *Ecology* 38: 256-262.

Gotte, S. W. 1988. "Nest site selection in the snapping turtle, mud turtle, and painted turtle." MS thesis. George Mason University, Fairfax, VA.

Gotte, S. W. 1992. "Testudines: *Chrysemys picta picta* (eastern painted turtle): predation." *Herpetological Review* 23:80.

Graham, S. 1973. "The first record of *Caretta caretta caretta* nesting on a Maryland beach." *Bulletin of the Maryland Herpetological Society* 9:24-26.

Grant, E. H. C., L. L. Bailey, J. L. Ware, and K. L. Duncan. 2008. "Prevalence of the amphibian pathogen *Batrachochytrium dendrobatidis* in stream and wetland amphibians in Maryland, USA." *Applied Herpetology* 5:233-241.

Gray, R. H. 1983. "Seasonal, annual and geographic variation in color morph frequencies of the cricket frog, *Acris crepitans*, in Illinois." *Copeia* 1983:300-311.

Gray, R. H., L. E. Brown, and L. Blackburn. 2005. "*Acris crepitans* Baird, 1854(b), northern cricket frog." In *Amphibian Declines: The Conservation Status of United States Species*, edited by M. J. Lannoo, 441-443. Berkeley: University of California Press.

Green, D. M., L. A. Weir, G. S. Casper, and M. J. Lannoo. 2013. *North American Amphibians: Distribution and Diversity*. Berkeley and Los Angeles: University of California Press.

Green, N. B., and T. K. Pauley. 1987. *Amphibians and Reptiles in West Virginia*. Pittsburgh: University of Pittsburgh Press.

Grizzell, R. A. 1949. "Hibernation sites of three snakes and a salamander." *Copeia* 1949:231-232.

Grogan, W. L., Jr. 1973. "A northern pine snake, *Pituophia m. melanoleucus*, from Maryland." *Bulletin of the Maryland Herpetological Society* 9:27-30.

Grogan, W. L., Jr. 1974a. "A new county record for the four-toed salamander, *Hemidactylium scutatum*, in Maryland." *Bulletin of the Maryland Herpetological Society* 10:32-33.

Grogan, W. L., Jr. 1974b. "Notes on *Lampropeltis calligaster rhombomaculata* and *Rana virgatipes*." *Bulletin of the Maryland Herpetological Society* 10:33-34.

Grogan, W. L., Jr. 1975. "A Maryland hibernaculum of northern brown snakes, *Storeria d. dekayi*." *Bulletin of the Maryland Herpetological Society* 11:27.

Grogan, W. L., Jr. 1981. "Two new reptile county records for Maryland." *Bulletin of the Maryland Herpetological Society* 17:110.

Grogan, W. L., Jr. 1985. "New distribution records for Maryland reptiles and amphibians." *Bulletin of the Maryland Herpetological Society* 21:74-75.

Grogan, W. L., Jr. 1994. "New herpetological distribution records from Maryland's Eastern Shore." *Bulletin of the Maryland Herpetological Society* 30:27-32.

Grogan, W. L., Jr. 2008. "*Storeria o. occipitomaculata* (northern red-bellied snake)." *Herpetological Review* 39:374.

Grogan, W. L., Jr., and P. G. Bystrak. 1973a. "The amphibians and reptiles of Kent Island, Maryland." *Bulletin of the Maryland Herpetological Society* 9:115-118.

Grogan, W. L., Jr., and P. G. Bystrak. 1973b. "Early breeding activity of *Rana sphencephala* and *Bufo woodhousei fowleri* in Maryland." *Bulletin of the Maryland Herpetological Society* 9:106.

Grogan, W. L., Jr., and M. T. Close. 2008. "Geographic distribution: *Cemophora coccinea copei* (northern scarlet snake)." *Herpetological Review* 39:370.

Grogan, W. L., Jr., and D. C. Forester. 1998. "New records of the milk snake, *Lampropeltis triangulum*, from the coastal plain of the Delmarva Peninsula, with comments on the status of *L. t. temporalis*." *Maryland Naturalist* 42:5-14.

Grogan, W. L., Jr., and E. C. Prince. 1971. "Notes on hatching mole snakes, *Lampropeltis calligaster rhombomaculata* Holbrook, in Maryland." *Bulletin of the Maryland Herpetological Society* 7:42.

Grogan, W. L., Jr., and J. J. White, Jr. 1973. "An additional corn snake, *Elaphe guttata guttata*, from western Maryland." *Bulletin of the Maryland Herpetological Society* 9:33-34.

Grogan, W. L., Jr., and G. Williams. 1973. "Notes on hatchling painted turtles, *Chrysemys p. picta*, from Maryland." *Bulletin of the Maryland Herpetological Society* 9:108-110.

Grosse, A. M., J. D. van Dijk, K. L. Holcomb, and J. C. Maerz. 2009. "Diamondback terrapin mortality in crab pots in a Georgia tidal marsh." *Chelonian Conservation and Biology* 8:98-100.

Groves, F. 1991. "Snakes alive." *The Sun* (1837-1991), Aug 5, 1991. ProQuest Historical Newspapers: *The Baltimore Sun*.

Groves, J. D. 1972. "Additional notes on the turtle, *Chrysemys scripta*, in Maryland." *Bulletin of the Maryland Herpetological Society* 8:52-53.

Groves, J. D. 1984. "The sea turtles of Maryland." In *Threatened and Endangered Plants and Animals of Maryland*, edited by A. Norden, D. Forester, and G. Fenwick, 352-358. Natural Heritage Program Special Publication 84-1. Annapolis: Maryland Department of Natural Resources.

Groves, J. D., and A. W. Norden. 1995. "Occurrence of the ground skink, *Scincella lateralis*, in Queen Anne's County, Maryland, and a comment on its reproductive cycle." *Bulletin of the Maryland Herpetological Society* 31:143-146.

Groves, M. F. 1940. "The reptiles and amphibians of Cherry Hill, Baltimore City, Maryland: reptiles." *Natural History Society of Maryland, Junior Division Bulletin* 4:1-6.

Groves, M. F. 1941. "An unusual feeding record of the keeled green snake, *Opheodrys aestivus*." *Bulletin of the Natural History Society of Maryland* 12:27.

Groves, M. F. 1946. "Oviposition in the eastern worm snake." *Maryland: A Journal of Natural History* 16:84.

Grubb, J. C. 1973. "Olfactory orientation in *Bufo woodhousei fowleri*, *Pseudacris clarki* and *Pseudacris streckeri*." *Animal Behaviour* 21: 726-732.

Gunzburger, M. S. 2006. "Reproductive ecology of the green treefrog (*Hyla cinerea*) in northwestern Florida." *American Midland Naturalist* 155:321-328.

Guttman, S. I., and A. A. Karlin. 1986. "Hybridization of cryptic species of two-lined salamanders (*Eurycea bislineata* complex)." *Copeia* 1986:96-108.

Hagood, S. 2009. "Genetic differentiation of selected eastern box turtle (*Terrapene carolina*) populations in fragmented habitats, and a comparison of road-based mortality rates to population size." PhD dissertation. University of Maryland, College Park.

Hagood, S., and M. J. Bartles. 2010. "Reducing road-based habitat fragmentation: an eastern box turtle (*Terrapene c. carolina*) pilot project." In *Proceedings of the 2009 International Conference on Ecology and Transportation*, edited by P. J. Wagner, D. Nelson, and E. Murray, 539-545. Raleigh, NC: Center for Transportation and the Environment.

Hailman, J. P., and R. G. Jaeger. 1974. "Phototactic responses to spectrally dominant stimuli and use of colour vision by adult anuran amphibians: a comparative survey." *Animal Behaviour* 22:757-795.

Hall, R. J. 1977. "A population analysis of two species of streamside salamanders, Genus *Desomgnathus*." *Herpetologica* 33:109-1134.

Hall, R. J., P. F. P. Henry, and C. M. Bunck. 1999. "Fifty-year trends in a box turtle population in Maryland." *Biological Conservation* 88:165-172.

Hall, R. J., and B. M. Mulhern. 1984. "Are anuran amphibians heavy metal accumulators?" In *Vertebrate Ecology and Systematics—A Tribute to Henry S. Fitch*, edited by R. A. Seigel, L. E. Hunt, J. L. Knight, L. Malaret, and N. L. Zuschlag, 123-133. Special Publication no. 10. Lawrence: Museum of Natural History, University of Kansas.

Hall, R. J., and D. P. Stafford. 1972. "Studies in the life history of Wehrle's salamander, *Plethodon wehrlei*." *Herpetologica* 28:300-309.

Hallgren-Scaffidi, L. 1986. "Habitat, home range and population study of the eastern box turtle (*Terrapene carolina*)." MS thesis. University of Maryland, College Park.

Hamilton, W. J., Jr. 1932. "The food and feeding habits of some eastern salamanders." *Copeia* 1932:83-86.

Hamilton, W. J., Jr. 1943. "Winter habits of the dusky salamander in central New York." *Copeia* 1943:192.

Hammerson, G. A. 2007. "*Cemophora coccinea*." The IUCN Red List of Threatened Species. Version 2015.3. Online at www .iucnredlist.org.

Harding, J. H. 1997. *Amphibians and Reptiles of the Great Lakes Region*. Ann Arbor: University of Michigan Press.

Harding, J. H., and T. J. Bloomer. 1979. "The wood turtle, *Clemmys insculpta*: a natural history." *Bulletin of the New York Herpetological Society* 15:9-26.

Hardy, D. L., Sr. 1992. "A review of first aid measures for pit viper bite in North America with an appraisal of extractor suction and stun gun electroshock." In *Biology of the Pitvipers*, edited by J. A. Campbell and E. D. Brodie, Jr., 405-414. Tyler, TX: Selva Press.

Hardy, J. D., Jr. 1962. "Comments on the Atlantic ridley turtle, *Lepidochelys olivacea kempi*, in the Chesapeake Bay." *Chesapeake Science* 3:217-220.

Hardy, J. D. 1969. "Records of the leatherback turtle, *Dermochelys coriacea coriacea* (Linnaeus), from the Chesapeake Bay." *Bulletin of the Maryland Herpetological Society* 5:92-96.

Hardy, J. D., Jr. 1972. "Reptiles of the Chesapeake Bay region." *Chesapeake Science* 13(suppl.):S128-S134.

Hardy, J. D., Jr. 2009. "Eastern narrow-mouth frog—*Gastrophryne carolinensis*." *Bulletin of the Maryland Herpetological Society* 45: 118-121.

Hardy, J. D., Jr., and R. J. Mansueti. 1962. "Checklist of the amphibians and reptiles of Calvert County, Maryland." University of Maryland Natural Resources Institute, Chesapeake Biological Laboratory Reference no. 62-34.

Hardy, L. M., and M. C. Lucas. 1991. "A crystalline protein is responsible for dimorphic egg jellies in the spotted salamander, *Ambystoma maculatum* (Shaw) (Caudata: Ambystomatidae)." *Comparative Biochemistry and Physiology* 100A:653-660.

Hardy, L. M., and L. R. Raymond. 1991. "Observations on the activity of the pickerel frog, *Rana palustris* (Anura: Ranidae), in northern Louisiana." *Journal of Herpetology* 25:220-222.

Harlan, R. 1827. "Genera of North American reptilia, and synopsis of the species." *Journal of the Academy of Natural Science* 5:317-372.

Harrell, T., W. Harrell, L. Pyne, and J. White. 2009. "*Farancia erytrogramma erytrogramma* (common rainbow snake)." *Catesbeiana* 29:95.

Harris, H. S., Jr. 1966a. "Additions to the distributional survey: Maryland and the District of Columbia—I." *Bulletin of the Maryland Herpetological Society* 2(4):24-26.

Harris, H. S., Jr. 1966b. "A checklist of the amphibians and reptiles of Patapsco State Park, Baltimore and Howard Counties, Maryland." *Bulletin of the Maryland Herpetological Society* 2(1):4-7.

Harris, H. S., Jr. 1968. "Additions to the distribution survey: Maryland and the District of Columbia—II." *Bulletin of the Maryland Herpetological Society* 4:85-86.

Harris, H. S., Jr. 1969a. "Additions to the distributional survey: Maryland and the District of Columbia—III." *Bulletin of the Maryland Herpetological Society* 5:81-82.

Harris, H. S., Jr. 1969b. "Distributional survey: Maryland and the District of Columbia." *Bulletin of the Maryland Herpetological Society* 5:97-161.

Harris, H. S., Jr. 1975. "Distributional survey (Amphibia/Reptilia): Maryland and the District of Columbia." *Bulletin of the Maryland Herpetological Society* 11:73-167.

Harris, H. S., Jr. 1995. "The Maryland Herpetological Society survey." *Bulletin of the Maryland Herpetological Society* 31:112-115.

Harris, H. S., Jr. 2004. "Miscellaneous comments on select Maryland amphibians and reptiles." *Bulletin of the Maryland Herpetological Society* 40:189-195.

Harris, H. S., Jr. 2005. Periodic species abundance in snakes. *Bulletin of the Maryland Herpetological Society* 42:25-29.

Harris, H. S., Jr. 2006. "*Pseudemys* and *Trachemys*, indigenous and feral, in Maryland." *Bulletin of the Maryland Herpetological Society* 42:184-187.

Harris, H. S., Jr. 2007a. "The distribution in Maryland, both recent and historic, of the timber rattlesnake, *Crotalus horridus*." *Bulletin of the Maryland Herpetological Society* 43:8-13.

Harris, H. S., Jr. 2007b. "The father of Delmarva herpetology and the pine snake." *Bulletin of the Maryland Herpetological Society* 43:4-7.

Harris, H. S., Jr. 2009. "The past history of documenting the distributions of amphibians and reptiles of Maryland and the District of Columbia." *Bulletin of the Maryland Herpetological Society* 45:14-16.

Harris, H. S., Jr. 2013. "The distribution of the northern scarlet snake, *Cemophora coccinea copei* (Jan, 1863), in Maryland and Delaware with a review of the known records." *Bulletin of the Maryland Herpetological Society* 49:40-46.

Harris, H. S., Jr. 2015. "Preliminary note on the amphibians and reptiles of the Severn Run Natural Environmental Area, Anne Arundel County, Maryland." *Bulletin of the Maryland Herpetological Society* 51:9-12.

Harris, H. S., Jr., and K. Crocetti. 2007. "The wood turtle, *Clemmys insculpta*, in Talbot/Queen Ann's County, Maryland?" *Bulletin of the Maryland Herpetological Society* 43:167-168.

Harris, H. S., Jr., and K. Crocetti. 2008. "The eastern spadefoot toad, *Scaphiopus holbrookii* in Maryland." *Bulletin of the Maryland Herpetological Society* 44:107-110.

Harris, H. S., Jr., R. Franz, and C. J. Stine. 1967. "Some problems in Maryland Herpetology." *Bulletin of the Maryland Herpetological Society* 3:61-62.

Harris, H. S., Jr., and D. Lyons. 1968a. "The first record of the green salamander, *Aneides aeneus* (Cope and Packard), in Maryland." *Journal of Herpetology* 1:106-107.

Harris, H. S., Jr., and D. Lyons. 1968b. "The green salamander, *Aneides aeneus* (Cope and Packard), in Maryland." *Bulletin of the Maryland Herpetological Society* 4:1-6.

Harris, R. N. 1980. "The consequences of within-year timing of breeding in *Ambystoma maculatum*." *Copeia* 1980:719-722.

Harris, R. N. 1981. "Intrapond homing behavior in *Notophthalmus viridescens*." *Journal of Herpetology* 15:355-356.

Harris, R. N., and D. E. Gill. 1980. "Communal nesting, brooding behavior, and embryonic survival of the four-toed salamander *Hemidactylium scutatum*." *Herpetologica* 36:141-144.

Hassinger, D. D., J. D Anderson, and G. H. Dalrymple. 1970. "The early life history and ecology of *Ambystoma tigrinum* and *Ambystoma opacum* in New Jersey." *American Midland Naturalist* 84:474-495.

Hay, W. P. 1902. "A list of the batrachians and reptiles of the District of Columbia and vicinity." *Proceedings of the Biological Society of Washington* 15:121-146.

Healy, W. R. 1970. "Reduction of neoteny in Massachusetts populations of *Notophthalmus viridescens*." *Copeia* 1970:578-581.

Healy, W. R. 1975. "Terrestrial activity and home range in efts of *Notophthalmus viridescens*." *American Midland Naturalist* 93:131-138.

Heatwole, H. 1961. "Habitat selection and activity of the wood frog, *Rana sylvatica* LeConte." *American Midland Naturalist* 66:301-313.

Heckscher, C. M. 1995. "Distribution and habitat associations of the eastern mud salamander, *Pseudotriton montanus montanus*, on the Delmarva Peninsula." *Maryland Naturalist* 39:11-14.

Hecnar, S. J., and R. T. M'Closkey. 1996. "Regional dynamics and the status of amphibians." *Ecology* 77:2091-2097.

Hedges, S. B. 1986. "An electrophoretic analysis of Holarctic hylid frog evolution." *Systematic Zoology* 35:1-21.

Henderson, R. W. 1970. "Feeding behavior, digestion, and water requirements of *Diadophis punctatus arnyi* Kennicott." *Herpetologica* 26:520-526.

Hennessee, L., M. J. Valentino, and A. M. Lesh. 2003. "Updating shore erosion rates in Maryland." Maryland Geological Survey, Coastal and Estuarine Geology File Report no. 03-05.

Henshaw, H. W. 1907. "An extension of the range of the wood tortoise." *Proceedings of the Biological Society of Washington* 20(14):65.

Heriot, T. 1588. "Briefe and true report of the new found land of Virginia." In *Electronic Texts in American Studies*, edited by P. Royster. Lincoln: University of Nebraska.

Herman, D. W. 1983. "*Cemophora coccinea copei* (northern scarlet snake): coloration and reproduction." *Herpetological Review* 14:119.

Herrmann, H. L., K. J. Babbitt, M. J. Baber, and R. G. Congalton. 2005. "Effects of landscape characteristics on amphibian distribution in a forest-dominated landscape." *Biological Conservation* 123:139–149.

Hershey, J. L., and D. C. Forester. 1980. "Sensory orientation in *Notophthalmus v. viridescens* (Amphibia: Salamandridae)." *Canadian Journal of Zoology* 58:266–276.

Highton, R. 1956. "The life history of the slimy salamander, *Plethodon glutinosus*, in Florida." *Copeia* 1956:75–93.

Highton, R. 1962a. "Geographic variation in the life history of the slimy salamander." *Copeia* 1962:597–613.

Highton, R. 1962b. "Revision of North American salamanders of the genus *Plethodon*." *Bulletin of the Florida State Museum* 6:235–367.

Highton, R. 1972. "Distributional interactions among eastern North American salamanders of the genus *Plethodon*." In *The Distributional History of the Biota of the Southern Appalachians. Part III: Vertebrates*, edited by P. C. Holt, 139–188. Virginia Polytechnic Institute and State University, Research Division Monograph 4. Blacksburg, VA: Virginia Polytechnic Institute and State University.

Highton, R. 1986. "*Plethodon hoffmani* Highton, valley and ridge salamander." *Catalogue of American Amphibians and Reptiles,* no. 392.

Highton, R. 1999. "Geographic protein variation and speciation in the salamanders of the *Plethodon cinereus* group with the description of two new species." *Herpetologica* 55:43–90.

Highton, R. 2005. "Declines of eastern North American woodland salamanders (*Plethodon*)." In *Amphibian Declines: The Conservation Status of United States Species*, edited by M. J. Lannoo, 34–54. Berkeley: University of California Press.

Highton, R. G., G. C. Maha, and L. R. Maxson. 1989. *Biochemical Evolution in the Slimy Salamanders of the* Plethodon Glutinosus *Complex in the Eastern United States*. Urbana: University of Illinois Press.

Highton, R., and R. B. Peabody. 2000. "Geographic protein variation and speciation in salamanders of the *Plethodon jordani* and *Plethodon glutinosus* complexes in the southern Appalachian Mountains with the description of four new species." In *The Biology of Plethodontid Salamanders*, edited by R. C. Bruce, R. G. Jaeger, and L. Houck, 31–93. New York: Kluwer Academic/Plenum Publishers.

Hildebrand, W. G. 2003. "New distributional record for the southern leopard frog in Frederick County, Maryland." *Bulletin of the Maryland Herpetological Society* 39:62–63.

Hildebrand, W. G. 2005. "Maryland's anuran populations: are they at risk from anthropomorphic impact I. Calling survey of Maryland anurans." *Bulletin of the Maryland Herpetological Society* 41(4):121–140.

Hildebrand, W. G. 2009. "Eastern spadefoot toad, *Scaphiopus holbrookii*." *Bulletin of the Maryland Herpetological Society* 45:47–50.

Hildebrand, W. G. 2011. "Mississippi map turtle, *Graptemys pseudogeographica kohnii*, documented in Frederick County Maryland." *Bulletin of the Maryland Herpetological Society* 47:51–52.

Hill, R. E., and S. P. Mackessy. 2000. "Characterization of venom (Duvernoy's secretion) from twelve species of colubrid snakes and partial sequence of four venom proteins." *Toxicon* 38:1663–1687.

Hillis, D. M. 1976. "Early breeding behavior of *Notophthalmus v. viridescens* in Allegany County, Maryland." *Bulletin of the Maryland Herpetological Society* 12:55.

Hillis, D. M. 1977. "Sex ratio, mortality rate, and breeding stimulus in a Maryland population of *Ambystoma maculatum*." *Bulletin of the Maryland Herpetological Society* 13:84–91.

Hillis, R. E., and E. D. Bellis. 1971. "Some aspects of the ecology of the hellbender, *Cryptobranchus alleganiensis alleganiensis*, in a Pennsylvania stream." *Journal of Herpetology* 5:121–126.

Holbrook, J. E. 1836. *North American Herpetology; or a Description of the Reptiles Inhabiting the United States*, vol. I. Philadelphia: J. Dobson.

Holbrook, J. E. 1838a. *North American Herpetology; or a Description of the Reptiles Inhabiting the United States*, vol. II. Philadelphia: J. Dobson.

Holbrook, J. E. 1838b. *North American Herpetology; or a Description of the Reptiles Inhabiting the United States*, vol. III. Philadelphia: J. Dobson.

Holbrook, J. E. 1840. *North American Herpetology; or a Description of the Reptiles Inhabiting the United States*, vol. IV. Philadelphia: J. Dobson.

Holbrook, J. E. 1842a. *North American Herpetology; or a Description of the Reptiles Inhabiting the United States*, vol. III. Philadelphia: J. Dobson.

Holbrook, J. E. 1842b. *North American Herpetology; or a Description of the Reptiles Inhabiting the United States*, vol. IV. Philadelphia: J. Dobson.

Holbrook, J. E. 1842c. *North American Herpetology; or a Description of the Reptiles Inhabiting the United States*, vol. V. Philadelphia: J. Dobson.

Holomuzki, J. R. 1982. "Homing behavior of *Desmognathus ochrophaeus* along a stream." *Journal of Herpetology* 16:307–309.

Holub, R. J., and T. J. Bloomer. 1977. "The bog turtle, *Clemmys muhlenbergii*: a natural history." *Bulletin of the New York Herpetological Society* 13:9–23.

Hom, C. L. 1987. "Reproductive ecology of female dusky salamanders, *Desmognathus fuscus* (Plethodontidae), in the southern Appalachians." *Copeia* 1987:768–777.

Hom, C. L. 1988. "Cover object choice by female dusky salamanders, *Desmognathus fuscus*." *Journal of Herpetology* 22:247–249.

Hoose, P. M. 1981. "*Building an Ark: Tools for the Preservation of Natural Diversity through Land Protection*." Covelo, CA: Island Press.

Hopkins, H. H. 1890. "Observations on the copperhead snake, *Ancistrodon contortrix*." *Transactions of the Maryland Academy of Sciences* 1:89–96.

Horchler, D. C. 2010. "Long-term growth and monitoring of the eastern hellbender (*Cryptobranchus alleganiensis alleganiensis*) in eastern West Virginia." PhD dissertation. Marshall University, Huntington, WV.

Hotchkin, P. E., C. D. Camp, and J. L. Marshall. 2001. "Aspects of the life history and ecology of the coal skink, *Eumeces anthracinus*, in Georgia." *Journal of Herpetology* 35:145–148.

Howard, M. 1923. "Some abstracts of old Baltimore County records." *Maryland Historical Magazine* 18:1–37.

Howard, R. D. 1988. "Sexual selection on male body size and mating behavior in American toads, *Bufo americanus*." *Animal Behaviour* 36:1796–1808.

Howard, R. D., and J. R. Young. 1998. "Individual variation in male vocal traits and female mating preferences in *Bufo americanus*." *Animal Behaviour* 55:1165–1179.

Howard, R. R., and E. D. Brodie. 1973. "A Batesian mimetic complex in salamanders: responses to avian predators." *Herpetologica* 29:33–41

Howden, H. F. 1946. "The brown king snake or mole snake (*Lampropeltis rhombomaculata*) in Maryland." *Maryland: A Journal of Natural History* 16:38–40.

Howze, J. M., and L. L. Smith. 2012. "Factors influencing eastern kingsnake diel activity." *Copeia* 2012:460–464.

Hulse, A. C., C. J. McCoy, and E. J. Censky. 2001. *Amphibians and Reptiles of Pennsylvania and the Northeast*. Ithaca, NY, and London: Cornell University Press.

Hunsinger, T. W., and M. J. Lannoo. 2005. "*Notophthalmus viridescens* (Rafinesque, 1820), eastern newt." In *Amphibian Declines: The Conservation Status of United States Species*, edited by M. J. Lannoo, 889–894. Berkeley: University of California Press.

Hunter, M. L., Jr., J. Albright, and J. Arbuckle, editors. 1992. "The amphibians and reptiles of Maine." *Maine Agricultural Experiment Station Bulletin*, no. 838.

Hurlbert, S. H. 1969. "The breeding migrations and interhabitat wandering of the vermilion-spotted newt, *Notophthalmus viridescens* (Rafinesque)." *Ecological Monographs* 39:465–488.

Hurlbert, S. H. 1970a. "The post-larval migration of the red-spotted newt *Notophthalmus viridescens* (Rafinesque)." *Copeia* 1970:515–528.

Hurlbert, S. H. 1970b. "Predator responses to the vermilion-spotted newt (*Notophthalmus viridescens*)." *Journal of Herpetology* 4:47–55.

Hutchins, M., J. B. Murphy, and N. Schlager. 2003. *Grzimek's Animal Life Encyclopedia*, 2nd ed., vol. 7: *Reptiles*. Farmington Hills, MI: Gale.

Ireland, P. H. 1979. "*Eurycea longicauda* (Green), long-tailed salamander." *Catalogue of American Amphibians and Reptiles*, no. 221.

Irwin, J. T., J. P. Costanzo, and R. E. Lee. 1999. "Terrestrial overwintering in the northern cricket frog, *Acris crepitans*." *Canadian Journal of Zoology* 77:1240–1246.

ISSG (Invasive Species Specialist Group). 2010. "Species profile: *Trachemys scripta elegans*." Global Invasive Species Database. Online at http://www.iucngisd.org/gisd/species.php?sc=71.

IUCN SSC Amphibian Specialist Group. 2014. "*Plethodon glutinosus*." The IUCN Red List of Threatened Species 2014: e.T59340A56365349. Online at http://dx.doi.org/10.2305/IUCN.UK.2014-1.RLTS.T59340A56365349.en.

Jackrel, S. L., and H. K. Reinert. 2011. "Behavioral responses of a dietary specialist, the queen snake (*Regina septemvittata*), to potential chemoattractants released by its prey." *Journal of Herpetology* 45:272–276.

Jackson, S. D., R. M. Richmond, T. F. Tyning, and C. W. Leahy, editors. 2010. *Massachusetts Herpetological Atlas 1992–1998*. Lincoln: Massachusetts Audubon Society, and Amherst: University of Massachusetts.

Jacobs, J. F. 1987. "A preliminary investigation of geographic generic variation and systematics of the two-lined salamander, *Eurycea bislineata* (Green)." *Herpetologica* 43:423–446.

Jaeger, E. C. 1972. *A Source-Book of Biological Names and Terms*, 3rd ed. Springfield, IL: Charles C. Thomas Publisher.

Jensen, J. B., C. D. Camp, W. Gibbons, and M. J. Elliott, editors. 2008. *Amphibians and Reptiles of Georgia*. Athens: University of Georgia Press.

Jensen, J. B., and C. Waters. 1999. "The 'spring lizard' bait industry in the state of Georgia, USA." *Herpetological Review* 30:20–21.

Jeyasuria, P., W. M. Roosenburg, and A. R. Place. 1994. "Role of P-450 aromatase in sex determination of the diamondback terrapin, *Malaclemys terrapin*." *Journal of Experimental Zoology* 270:95–111.

Johnson, B. K., and J. L. Christiansen. 1976. "The food and food habits of Blanchard's cricket frog, *Acris crepitans blanchardi* (Amphibia, Anura, Hylidae) in Iowa." *Journal of Herpetology* 10:63–74.

Johnson, C. F. 1963. "Additional evidence of sterility between call-types in the *Hyla versicolor* complex." *Copeia* 1963:139–143.

Johnson, K. A. 2003. "Abiotic factors influencing the breeding, movement, and foraging of the eastern spadefoot (*Scaphiopus holbrookii*) in West Virginia." MS thesis. Marshall University, Huntington, WV.

Johnson, M. J. 1984. "The distribution and habitat of the pickerel frog in Wisconsin." MS thesis. University of Wisconsin, Stevens Point.

Johnson, R. H., and M. van Deusen. 1979. "Herpetologic survey of the proposed expansion site for Vienna steam-electric station, Vienna, Dorchester County, Maryland." Chesapeake Bay Institute, Johns Hopkins University Special Report no. 76.

Johnson, R. H., and M. van Deusen. 1980. "Reptiles and amphibians in the vicinity of Vienna, Maryland." *Bulletin of the Maryland Herpetological Society* 16:70–76.

Johnson, T. R. 1987. *The Amphibians and Reptiles of Missouri*. St. Louis: Missouri Department of Conservation.

Johnston, C., transcriber. 1922. "News from the *Maryland Gazette*." *Maryland Historical Magazine* 17:364–379.

Jones, H. 1699. "Part of a letter from the Reverend Mr. Hugh Jones to the Reverend Dr. Benjamin Woodroofe, F.R.S. concerning several observables in Maryland." *Philosophical Transactions (Royal Society of London)* 21:436–442.

Jones, R. L. 1986. "Reproductive biology of *Desmognathus fuscus* and *Desmognathus santeetlah* in the Unicoi Mountains." *Herpetologica* 42:323–334.

Jordan, G. T., and M. Kaups. 1989. *The American Backwoods Frontier: An Ethnic and Ecological Interpretation*. Baltimore: Johns Hopkins University Press.

Jordan, R. 1970. "Death-feigning in a captive red-bellied snake, *Storeria occipitomaculata* (Storer)." *Herpetologica* 26:466–468.

Juterbock, J. E. 1986. "The nesting behavior of the dusky salamander, *Desmognathus fuscus* I: nesting phenology." *Herpetologica* 42:457–471.

Kats, L. B., J. W. Petranka, and A. Sih. 1988. "Antipredator defenses and the persistence of amphibian larvae with fishes." *Ecology* 69:1865–1870.

Kaufmann, J. H. 1992. "The social behavior of wood turtles, *Clemmys insculpta*, in central Pennsylvania." *Herpetological Monographs* 6:1–25.

Kearney, R. F. 2003. *Partners in Flight Landbird Conservation Plan*

Physiographic Area 10: Mid-Atlantic Piedmont. The Plains, VA: American Bird Conservancy.

Keim, T. D. 1914. "Amphibian and reptiles of Jennings Maryland." *Copeia* 2:2.

Keller, M. J., and H. C. Gerhardt. 2001. "Polyploidy alters advertisement call structure in gray treefrogs." *Proceedings of the Royal Society of London Series B* 268:341-345.

Kellner, A., and D. M. Green. 1995. "Age structure and age at maturity in Fowler's toads, *Bufo woodhousii fowleri*, at their northern range limit." *Journal of Herpetology* 29:485-489.

Kelly, E. B. 1949. "Howard A. Kelly, naturalist (1858-1943)." *Maryland Naturalist* 19:26-31.

Kelly, H. A. 1926. "Snakes and snake bite." *Hygeia: A Journal of Individual and Community Health* 4:32-38.

Kelly, H. A., A. W. Davis, and H. C. Robertson. 1936. *Snakes of Maryland.* Baltimore: Natural History Society of Maryland.

Kenney, G., and C. Stearns. 2015. "Recovery plan for New York state populations of the northern cricket frog (*Acris crepitans*)." New York State Department of Environmental Conservation. Online at http://www.dec.ny.gov/docs/wildlife_pdf/crickfrogrecplan15.pdf.

Kephart, D. G., and S. J. Arnold. 1982. "Garter snake diets in a fluctuating environment: a seven-year study." *Ecology* 63:1232-1236.

Kerney, R., E. Kim, R. P. Hangarter, A. A. Heiss, C. D. Bishop, and B. K. Hall. 2011. "Intracellular invasion of green algae in a salamander host." *Proceedings of the National Academy of Sciences of the United States of America* 108:6497-6502.

Keyler, D. E. 2008. "Timber rattlesnake (*Crotalus horridus*) envenomations in the upper Mississippi Valley." In *The Biology of Rattlesnakes*, edited by W. K. Hayes, K. R. Beaman, M. D. Cardwell, and S. P. Bush, 569-580. Loma Linda, CA: Loma Linda University Press.

Kiester, A. R., and L. L. Willey. 2015. "*Terrapene carolina* (Linnaeus 1758)—eastern box turtle, common box turtle." In *Conservation Biology of Freshwater Turtles and Tortoises: A Compilation Project of the IUCN/SSC Tortoise and Freshwater Turtle Specialist Group*, edited by A. G. J. Rhodin, P. C. H. Pritchard, P. P. van Dijk, R. A. Saumure, K. A. Buhlmann, J. B. Iverson, and R. A. Mittermeier, 1-25. Chelonian Research Monographs 5(8):085. Lunenburg, MA: Chelonian Research Foundation, and Ojai, CA: Turtle Conservancy.

Kimmel, T. 2007. *Sea Turtle Tagging and Health Assessment Study in the Maryland Portion of the Chesapeake Bay.* Final report. Oxford: Maryland Department of Natural Resources, Cooperative Oxford Laboratory.

King, R. B. 1993. "Determinants of offspring number and size in the brown snake, *Storeria dekayi.*" *Journal of Herpetology* 27:175-185.

King, R. B., S. Hauff, and J. B. Phillips. 1994. "Physiological color change in the green treefrog: responses to background brightness and temperature." *Copeia* 1994:422-432.

Kinneary, J. J. 1993. "Salinity relations of *Chelydra serpentina* in a Long Island estuary." *Journal of Herpetology* 27:441-446.

Kirkley, D. 1949. "Small game hunters: naturalists exploring the shore found snakes, salamanders, and skinks—but the 'big ones' got away." *Sunday Sun Magazine*, Brown Section. *The Sun* (1837-1991), August 28. ProQuest Historical Newspapers: *The Baltimore Sun.*

Kirkwood, F. C. 1895. "A list of birds of Maryland." *Transactions of the Maryland Academy of Sciences* 1:241-374.

Kleopfer, J. D., J. Gallegos, D. Stolley, and J. C. Mitchell. 2006. "*Chelonia mydas* (green seaturtle): nesting." *Herpetological Review* 37:453.

Klimkiewicz, M. K. 1972. "Reptiles of Mason Neck." *Atlantic Naturalist* 27:20-25.

Klimstra, W. D. 1959. "Foods of the racer, *Coluber constrictor*, in southern Illinois." *Copeia* 1959:210-214.

Klingel, G. C. 1929. "Lizard hunting in the black republic." *Natural History: Journal of the American Museum of Natural History* 29:451-464.

Klingel, G. C. 1930. *Department of Herpetology.* First annual report of the Natural History Society of Maryland for the year 1929. Baltimore: Natural History Society of Maryland.

Klingel, G. C. 1932a. "Frog photography." *Bulletin of the Natural History Society of Maryland* 2:85-86.

Klingel, G. C. 1932b. "Shipwrecked on Inagua." *Natural History: Journal of the American Museum of Natural History* 32:42-55.

Klingel, G. C. 1944. "Notes from field and laboratory." *Maryland: A Journal of Natural History* 14:73.

Knutson, M. G., W. B. Richardson, D. M. Reineke, B. R. Gray, J. R. Parmelee, and S. E. Weick. 2004. "Agricultural ponds support amphibian ponds." *Ecological Applications* 14:669-684.

Kramer, D. C. 1973. "Movements of western chorus frogs *Pseudacris triseriata triseriata* tagged with Co 60." *Journal of Herpetology* 7:231-235.

Kramer, D. C. 1974. "Home range of the western chorus frog *Pseudacris triseriata triseriata.*" *Journal of Herpetology* 8:245-246.

Kraus, F. 2009. *Alien Reptiles and Amphibians: A Scientific Compendium and Analysis.* Dordrecht, The Netherlands: Springs.

Krenz, J. D., and D. E. Scott. 1994. "Terrestrial courtship affects mating locations in *Ambystoma opacum.*" *Herpetologica* 50:46-50.

Krzysik, A. J. 1979. "Resource allocation, coexistence, and niche structure of a streambank salamander community." *Ecological Monographs* 49:173-194.

Krzysik, A. J. 1980. "Trophic aspects of brooding behavior in *Desmognathus fuscus fuscus.*" *Journal of Herpetology* 14:426-428.

LaBranche, J., M. McCoy, D. Clearwater, P. Turgeon, and G. Setzer. 2003. *Maryland State Wetlands Conservation Plan.* Baltimore: Maryland Department of the Environment, Nontidal Wetlands and Waterways Division.

Lancaster, D. L., and S. E. Wise. 1996. "Differential response by the ringneck snake, *Diadophis punctatus*, to odors of tail-autotomizing prey." *Herpetologica* 52:98-108.

Landreth, H. F., and D. E. Ferguson. 1968. "The sun compass of Fowler's toad, *Bufo woodhousei fowleri.*" *Behaviour* 30:27-43.

Lang, E., C. Gerhardt, and C. Davidson. 2009. *The Frogs and Toads of North America.* Boston: Houghton Mifflin Harcourt.

Larsen, K. W., P. T. Gregory, and R. Antoniak. 1993. "Reproductive ecology of the common garter snake *Thamnophis sirtalis* at the northern limit of its range." *American Midland Naturalist* 129:336-345.

Lazell, J. D., Jr., and P. J. Auger. 1981. "Predation on diamondback terrapin (*Malaclemys terrapin*) eggs by dunegrass (*Ammophila breviligulata*)." *Copeia* 1981:723-724.

Lee, D. S. 1972a. "List of the amphibians and reptiles of Assateague Island." *Bulletin of the Maryland Herpetological Society* 8:90-95.

Lee, D. S. 1972b. "*Necturus* in Maryland." *Bulletin of the Maryland Herpetological Society* 8:66.

Lee, D. S. 1973a. "Additional reptiles and amphibians from Assateague Island." *Bulletin of the Maryland Herpetological Society* 9: 110-111.

Lee, D. S. 1973b. "An annotated list of amphibians, reptiles and mammals of Irish Grove Sanctuary, Somerset County, Maryland." *Maryland Birdlife* 29:143-149.

Lee, D. S. 1973c. "Another record for the queen snake in southern Maryland." *Bulletin of the Maryland Herpetological Society* 9: 107.

Lee, D. S. 1973d. "Seasonal breeding distributions for selected Maryland and Delaware amphibians." *Bulletin of the Maryland Herpetological Society* 9:101-104.

Lee, D. S. 1988. "Wm. H. Fisher's Mammals of Maryland: a previously unknown early compilation of the state's fauna." *Maryland Naturalist* 32:9-37.

Lee, D. 1993. "Reflections on Roger Conant's contributions to regional herpetology." *Maryland Naturalist* 37:1-6.

Lee, D. S. 2012. "The future of map turtles: will the mutts take over?" *Bulletin of the Chicago Herpetological Society* 47:1-4.

Lee, D. S., and A. W. Norden. 1973. "A food study of the green salamander, *Aneides aeneus*." *Journal of Herpetology* 7:53-54.

Lee, D. S., and A. W. Norden. 1996. "The distribution, ecology and conservation of bog turtles, with special emphasis on Maryland." *Maryland Naturalist* 40:7-46.

Lemay, L., and A. G. Marsiglia. 1952. "The coal skink, *Eumeces anthracinus* (Baird), in Maryland." *Copeia* 1952:193.

Lemmon, E. M., A. R. Lemmon, J. T. Collins, J. A. Lee-Yaw, and D. Cannatella. 2007. "Phylogeny-based delimitation of species boundaries and contact zones in the trilling chorus frogs (*Pseudacris*)." *Molecular Phylogenetics and Evolution* 44:1068-1082.

Lentz, J. 2004a. *Home Range and Habitat Preferences of the Eastern Box Turtle at the Jug Bay Wetlands Sanctuary*. Technical Report of the Jug Bay Wetlands Sanctuary. Lothian, MD: Jug Bay Wetlands Sanctuary.

Lentz, J. 2004b. "Home range and habitat preferences of *Terrapene carolina carolina* at the Jug Bay Wetlands Sanctuary." Senior thesis. Hamilton College, Clinton, NY.

Leviton, A. 1971. *Reptiles and Amphibians of North America*. New York: Doubleday & Company.

Lindeman, P. V. 2013. *The Map Turtle and Sawback Atlas: Ecology, Evolution, Distribution, and Conservation*. Norman: University of Oklahoma Press.

Linebaugh, S. 1995. *Basking Patterns of Aquatic Turtles*. Technical Report of the Jug Bay Wetlands Sanctuary. Lothian, MD: Jug Bay Wetlands Sanctuary.

Linzey, D. W., and M. J. Clifford. 1981. *Snakes of Virginia*. Charlottesville: University Press of Virginia.

Lister, T. W., J. L. Perdue, C. J. Barnett, B. J. Butler, S. J. Crocker, G. M. Domke, D. Griffith, M. A. Hatfield, C. M. Kurtz, A. J. Lister, R. S. Morin, W. K. Moser, M. D. Nelson, C. H. Perry, R. J. Piva, R. Riemann, R. Widmann, and C. W. Woodall. 2011. *Maryland's Forests 2008*. Resource Bulletin NRS-58. Newton Square, PA: US Department of Agriculture, Forest Service, Northern Research Station.

Littleford, R. A. 1945. "DeKay's snake in Maryland." *Copeia* 1945:50.

Loss, S. R., T. Will, and P. P. Marra. 2013. "The impact of free-ranging domestic cats on wildlife of the United States." *Nature Communications* 4:1396.

Lovich, J. E., C. H. Ernst, and R. T. Zappalorti. 1998. "Geographic variation in growth and sexual size dimorphism of bog turtles (*Clemmys muhlenbergii*)." *American Midland Naturalist* 139: 69-78.

Lovich, J. E., D. W. Herman, and K. M. Fahey. 1992. "Seasonal activity and movements of bog turtles (*Clemmys muhlenbergii*) in North Carolina." *Copeia* 1992:1107-1111.

Lutcavage, M., and J. A. Musick. 1985. "Aspects of the biology of sea turtles in Virginia." *Copeia* 1985:449-456.

Lynch, J. F. 1980. "Amphibians and reptiles of the Chesapeake Bay Center for Environmental Studies." Unpublished typed list.

Macaulay, P., and W. R. Fisher. 1836. "Miscellaneous intelligence 49: Maryland Academy of Science and Literature." *American Journal of Science and Arts* 30:192-194.

Madison, D. M. 1997. "The emigration of radio-implanted spotted salamanders, *Ambystoma maculatum*." *Journal of Herpetology* 31: 542-551.

Mansueti, R. 1939a. "A collection of amphibians and reptiles made in Frederick County." *Natural History Society of Maryland, Junior Division Bulletin* 3:53-55.

Mansueti, R. 1939b. "Reptiles noted during 1938 in and around the Patapsco State Park." *Natural History Society of Maryland, Junior Division Bulletin* 3:5-11.

Mansueti, R. 1940. "Amphibians." *Natural History Society of Maryland, Junior Division Bulletin* 4:6-9.

Mansueti, R. 1941a. "A descriptive catalogue of the amphibians and reptiles from in and around Baltimore City, Maryland." *Proceedings of the Natural History Society of Maryland*, no. 7.

Mansueti, R. 1941b. "A descriptive catalogue of the amphibians and reptiles from in and around Baltimore City, Maryland, within a radius of miles of twenty miles." *Proceedings of the Natural History Society of Maryland*, no. 8.

Mansueti, R. 1941c. "The herpetofauna of the Patapsco State Park, Maryland." *Natural History Society of Maryland, Junior Division Bulletin* 5:7-17.

Mansueti, R. 1942. "Notes on the herpetology of Calvert County, Maryland." *Bulletin of the Natural History Society of Maryland* 12: 33-43.

Mansueti, R. 1947. "The spadefoot toad in Maryland." *Maryland: A Journal of Natural History* 17:7-9, 12-14.

Mansueti, R. 1953. "Notes on Eastern Shore, Maryland, turtles and terrapins." Chesapeake Biological Laboratory. Mimeo.

Mansueti, R. 1958. "The Cranesville Pine Swamp." *Atlantic Naturalist* 13:72-84.

Mansueti, R., and R. Simmons. 1943. "The four-toed salamander in the Baltimore area." *Bulletin of the Natural History Society of Maryland* 13:81.

Mansueti, R., and D. H. Wallace. 1960. "Notes on the soft-shell turtle (*Trionyx*) in Maryland waters." *Chesapeake Science* 1:71-72.

Manville, R. H. 1968. *Natural History of Plummers Island, Maryland. XX. Annotated List of the Vertebrates*. Special publication. Washington, DC: Washington Biologists Field Club.

Marchand, M., M. M. Quinlan, and C. W. Swarth. 2003. "Movement patterns and habitat use of eastern box turtles at the Jug Bay Wetlands Sanctuary, Maryland." In *Conservation and Ecology of Turtles of the Mid-Atlantic Region*, edited by C. W. Swarth,

W. M. Roosenburg, and E. Kiviat, 55–62. Salt Lake City, UT: Bibliomania.

Marsiglia, A. G. 1950. "New county records for the slimy salamander in Maryland." *Maryland Naturalist* 20:57–59.

Martin, P. 1995. "Reptiles in Hartford County Maryland: distribution and behavior of reptiles at Camp Moshava." *Bulletin of the Maryland Herpetological Society* 31:102–105.

Martin, W. H. 1992. "Phenology of the timber rattlesnake (*Crotalus horridus*) in an unglaciated section of the Appalachian Mountains." In *Biology of the Pitvipers*, edited by J. A. Campbell and E. D. Brodie, Jr., 259–277. Tyler, TX: Selva Press.

Martin, W. H. 1993. "Reproduction of the timber rattlesnake (*Crotalus horridus*) in the Appalachian Mountains." *Journal of Herpetology* 27:133–143.

Martin, W. H. 2002. "Life history constraints on the timber rattlesnake (*Crotalus horridus*) at its climatic limits." In *Biology of the Vipers*, edited by G. W. Schuett, M. Hoggren, M. E. Dorcas, and H. W. Greene, 285–306. Eagle Mountain, UT: Eagle Mountain Publishing.

Martin, W. H., W. S. Brown, E. Possardt, and J. B. Sealy. 2008. "Biological variation, management units, and a conservation action plan for the timber rattlesnake (*Crotalus horridus*)." In *The Biology of Rattlesnakes*, edited by W. K. Hayes, K. R. Beaman, M. D. Cardwell, and S. P. Bush, 447–462. Loma Linda, CA: Loma Linda University Press.

Martof, B. S., W. M. Palmer, J. R. Bailey, J. R. Harrison III, and J. Dermid. 1980. *Amphibians and Reptiles of the Carolinas and Virginia*. Chapel Hill: University of North Carolina Press.

Marvin, G. A. 1998. *Plethodon kentucki* and *Plethodon glutinosus*: evidence of interspecific interference competition. *Canadian Journal of Zoology* 76:92–103.

Marye, W. B. 1921. "The Baltimore County 'Garrison' and the Old Garrison Roads." *Maryland Historical Magazine* 16:105–149.

Maryland Academy of Sciences. 1928. "The museum." *Maryland Academy of Sciences Bulletin* 7:60–62.

Maryland Conservationist. 1932. "A monster of the tropical seas captured in Maryland waters." *Maryland Conservationist* 9(3):21–22.

Maryland Herpetological Society. 1965. "Maryland Herpetological Society." *Maryland Herpetological Society Bulletin* 1:2.

Maryland Historical Magazine. 1908. "A Maryland sigurd." *Maryland Historical Magazine* 3:279.

MASL (Maryland Academy of Science and Literature). 1837a. "Article X. Directions for preparing specimens of natural history." *Transactions of the Maryland Academy of Science and Literature* 1:148–156.

MASL (Maryland Academy of Science and Literature). 1837b. "Miscellaneous intelligence 2: Proceedings of the Maryland Academy of Science and Literature, 1936." *American Journal of Science and Arts* 31:395–399.

Massal, L. R., J. W. Snodgrass, and R. E. Casey. 2007. "Nitrogen pollution of stormwater ponds: potential for toxic effects on amphibian embryos and larvae." *Applied Herpetology* 4:19–29.

Matson, T. O. 1990. "Erythrocyte size as a taxonomic character in the identification of Ohio *Hyla chrysoscelis* and *H. versicolor*." *Herpetologica* 46:457–462.

Matsuda, B. M., D. M. Green, and P. T. Gregory. 2006. *Amphibians and Reptiles of British Columbia*. Victoria, BC: Royal British Columbia Museum.

Matthews, S. 2007. *The Effects of Weather, Rain Events, Humidity, and Temperature on the Seasonal Movements of Female Eastern Box Turtles*. Technical Report of the Jug Bay Wetlands Sanctuary. Lothian, MD: Jug Bay Wetlands Sanctuary.

McAlister, W. H. 1963. "A post-breeding concentration of the spring peeper." *Herpetologica* 19:293.

McAtee, W. L. 1918. "A sketch of the natural history of the District of Columbia together with an indexed edition of the U.S. Geological Survey's 1917 map of Washington and Vicinity." *Bulletin of the Biological Society of Washington* 1:1–142.

McCann, J., P. Bendel, and D. Limpert. 2002. *A Survey of the Mammals Occurring at the Cove Point Liquefied Natural Gas Terminal Property, Calvert County, Maryland: Final Report*. Prince Frederick, MD: Cove Point Natural Heritage Trust.

McCauley, R. H., Jr. 1939a. "Differences in the young of *Eumeces fasciatus* and *Eumeces laticeps*." *Copeia* 1939:93–95.

McCauley, R. H., Jr. 1939b. "An extension of the range of *Abastor erythrogrammus*." *Copeia* 1939:54.

McCauley, R. H., Jr. 1940. "A distributional study of the reptiles of Maryland and the District of Columbia." PhD dissertation. Cornell University, Ithaca, NY.

McCauley, R. H., Jr. 1941. "*Diadophis punctatus* in Maryland." *Copeia* 1941:55.

McCauley, R. H., Jr. 1945. *The Reptiles of Maryland and the District of Columbia*. Hagerstown, MD: Published by the author.

McCauley, R. H., Jr., and C. S. East. 1940. "Amphibians and reptiles from Garrett County, Maryland." *Copeia* 1940:120–123.

McCauley, R. H., and R. Mansueti. 1943. "*Clemmys muhlenbergii* in Maryland." *Copeia* 1943:197.

McCauley, R. H., Jr., and R. Mansueti. 1944. "Notes on a Muhlenberg turtle." *Maryland: A Journal of Natural History* 14:68–69.

McClellan, W. H. 1916. "Woodstock College snake collection." *Woodstock Letters* 45:332–338.

McClellan, W. H., R. Mansueti, and F. Groves. 1943. "The lizards of central and southern Maryland." *Proceedings of the Natural History Society of Maryland*, no. 8.

McCoy, C. J. 1992. "Rediscovery of the mud salamander (*Pseudotriton montanus*, Amphibia, Plethodontidae) in Pennsylvania, with restriction of the type-locality." *Journal of the Pennsylvania Academy of Sciences* 66:92–93.

McDaniel, V. R., and J. E. Gardner. 1977. "Cave fauna of Arkansas: vertebrate taxa." *Proceedings of the Arkansas Academy of Science* 31:68–71.

McGregor, J. H., and W. R. Teska. 1989. "Olfaction as an orienting mechanism in migrating *Ambystoma maculatum*." *Copeia* 1989:779–781.

McKelvy, A. D., and F. T. Burbrink. 2016. "Ecological divergence in the yellow-bellied kingsnake (*Lampropeltis calligaster*) at two North American biodiversity hotspots." *Molecular Phylogenetics and Evolution* 106:61–72.

McLeod, R. F. 1995. "Effect of timber harvest and prescribed burning on the distribution and abundance of reptiles and amphibians at Remington Farms, Maryland." MS thesis. Frostburg State University, Frostburg, MD.

McLeod, R. F., and E. J. Gates. 1998. "Response of herpetofaunal communities to forest cutting and burning at Chesapeake Farms, Maryland." *American Midland Naturalist* 139:164–177.

MDNR (Maryland Department of Natural Resources). 2005. *Mary-*

land Wildlife Diversity Conservation Plan. Annapolis: Maryland Department of Natural Resources.

MDNR (Maryland Department of Natural Resources). 2015. *Maryland State Wildlife Action Plan 2015-2025*. Annapolis: Maryland Department of Natural Resources.

MDNR (Maryland Department of Natural Resources). 2016a. *DNR Owned Land Acreage: Fiscal Year 2016*. Annapolis: Maryland Department of Natural Resources.

MDNR (Maryland Department of Natural Resources). 2016b. *List of Rare, Threatened, and Endangered Animals of Maryland*. Annapolis: Maryland Department of Natural Resources.

MDP (Maryland Department of Planning). 1973. *Natural Soil Groups Technical Report*. Annapolis: Maryland Department of Planning.

MDP (Maryland Department of Planning). 2010. *Land Use/Land Cover Layer Update*. Annapolis: Maryland Department of Planning.

MDP (Maryland Department of Planning). 2016. *Maryland's Slow Population Growth Continues through 2016*. Annapolis: Maryland Department of Planning.

Meanley, B. 1951. "Carpenter frog, *Rana virgatipes*, on the coastal plain of Maryland." *Proceedings of the Biological Society of Washington* 64:59.

Meeks, R. L., and G. R. Ultsch. 1990. "Overwintering behavior of snapping turtles." *Copeia* 1990:880-884.

Melville, R. V. 1995. *Towards Stability in the Names of Animals: A History of the International Commission on Zoological Nomenclature 1895-1995*. London: International Trust for Zoological Nomenclature, The Natural History Museum.

Merovich, C. E., and J. H. Howard. 2000. "Amphibian use of constructed ponds on Maryland's Eastern Shore." *Journal of the Iowa Academy of Science* 107:151-159.

Meshaka, W. E., and J. T. Collins. 2010. *A Pocket Guide to Pennsylvania Frogs and Toads*. Harrisburg: Pennsylvania Historical and Museum Commission.

MGS (Maryland Geological Survey). 2004. *Highest and Lowest Elevations in Maryland's Counties*. Baltimore: Maryland Geological Survey.

MGS (Maryland Geological Survey). 2015. "Maryland geology." Online at http://www.mgs.md.gov/geology.

Miller, B. T., and M. L. Niemiller. 2011. "Slimy salamander complex." In *The Amphibians of Tennessee*, edited by M. L. Niemiller and R. G. Reynolds, 192-195. Knoxville: University of Tennessee Press.

Miller, F. P. 1967. "Maryland soils." *University of Maryland Cooperative Extension Service, Extension Bulletin*, no. 212.

Miller, R. [W.] 1977. "A new county record for the eastern spadefoot toad, *Scaphiopus h. holbrooki*, in Maryland." *Bulletin of the Maryland Herpetological Society* 13:118.

Miller, R. [W.] 1979. "Miscellaneous distributional records for Maryland amphibians and reptiles." *Bulletin of the Maryland Herpetological Society* 15:56-58.

Miller, R. [W.] 1980. "Distributional records for Maryland herpetofauna." *Bulletin of the Maryland Herpetological Society* 16:99-105.

Miller, R. W. 1982. "Distributional records for Maryland herpetofauna, II." *Bulletin of the Maryland Herpetological Society* 18:161-164.

Miller, R. [W.] 1984a. "Distributional records for Maryland herpetofauna, III." *Bulletin of the Maryland Herpetological Society* 20: 38-45.

Miller, R. W. 1984b. "Notes on the distribution of *Eurycea longicauda* in Maryland." *Bulletin of the Maryland Herpetological Society* 45:109-110.

Miller, R. W. 1993. "Comments on the distribution of *Clemmys insculpta* on the Coastal Plain of Maryland." *Herpetological Review* 24:90-93.

Miller, R. W. 2009a. "Deletion of *Eurycea guttolineata* from the herpetofauna of Maryland." *Bulletin of the Maryland Herpetological Society* 45:109-110.

Miller, R. W. 2009b. "The reptiles of Maryland and the District of Columbia": missing specimen lists. *Smithsonian Herpetological Information Service*, no. 140.

Miller, R. W. 2010. "Notes on the distribution of *Ambystoma tigrinum* in Maryland and Virginia." *Bulletin of the Chicago Herpetological Society* 45:157-160.

Miller, R. W. 2011. "Notes on the distribution of *Pseudotriton montanus* in Maryland." *Bulletin of the Chicago Herpetological Society* 46:44-47.

Miller, R. W. 2013. "Notes on the distribution of *Scaphiopus holbrookii* in Maryland." *Bulletin of the Chicago Herpetological Society* 48:166-171.

Miller, R. W. 2014. "Additional notes on the distribution of *Pseudotriton montanus* in Maryland, with comments on its conservation status." *Bulletin of the Chicago Herpetological Society* 49: 45-48.

Miller, R. W. 2015a. "Comments on the herpetological publications of Romeo J. Mansueti, with special reference to Baltimore, Maryland. Part 1: introductions and amphibians." *Bulletin of the Chicago Herpetological Society* 50:142-153.

Miller, R. W. 2015b. "Comments on the herpetological publications of Romeo J. Mansueti, with special reference to Baltimore, Maryland. Part 2: reptiles." *Bulletin of the Chicago Herpetological Society* 50:89-208.

Miller, R. W. 2015c. "Notes on the distribution of *Necturus maculosus* in Maryland and the birth of reenactment herpetology." *Bulletin of the Chicago Herpetological Society* 50:51-54.

Miller, R. W. 2016. "Notes on the distribution of *Cemophora coccinea* in Maryland." *Bulletin of the Chicago Herpetological Society* 51:132-138.

Miller, R. [W.], and G. Grall. 1978. "Reproductive data on *Lampropeltis triangulum temporalis* from Maryland." *Bulletin of the Maryland Herpetological Society* 14:36-38.

Miller, R. W., and J. D. Zyla. 1992. "Additional rainbow snakes, *Farancia erytrogramma*, from Charles County, Maryland." *Bulletin of the Maryland Herpetological Society* 28:99-101.

Minton, S. A., Jr. 1972. *Amphibians and Reptiles of Indiana*. Monograph no. 3. Indianapolis: Indiana Academy of Science.

Minton, S. A., Jr. 2001. *Amphibians and Reptiles of Indiana*, 2nd ed. Indianapolis: Indiana Academy of Science.

Minton, S. A., Jr., and H. B. Bechtel. 1958. "Another Indiana record of *Cemophora coccinea* and a note on egg-eating." *Copeia* 1958:47.

Minton, S. A., Jr., and J. E. Minton. 1948. "Notes on a herpetological collection from the middle Mississippi Valley." *American Midland Naturalist* 40:378-390.

MISC (Maryland Invasive Species Council). 2005. "Invasive species of concern in Maryland: vertebrates." Online at http://www.mdinvasivesp.org/list_vertebrates.html.

Mitchell, J. C. 1989. "An historical review of the Fairfax County, Virginia, bog turtle record." *Catesbeiana* 9:3-7.

Mitchell, J. C. 1994. *The Reptiles of Virginia*. Washington, DC: Smithsonian Institution Press.

Mitchell, J. C. 2005. "*Rana virgatipes* Cope, 1891, carpenter frog." In *Amphibian Declines: The Conservation Status of United States Species*, edited by M. J. Lannoo, 595-596. Berkeley: University of California Press.

Mitchell, J. C. 2008. *Inventory of Amphibians and Reptiles of Thomas Stone National Historic Site: Natural Resource Technical Report*. Technical report NPS/NER/NRTR—2008/111. Philadelphia: National Park Service.

Mitchell, J. C. 2013. "Emmett Reid Dunn and the early history of herpetology in Virginia." *Banisteria* 41:27-39.

Mitchell, J. C., and J. M. Anderson. 1994. "Amphibians and reptiles of Assateague and Chincoteague Islands." *Virginia Museum of Natural History Special Publication*, no. 2.

Mitchell, J., and R. A. Beck. 1992. "Free-ranging domestic cat predation on native vertebrates in rural and urban Virginia." *Virginia Journal of Science* 43:197-207.

Mitchell, J., and W. Gibbons. 2010. *Salamanders of the Southeast*. Athens: University of Georgia Press.

Mitchell, J. C., T. K. Pauley, D. I. Withers, S. M. Roble, B. T. Miller, A. L. Braswell, P. V. Cupp, Jr., and C. S. Hobson. 1999. "Conservation status of the southern Appalachian herpetofauna." *Virginia Journal of Science* 50:13-35.

Mitchell, J. C., and K. K. Reay. 1999. *Atlas of Amphibians and Reptiles in Virginia*. Richmond: Virginia Department of Game and Inland Fisheries.

Mitchell, J. C., B. W. Steury, K. A. Buhlmann, and P. P. van Dijk. 2007. "Chinese softshell turtle (*Pelodiscus sinensis*) in the Potomac River and notes on eastern spiny softshells (*Apalone spinifera*) in northern Virginia." *Banisteria* 30:41-43.

Mohr, C. E. 1943. "The eggs of the long-tailed salamander, *Eurycea longicauda* (Green)." *Proceedings of the Pennsylvania Academy of Science* 17:86.

Mohr, C. E. 1944. "A remarkable salamander migration." *Proceedings of the Pennsylvania Academy of Science* 18:51-54.

Molines, K. 1997. "The marbled salamander at Jug Bay." *Audubon Naturalist News* 23(9):9-11.

Molines, K. 2008. *Development of an Identification Process Using Digital Photographs: Marbled Salamander Migration Study*. Final report. Lothian, MD: Jug Bay Wetlands Sanctuary.

Molines, K., and C. Swarth. 1997. "Marbled salamander migration patterns in a Coastal Plain forested wetland." Abstract. Conference on Conservation Ecology of Amphibians in the Mid-Atlantic Region, Lothian, MD.

Molines, K., and C. Swarth. 1999. "Aspects of the migratory activity of marbled salamanders (*Ambystoma opacum*)." Abstract. Canadian Amphibian and Reptile Conservation Network International Symposium on Conservation Biology of Amphibians and Reptiles and Herpetological Research. Quebec City, QC, Canada.

Moll, D., and E. O. Moll. 2004. *The Ecology, Exploitation, and Conservation of River Turtles*. New York: Oxford University Press.

Monahan, W., and K. Gallo. 2014. "Inventory and monitoring of park biodiversity." *Park Science* 31:18-19.

Montague, J. R., and J. W. Poinski. 1978. "Note on the brooding behavior of *Desmognathus fuscus fuscus* (Raf.) (Amphibia, Urodela, Plethodontidae) in Columbiana County, Ohio." *Journal of Herpetology* 12:104.

Moore, J. A. 1939. "Temperature tolerance and rates of development in the eggs of Amphibia." *Ecology* 20:459-478.

Moore, J. M., and M. Ouellet. 2014. "A review of colour phenotypes of the eastern red-backed salamander, *Plethodon cinereus*, in North America." *Canadian Field-Naturalist* 128:250-259.

Moore, J. M., and M. Ouellet. 2015. "Questioning the use of amphibian colour morph as an indicator of climate change." *Global Change Biology* 21:566-571.

Moore, M. J. C., and R. A. Seigel. 2006. "No place to nest or bask: effects of human disturbance on the nesting and basking habits of yellow-blotched map turtles (*Graptemys flavimaculata*)." *Biological Conservation* 130:386-393.

Morris, J. G. 1907. "Maryland Academy of Science and Literature." *Maryland Historical Magazine* 2:259-269.

Morris, M. A. 1974. "Observations on a large litter of the snake *Storeria dekayi*." *Transactions of the Illinois State Academy of Science* 67:359-360.

Morrow, J. L., J. H. Howard, S. A. Smith, and D. K. Poppel. 2001. "Habitat selection and habitat use by the bog turtle (*Clemmys muhlenbergii*) in Maryland." *Journal of Herpetology* 35:545-552.

Mount, R. H. 1975. *The Reptiles and Amphibians of Alabama*. Auburn, AL: Auburn University Agricultural Experiment Station.

MSA (Maryland State Archives). 2015. "Maryland manual: Maryland at a glance: population." Online at http://msa.maryland.gov/msa/mdmanual/01glance/html/pop.html.

MTSG (Marine Turtle Specialist Group). 1996. "*Lepidochelys kempii*." The IUCN Red List of Threatened Species 1996: e.T11533A3292342. Online at http://dx.doi.org/10.2305/IUCN.UK.1996.RLTS.T11533A3292342.en.

Müller-Schwarze, D. 2006. *Chemical Ecology of Vertebrates*. Cambridge: Cambridge University Press.

Musick, J. A. 1972. "Herptiles of the Maryland and Virginia Coastal Plain." In *A Checklist of the Biota of Lower Chesapeake Bay with Inclusions from the Upper Bay and the Virginian Sea*, edited by M. L. Wass et al., 213-242. Special Scientific Report no. 65. Gloucester Point: Virginia Institute of Marine Science.

NAOAC (North American Ornithological Atlas Committee). 1990. *Handbook for Atlasing American Breeding Birds*. Woodstock: Vermont Institute of Natural Science.

Nazdrowicz, N. H. 2009. "Geographic distribution: *Hemidactylium scutatum* (four-toed salamander)." *Herpetological Review* 40:106.

Nazdrowicz, N. H. 2015. "Ecology of the eastern long-tailed salamander (*Eurycea longicauda longicauda*) associated with springhouses." PhD dissertation. University of Delaware, Newark.

NCDC (National Climate Data Center). 2015. *Climate at a Glance: Maryland*. Asheville, NC: National Oceanic and Atmospheric Administration.

Neill, W. T. 1964. "Taxonomy, natural history, and zoogeography of the rainbow snake, *Farancia erytrogramma* (Palisot de Beauvois)." *American Midland Naturalist* 71:257-295.

Nelson, D. H., and J. W. Gibbons. 1972. "Ecology, abundance, and seasonal activity of the scarlet snake, *Cemophora coccinea*." *Copeia* 1972:582-584.

Nemuras, K., and T. Sparhawk. 1966. "Additional records of *Pseudemys rubriventris* and *Pseudemys scripta elegans* in Anne Arundel County, Maryland." *Bulletin of the Maryland Herpetological Society* 2(3):6-8.

Netting, M. G. 1938. "The occurrence of the eastern tiger salamander, *Ambystoma tigrinum tigrinum* (Green), in Pennsylvania and near-by states." *Annals of the Carnegie Museum* 27:159-166.

Netting, M. G. 1946. "Jefferson's salamander in Maryland." *Maryland: A Journal of Natural History* 16:60-61.

Netting, M. G., and N. Richmond. 1932. "The green salamander, *Aneides aeneus*, in northern West Virginia." *Copeia* 1932:101-102.

New York Daily Times. 1854. "Bite of a rattlesnake—drunkenness a remedy." *New York Daily Times* (1851-1857), June 12. ProQuest Historical Newspapers: *The New York Times*.

NHSM (Natural History Society of Maryland) Trustees. 1930. *First Annual Report of the Natural History Society of Maryland for the Year 1929*. Baltimore: Natural History Society of Maryland.

NHSM (Natural History Society of Maryland) Trustees. 1937. *Annual Report of the Natural History Society of Maryland for the Year 1936.* Baltimore: Natural History Society of Maryland.

Nickerson, M. A., and C. E. Mays. 1973. *The Hellbenders: North American "Giant Salamanders."* Milwaukee, WI: Milwaukee Public Museum Press.

Niederberger, A. J. 1993. "Aspects of the ecology of the wood turtle, *Clemmys insculpta* (Le Conte), in West Virginia." MS thesis. Marshall University, Huntington, WV.

Niemiller, M. L. 2011. "Allegheny mountain dusky salamander." In *The Amphibians of Tennessee*, edited by M. L. Niemiller and R. G. Reynolds, 128-130. Knoxville: University of Tennessee Press.

NJENSP (New Jersey Endangered and Nongame Species Program). 2002. *New Jersey's Herp Atlas Project: Herp Atlas Volunteer Training Manual*. Trenton: New Jersey Department of Environmental Protection.

NMFS (National Marine Fisheries Service). 2016. "Kemp's ridley turtle (*Lepidochelys kempii*)." Online at http://www.nmfs.noaa.gov/pr/species/turtles/kempsridley.html.

NMFS (National Marine Fisheries Service) and USFWS (US Fish and Wildlife Service). 1991. *Recovery Plan for the U.S. Population of Atlantic Green Turtle* (Chelonia mydas). Washington, DC: National Marine Fisheries Service.

NOAA (National Oceanic and Atmospheric Administration). 2017a. "Climate at a Glance: U.S. Time Series, Average Temperature." National Centers for Environmental Information. Published January 2017. Online at http://www.ncdc.noaa.gov/cag.

NOAA (National Oceanic and Atmospheric Administration). 2017b. "Climate at a Glance: U.S. Time Series, Precipitation." National Centers for Environmental Information. Published January 2017. Online at http://www.ncdc.noaa.gov/cag.

Noble, G. K. 1931. "The *basilisk*." *Natural History: Journal of the American Museum of Natural History* 31:93-100.

Noble, G. K., and G. Evans. 1932. "Observations and experiments on the life history of the salamander, *Desmognathus fuscus fuscus* (Rafinesque)." *American Museum Novitates*, no. 533.

Noble, G. K., and G. Hassler. 1936. "Three Salientia of geographic interest from southern Maryland." *Copeia* 1936:63-64.

Norden, A. 1971. "A corn snake, *Elaphe guttata guttata*, from western Maryland." *Bulletin of the Maryland Herpetological Society* 7:25-27.

Norden, A. W. 1999. "Flood induced winter mortality of wood turtles (*Clemmys insculpta* Le Conte) in Maryland." *Maryland Naturalist* 43:18-19.

Norden, A. 2005. "The reptiles and amphibians of Cove Point, Calvert County, Maryland." *Bulletin of the Maryland Herpetological Society* 41:1-30.

Norden, A., and M. Browning. 2013. "Another yellowbellied slider (*Trachenys scripta scripta*) from Maryland." *Bulletin of the Maryland Herpetological Society* 49:20-21.

Norden, A. W., D. C. Forester, and G. H. Fenwick, editors. 1984. *Threatened and Endangered Plants and Animals of Maryland*. Natural Heritage Program Special Publication 84-1. Annapolis: Maryland Department of Natural Resources.

Norden, A. W., and B. B. Norden. 1989. "The Mediterranean gecko (*Hemidactylus turcicus*) in Baltimore, Maryland." *Maryland Naturalist* 33:57-58.

Norden, A. W., and B. B. Norden. 1998. "Additional Atlantic Kemp's ridley sea turtles, *Lepidochelys kempi* (Garman), from the Maryland portion of the Chesapeake Bay." *Maryland Naturalist* 42:17-19.

Norden, A. W., T. D. Schofield, and J. J. Evans. 1998. "Sea turtle strandings from Maryland waters reported to the National Aquarium in Baltimore, 1990 through 1997." *Maryland Naturalist* 42:20-23.

Norden, A., and J. Zyla. 1989. "The wood turtle, *Clemmys insculpta*, on the Maryland coastal plain." *Maryland Naturalist* 33:37-41.

Norman, J. 1939. "Amphibians and reptiles found around the Curtis Wright Airport, Bonnie View Golf Club, and Western Run Parkway." *Natural History Society of Maryland, Junior Division Bulletin* 3:58-63.

Norman, J. E. 1949. "Maryland turtles." *Maryland Naturalist* 19:13-16.

Norman, W. 1939a. "Amphibians and reptiles noted around the McMahon Quarry during 1939." *Natural History Society of Maryland, Junior Division Bulletin* 3:40-42.

Norman, W. 1939b. "Record of wood turtle from eastern shore, Maryland." *Natural History Society of Maryland, Junior Division Bulletin* 3:64.

"Note to Miss Davis." 1935. "Note to Miss Davis regarding the absence of specific reports of brown snake, *Pitamophis striatulus*, in the records of William Fisher, and to exclude the species from the Snakes of Maryland draft." Dr. Howard Kelly file. Natural History Society of Maryland Archives, Baltimore.

Nuebert, S. 1996. *Mammalian Predation on Turtle Nests*. Technical Report of the Jug Bay Wetlands Sanctuary. Lothian, MD: Jug Bay Wetlands Sanctuary.

Oldham, J. C., and H. M. Smith. 1991. "The generic status of the smooth green snake, *Opheodrys vernalis*." *Bulletin of the Maryland Herpetological Society* 27:201-215.

Oliver, J. A. 1955. *North American Amphibians and Reptiles*. Princeton, NJ: D. Van Nostrand Company.

Organ, J. A. 1961. "Studies of the local distribution, life history, and population dynamics of the salamander genus *Desmognathus* in Virginia." *Ecological Monographs* 31:189-220.

Orr, J. M. 2006. "Microhabitat use by the eastern worm snake, *Carphophis amoenus*." *Herpetological Bulletin* 97:29-35.

Orr, L. P. 1989. "*Desmognathus ochrophaeus.*" In *Salamanders of Ohio*, edited by R. A. Pfingsten and F. L. Downs, 181-189. Ohio Biological Survey Bulletin, New Series 7, no. 2. Columbus: College of Biological Sciences, Ohio State University.

Orser, P. N., and D. J. Shure. 1972. "Effects of urbanization on the salamander *Desmognathus fuscus fuscus.*" *Ecology* 53:1148-1154.

Orser, P. N., and D. J. Shure. 1975. "Population cycles and activity patterns of the dusky salamander, *Desmognathus fuscus fuscus.*" *American Midland Naturalist* 93:403-410.

Ortenburger, A. I., and R. D. Ortenburger. 1933. "Howard Atwood Kelly." *Publications of the Oklahoma Biological Survey* 5:9-13.

Otto, C. R. V. 2006. "Influences of wetland and landscape characteristics on the distribution of carpenter frogs in the northern extent of their range." MS thesis. Towson University, Towson, MD.

Otto, C. R. V., D. C. Forester, and J. W. Snodgrass. 2007a. "Influences of wetland and landscape characteristics on the distribution of carpenter frogs." *Wetlands* 27:261-269.

Otto, C. R. V., J. W. Snodgrass, D. C. Forester, J. C. Mitchell, and R. W. Miller. 2007b. "Climatic variation and the distribution of an amphibian polyploid complex." *Journal of Animal Ecology* 76:1053-1061.

Pais, R. C., S. A. Bonney, and W. C. McComb. 1988. "Herpetofaunal species richness and habitat associations in an eastern Kentucky forest." *Proceedings of the Annual Conference of the Southeastern Association of Fish and Wildlife Agencies* 42:448-455.

Pallin, L., E. M. Adams, H. F. Goyert, A. S. Friedlaender, and D. W. Johnston. 2015. "Density modeling for marine mammals and sea turtles with environmental covariates." In *Wildlife Densities and Habitat Use across Temporal and Spatial Scales on the Mid-Atlantic Outer Continental Shelf: Final Report to the Department of Energy EERE Wind & Water Power Technologies Office*, edited by K. A. Williams, E. E. Connelly, S. M. Johnson, and I. J. Stenhouse, chap. 15. Report BRI 2015-11. Portland, ME: Biodiversity Research Institute.

Palmer, W. M., and A. L. Braswell. 1995. *Reptiles of North Carolina.* Raleigh: University of North Carolina Press.

Palmer, W. M., and G. Tregembo. 1970. "Notes on the natural history of the scarlet snake *Cemophora coccinea copei* Jan in North Carolina." *Herpetologica* 26:300-302.

Parks, M. 1998. *Basking Ecology of Red-bellied Turtles.* Technical Report of the Jug Bay Wetlands Sanctuary. Lothian, MD: Jug Bay Wetlands Sanctuary.

PARS. 2017. "Pennsylvania amphibian & reptile survey." Online at https://www.paherpsurvey.org.

Parsons, J. J. 1962. *The Green Turtle and Man.* Gainesville: University of Florida Press.

Pauley, T. K., and W. H. England. 1969. "Time of mating and egg deposition in the salamander, *Plethodon wehrlei* Fowler and Dunn, in West Virginia." *Proceedings of the West Virginia Academy of Science* 41:155-160.

Pauley, T. K., and M. B. Watson. 2005. "*Eurycea cirrigera* (Green, 1830), southern two-lined salamander." In *Amphibian Declines: The Conservation Status of United States Species*, edited by M. J. Lannoo, 740-743. Berkeley: University of California Press.

Pauley, T. K, M. K. Watson, and J. C. Mitchell. 2005. *Reptile and Amphibian Inventories in Eight Parks in the National Capital Region.* Final report TIC no. NCRO-D-61. Washington, DC: National Park Service, National Capitol Region Network, Center for Urban Ecology.

Pearson, P. G. 1955. "Population ecology of spadefoot toads, *Scaphiopus h. holbrookii* (Harlan)." *Ecological Monographs* 25:233-267.

Peel, M. C., B. L. Finlayson, and T. A. Mcmahon. 2007. "Updated world map of the Köppen-Geiger' climate classification." *European Geosciences Union, Hydrology and Earth System Sciences Discussions* 4:439-473.

Perkins, D. W. 1885. *Guide Book through the World's Industrial and Cotton Centennial Exposition at New Orleans.* Harrisburg, PA: Lane S. Hart Printers and Binder.

Perring, E. H., and S. M. Walters. 1962. *Atlas of the British Flora.* London: T. Nelson.

Peterson, E. E., and N. S. Urquhart. 2006. "Predicting water quality impaired stream segments using landscape-scale data and a regional geostatistical model: a case study in Maryland." *Environmental Monitoring and Assessment* 21:613-636.

Petiver, J. 1698. Remarks by Mr. James Petiver, apothecary, and fellow of the Royal Society, on some animals, plants, etc. sent to him from Maryland, by the Reverend Mr. Hugh Jones. *Proceedings of the Royal Society of London, Philosophical Transactions of the Royal Society* 20:393-406.

Petiver, J. 1767. "Gazophylacii Naturae & Artis Decas Prima." In *Jacobi Petiveri Opera, Historiam Naturalem Spectantia . . .* London: Printed for John Millan, Bookseller, near White Hall.

Petranka, J. W. 1998. *Salamanders of the United States and Canada.* Washington, DC, and London: Smithsonian Institution Press.

Phillips, C. A. 1994. "Geographic distribution of mitochondrial DNA variants and the historical biogeography of the spotted salamander, *Ambystoma maculatum.*" *Evolution* 48:597-607.

Phillips, J. B. 1987. "Laboratory studies of homing orientation in the eastern red-spotted newt, *Notophthalmus viridescens.*" *Journal of Experimental Biology* 131:215-229.

Piersol, W. H. 1910. "Spawn and larva of *Ambystoma jeffersonianum.*" *American Naturalist* 44:732-738.

Platania, S. 1976. "A comment on the distribution of *Gyrinophilus porphyriticus* in western Maryland." *Bulletin of the Maryland Herpetological Society* 12:101.

Platt, D. R. 1969. "Natural history of the hognose snakes, *Heterodon platyrhinos* and *Heterodon nasicus.*" *University of Kansas Publication, Museum of Natural History* 18:253-420.

Platz, J. E. 1989. "Speciation within the chorus frog *Pseudacris triseriata*: morphometric and mating call analysis of the boreal and western subspecies." *Copeia* 1989:704-712.

Platz, J. E., and D. C. Forester. 1988. "Geographic variation in mating call among the four subspecies of the chorus frog: *Pseudacris triseriata.*" *Copeia* 1988:1062-1066.

Plenderleith, T. L. 2009. "The distribution and movement of the northern dusky salamander (*Desmognathus fuscus*) within a third-order stream system." MS thesis. Towson University, Towson, MD.

Plummer, M. V. 1991. "Patterns of feces production in free-living green snakes, *Opheodrys aestivus.*" *Journal of Herpetology* 25:222-226.

Plummer, M. W., and N. E. Mills. 1996. "Observations on trailing and mating behavior in hognose snakes (*Heterodon platirhinos*)." *Journal of Herpetology* 30:80–82.

Polley, K. P. 1989. "A *Terrapene carolina triunguis* from Maryland." *Bulletin of the Maryland Herpetological Society* 25:58–59.

Pollio, C. A., and S. L. Kilpatrick. 2002. "Status of *Pseudacris feriarum* in Prince William County, Virginia." *Bulletin of the Maryland Herpetological Society* 38:55–61.

Poole, K. G., and C. G. Murphy. 2006. "Preferences of female barking treefrogs, *Hyla gratiosa*, for larger males: univariate and composite tests." *Animal Behavior* 73:513–524.

Pope, C. H. 1924. "Notes on North Carolina salamanders with especial reference to the egg-laying habits of *Leurognathus* and *Desmognathus*." *American Museum Novitates* 153:1–15.

Pope, C. H. 1928. "Some plethodontid salamanders from North Carolina and Kentucky, with a description of a new race of *Leurognathus*." *American Museum Novitates* 306:1–19.

Pope, C. H. 1944. *Amphibians and Reptiles of the Chicago Area*. Chicago: Chicago Natural History Museum Press.

Posey, C. R. 1973. "An observation on the feeding habits of the eastern king snake." *Bulletin of the Maryland Herpetological Society* 9:105.

Powell, R., R. Conant, and J. T. Collins. 2016. *Peterson Field Guide to Reptiles and Amphibians of Eastern and Central North America*, 4th ed. Boston and New York: Houghton Mifflin Harcourt.

Price, S. J., and M. E. Dorcas. 2011. "The Carolina Herp Atlas: an online, citizen-science approach to document amphibian and reptile occurrences." *Herpetological Conservation and Biology* 6:287–296.

Prince, E. C. 1957. "A typical Maryland habitat for the slimy salamander." *Maryland Naturalist* 27:5.

Prince, E. C., R. Duppstasdt, and D. Lyons. 1955. "An annotated list of amphibians and reptiles from the Broad Creek–Deep Run area, Harford County, Maryland." *Maryland Naturalist* 25:9–12.

Pritchard, P. C. H. 1990. "Kemp's ridleys are rarer than we thought." *Marine Turtle Newsletter* 49:1–3.

Pritchard, P. C. H., and R. Marquez M. 1973. *Kemp's Ridley Turtle or Atlantic Ridley*, Lepidochelys kempi. IUCN Monograph no. 2, Marine Turtle Series. Morges, Switzerland: International Union for Conservation of Nature.

Proffitt, K. 2001. "The status of the wood turtle, *Clemmys insculpta*, on the eastern edge of the range in Maryland." MS thesis. Towson University, Towson, MD.

Pyron, R. A., and F. T. Burbrink. 2009. "Systematics of the common kingsnake (*Lampropeltis getula*; Serpentes: Colubridae) and the burden of heritage in taxonomy." *Zootaxa* 2241:22–32.

Quinlan, M., C. Swarth, and M. Marchand. 2003. "Characteristics of a high density eastern box turtle population on Maryland's Coastal Plain." Abstract. In *Conservation and Ecology of Turtles of the Mid-Atlantic Region*, edited by C. W. Swarth, W. M. Roosenburg, and E. Kiviat, 110. Salt Lake City, UT: Bibliomania.

Rabon, D. R., Jr., S. A. Johnson, R. Boettcher, M. Dodd, M. Lyons, S. Murphy, S. Ramsey, S. Roff, and K. Stewart. 2003. "Confirmed leatherback turtle (*Dermochelys coriacea*) nests from North Carolina, with a summary of leatherback nesting activities north of Florida." *Marine Turtle Newsletter* 101:4–8.

Radzio, T. A., and W. M. Roosenburg. 2005. "Diamondback terrapin mortality in the American eel pot fishery and evaluation of a bycatch reduction device." *Estuaries* 28:620–626.

Radzio, T. A., J. A. Smolinsky, and W. M. Roosenburg. 2013. "Low use of required terrapin bycatch reduction devices in a recreational crab pot fishery." *Herpetological Conservation and Biology* 8:222–227.

Rambo, K. 1992. "A specimen of the eastern spiny soft-shelled turtle, Apalone spinifer spinifer [*sic*] (LeSueur), from St. Mary's County, Maryland." *Maryland Naturalist* 36:3–4.

Rand, A. S. 1952. "Jumping ability of certain anurans with notes on endurance." *Copeia* 1952:15–20.

Randolph, M. R. 1984. "A survey of the order Caudata in Catoctin Mountain Park." Unpublished report to Catoctin Mountain Park.

Reading, C. J., L. M. Luiselli, G. C. Akani, X. Bonnet, G. Amori, J. M. Ballouard, E. Filippi, G. Naulleau, D. Pearson, and L. Rugiero. 2010. "Are snake populations in widespread decline?" *Biology Letters* 6:777–780.

Rebel, T. P. 1974. *Sea Turtles and the Turtle Industry of the West Indies, Florida, and the Gulf of Mexico*. Coral Gables, FL: University of Miami Press.

Redner, M. 2005. "*Rana palustris* LeConte, 1825, pickerel frog." In *Amphibian Declines: The Conservation Status of United States Species*, edited by M. J. Lannoo, 568–570. Berkeley: University of California Press.

Reed, C. F. 1956a. "Contributions to the herpetology of Maryland and Delmarva, 2: the herpetofauna of Harford County, Maryland." *Journal of the Washington Academy of Sciences* 46:58–60.

Reed, C. F. 1956b. *Contributions to the Herpetology of Maryland and Delmarva, 5: Bibliography to the Herpetology of Maryland, Delmarva, and the District of Columbia*. Baltimore: Privately published by Reed Herpetorium.

Reed, C. F. 1956c. *Contributions to the Herpetology of Maryland and Delmarva, 6: An Annotated Check List of the Lizards of Maryland and Delmarva*. Baltimore: Privately published by Reed Herpetorium.

Reed, C. F. 1956d. *Contributions to the Herpetology of Maryland and Delmarva, 7: An Annotated Check List of the Turtles of Maryland and Delmarva*. Baltimore: Privately published by Reed Herpetorium.

Reed, C. F. 1956e. *Contributions to the Herpetology of Maryland and Delmarva, 8: An Annotated Check List of the Snakes of Maryland and Delmarva*. Baltimore: Privately published by Reed Herpetorium.

Reed, C. F. 1956f. *Contributions to the Herpetology of Maryland and Delmarva, 9: An Annotated Check List of the Frogs and Toads of Maryland and Delmarva*. Baltimore: Privately published by Reed Herpetorium.

Reed, C. F. 1956g. *Contributions to the Herpetology of Maryland and Delmarva, 11: An Annotated Herpetofauna of the Del-Mar-Va Peninsula, Including Many New or Additional Localities*. Baltimore: Privately published by Reed Herpetorium.

Reed, C. F. 1956h. "Distribution of the wood turtle, *Clemmys insculpta*, in Maryland." *Herpetologica* 12:80.

Reed, C. F. 1956i. "*Hyla cinerea* in Maryland, Delaware, and Virginia, with notes on the taxonomic status of *Hyla cinerea evittata*." *Journal of the Washington Academy of Sciences* 46:328–332.

Reed, C. F. 1956j. "The spadefoot toad in Maryland." *Herpetologica* 12:294-295.

Reed, C. F. 1957a. *Contributions to the Herpetology of Maryland and Delmarva, 10: An Annotated Check List of the Salamanders of Maryland and Delmarva*. Baltimore: Privately published by Reed Herpetorium.

Reed, C. F. 1957b. "Contributions to the herpetology of Maryland and Delaware, 12: the herpetofauna of Anne Arundel County, MD." *Journal of the Washington Academy of Sciences* 47:64-66.

Reed, C. F. 1957c. "Contributions to the herpetology of Maryland and Delaware, 15: the herpetofauna of Somerset County, Md." *Journal of the Washington Academy of Sciences* 47:127-128.

Reed, C. F. 1957d. "*Rana virgatipes* in southern Maryland, with notes upon its range from New Jersey to Georgia." *Herpetologica* 13:137-138.

Reed, C. F. 1958a. "Contributions to the herpetology of Maryland and Delmarva, 13: Piedmont herpetofauna on coastal Delmarva." *Journal of the Washington Academy of Sciences* 48:95-99.

Reed, C. F. 1958b. "Contributions to the herpetology of Maryland and Delmarva, 17: southeastern herptiles with northern limits on coastal Maryland, Delmarva, and New Jersey." *Journal of the Washington Academy of Sciences* 48:28-32.

Reger, J. P., and E. T. Cleaves. 2008. *Explanatory Text for the Physiographic Map of Maryland*. Version MDPHYS2003.2. Baltimore: Maryland Geological Survey.

Resetarits, W. J., Jr. 1986. "Ecology of cave use by the frog, *Rana palustris*." *American Midland Naturalist* 116:256-266.

Richards-Dimitrie, T. M. 2011. "Spatial ecology and diet of Maryland endangered northern map turtles (*Graptemys geographica*) in an altered river system: implications for conservation and management." MS thesis. Towson University, Towson, MD.

Richards-Dimitrie, T., S. E. Gresens, S. A. Smith, and R. A. Seigel. 2013. "Diet of northern map turtles (*Graptemys geographica*): sexual differences and potential impacts of an altered river system." *Copeia* 2013:477-484.

Richmond, N. D. 1945. "The habits of the rainbow snake in Virginia." *Copeia* 1945:28-30.

Richmond, N. D. 1956. "Autumn mating of the rough green snake." *Herpetologica* 12:325.

Rider, D. R. 2015. "Forestry and tree planting in Maryland." *Tree Planters Notes*, no. 2.

Robbins, C. S., and E. A. T. Blom, editors. 1996. *Atlas of the Breeding Birds of Maryland and the District of Columbia*. Pittsburgh, PA: University of Pittsburgh Press.

Robertson, H. C. 1939. "A preliminary report on the reptiles and amphibians of Catoctin National Park, Maryland." *Bulletin of the Natural History Society of Maryland* 9:88-93.

Robertson, H. C. 1947. "Notes on the green turtle in marine waters of Maryland." *Maryland: A Journal of Natural History* 17:29-32.

Robertson, M. P., G. S. Cumming, and B. F. N. Erasmus. 2010. "Getting the most out of atlas data." *Diversity and Distributions* 16:363-375.

Robinson, K. M., and G. G. Murphy. 1978. "The reproductive cycle of the eastern spiny softshell turtle (*Trionyx spiniferus spiniferus*)." *Herpetologica* 34:137-140.

Robison, R. F. 2010. "Howard Atwood Kelly (1858-1943): founding professor of gynecology at Johns Hopkins Hospital & pioneer American radium therapist." *Journal of Oncology* 60:21e-35e.

Rochford, M. R., J. M. Lemm, K. L. Krysko, L. A. Somma, R. W. Hansen, and F. J. Mazzotti. 2015. "Spreading holiday spirit and northwestern salamanders, *Ambystoma gracile* (Baird 1859) (Caudata: Ambystomatidae), across the USA." *IRCF Reptiles and Amphibians* 22:126-127.

Roeder, W. G. 1966. "Herpetofauna conservation: can it succeed?" *Bulletin of the Maryland Herpetological Society* 2(1):2.

Roosenburg, W. M. 1991. "The diamondback terrapin: habitat requirements, population dynamics, and opportunities for conservation." In *New Perspectives in the Chesapeake System: A Research and Management Partnership: Proceedings of a Conference, December 4-6, 1990, Baltimore, MD*, edited by J. A. Mihirsky and A. Chaney, 227-234. Chesapeake Research Consortium Publication no. 137. Solomons, MD: Chesapeake Research Consortium.

Roosenburg, W. M. 1992. "Life history consequences of nest site choice by the diamondback terrapin, *Malaclemys terrapin*." PhD dissertation. University of Pennsylvania, Philadelphia.

Roosenburg, W. M., and K. C. Kelley. 1996. "The effect of egg size and incubation temperature on growth in the turtle, *Malaclemys terrapin*." *Journal of Herpetology* 30:198-204.

Roosenburg, W. M., D. M. Spontak, S. P. Sullivan, E. L. Matthews, M. L. Heckman, R. J. Trimbath, R. P. Dunn, E. A. Dustman, L. Smith, and L. J. Graham. 2014. "Nesting habitat creation enhances recruitment in a predator free environment: *Malaclemys* nesting at the Paul S. Sarbanes ecosystem restoration project." *Restoration Ecology* 22:815-823.

Ross, J. P., S. Beavers, D. Mundell, and M. Airth-Kindree. 1989. *The Status of Kemp's Ridley*. Washington, DC: Center for Marine Conservation.

Rossman, D. A. 1963. "The Colubrid snake genus *Thamnophis*: a revision of the *Sauritis* group." *Bulletin of the Florida State Museum, Biological Science* 77:99-178.

Rossman, D. A., N. B. Ford, and R. A. Seigel. 1996. *The Garter Snakes: Evolution and Ecology*. Norman: University of Oklahoma Press.

Rossman, D. A., and P. A. Myer. 1990. "Behavioral and morphological adaptations for snail extraction in the North American brown snakes (genus *Storeria*)." *Journal of Herpetology* 24:434-438.

Roth, N. E., M. T. Southerland, G. Mercurio, J. C. Chaillou, P. F. Kazyak, S. S. Stranko, A. P. Prochaska, D. G. Heimbuch, and J. C. Seibel. 1999. *The State of the Streams: 1995-1997 Maryland Biological Stream Survey Results*. Annapolis: Maryland Department of Natural Resources.

Rothblum, L., and T. A. Jenssen. 1978. "Display repertoire analysis of *Sceloporus undulatus hyacinthinus* (Sauria: Iguanidae) from south-western Virginia." *Animal Behaviour* 26:130-137.

Ruane, S., R. W. Bryson, Jr., R. A. Pyron, and F. T. Burbrink. 2014. "Coalescent species delimitation in milksnakes (genus *Lampropeltis*) and impacts of phylogenetic comparative analyses." *Systematic Biology* 63:231-250.

Rubbo, M. J., V. R. Townsend, Jr., S. D. Smyers, and R. G. Jaeger. 2001. "The potential for invertebrate-vertebrate intraguild predation: the predatory relationship between wolf spiders (*Gladicosa pulchra*) and ground skinks (*Scincella lateralis*)." *Canadian Journal of Zoology* 79:1465-1471.

Ruhe, B. M., and R. Koval. 2009. Acris crepitans: *Status in Pennsyl-*

vania. Pennsylvania Biological Survey, Amphibian and Reptile Technical Committee Report. Altoona: Pennsylvania Biological Survey.

Sauer, J. R., W. A. Link, J. E. Fallon, K. L. Pardieck, and D. J. Ziolkowski, Jr. 2013. "The North American Breeding Bird Survey 1966-2011: summary analysis and species accounts." *North American Fauna* 79:1-32.

Saul, D. 1966. "Conservation laws for herpetology." *Bulletin of the Maryland Herpetological Society* 2(1):3.

Saunders, C., and L. H. Benedict. 2008. "Status of the rainbow snake, *Farancia erytrogramma*, in southern Maryland." *Bulletin of the Maryland Herpetological Society* 44:122-135.

Savva, Y., C. W. Swarth, J. Gupchup, and K. Szlavecz. 2010. "Thermal environments of overwintering eastern box turtles (*Terrapene carolina carolina*)." *Canadian Journal of Zoology* 88:1086-1094.

Say, T. 1825. "Descriptions of three new species of *Coluber*, inhabiting the United States." *Journal of the Academy of Natural Sciences of Philadelphia* 4:237-241.

Sayler, A. 1966. "The reproductive ecology of the red-backed salamander, *Plethodon cinereus*, in Maryland." *Copeia* 1966:183-193.

Scarpulla, E. J. 1989. "First records for the leatherback turtle (*Dermochelys coriacea*) along Maryland's Atlantic coast." *Maryland Naturalist* 33:59-60.

Schaaf, R. T., and P. W. Smith. 1970. "Geographic variation in the pickerel frog." *Herpetologica* 26:240-254.

Scharf, J. T. 1882. *History of Western Maryland: Being a History of Frederick, Montgomery, Carroll, Washington, Allegany, and Garrett Counties from the Earliest Period to the Present Day; Including Biographical Sketches of the Representative Men*, vol. I. Philadelphia: Everts.

Schlesinger, M. D., J. A. Feinberg, N. H. Nazdrowicz, J. D. Kleopfer, J. Beane, J. F. Bunnell, J. Burger, E. Corey, K. Gipe, J. W. Jaycox, E. Kiviat, J. Kubel, D. Quinn, C. Raithel, S. Wenner, E. L. White, B. Zarate, and H. B. Shaffer. 2017. "Distribution, identification, landscape setting, and conservation of *Rana kauffeldi* in the northeastern U.S." Report to the Wildlife Management Institute for grant RCN 2013-03.

Schmidt, M. F., Jr. 2010. *Maryland's Geology*. Atglen, PA: Schiffer Publishing.

Schreiber, J. F. 1952. "Ecology of a Bare Hills cave, Bare Hill, Maryland." *Maryland Naturalist* 22:9-18.

Schuett, G. W., and R. E. Gatten, Jr. 1980. "Thermal preference in snapping turtles (*Chelydra serpentina*)." *Copeia* 1980:149-152.

Schwalbe, C. R., and P. C. Rosen. 1988. "Preliminary report on effect of bullfrogs on wetland herpetofauna in southeastern Arizona." In *Proceedings of the Symposium on Management of Amphibians, Reptiles, and Small Mammals in North America*, edited by R. C. Szaro, K. E. Severson, and D. R. Patton, 166-173. General Technical Report RM-166. Fort Collins, CO: US Department of Agriculture Forest Service.

Schwartz, A. 1954. "The salamander *Aneides aeneus* in South Carolina." *Copeia* 154:296-298.

Schwartz, F. J. 1961. *Maryland Turtles*. Educational Series no. 50. Solomons, MD: Maryland Department of Research and Education.

Schwartz, F. J. 1967. *Maryland Turtles*. Natural Resources Institute, Education Series no. 79. Solomons, MD: Natural Resources Institute, University of Maryland.

Schwartz, F. J. 1978. "Behavioral and tolerance responses to cold water temperatures by three species of sea turtles (Reptilia, Cheloniidae) in North Carolina." *Florida Marine Resources Publication* 33:16-18.

Schwartz, F. J., and B. L. Dutcher. 1961. "A record of the Mississippi map turtle, *Graptemys kohni*, in Maryland." *Chesapeake Science* 2:100-101.

Scott, J. 1801. *A Geographical Description of the States of Maryland and Delaware*. Philadelphia: Kimber, Conrad, and Company.

Scribner, S. J., and P. J. Weatherhead. 1995. "Locomotion and antipredator behaviour in three species of semi-aquatic snakes." *Canadian Journal of Zoology* 73:321-329.

Seibert, H. C., and R. A. Brandon. 1960. "The salamanders of southeastern Ohio." *Ohio Journal of Science* 60:291-303.

Seigel, R. A. 2005. "The importance of population demography in the conservation of box turtles: what do we know and what do we need to learn?" In *Summary of the Eastern Box Turtle Regional Conservation Workshop*, edited by C. W. Swarth and S. Hagood, 6-7. Washington, DC: The Humane Society of the United States.

Seigel, R. A., and H. S. Fitch. 1985. "Annual variation in reproduction in snakes in a fluctuating environment." *Journal of Animal Ecology* 54:497-505.

Selleck, J. 2014. "Cameras and cell phones at the bioblitz." *Park Science* 31:102-103.

Semlitsch, R. D. 1980. "Geographic and local variation in population parameters of the slimy salamander *Plethodon glutinosus*." *Herpetologica* 36:6-16.

Semlitsch, R. D. 1988. "Allotopic distribution of two salamanders: effects of fish predation and competitive interactions." *Copeia* 1988:290-298.

Sever, D. M. 2005. "*Eurycea bislineata* (Green, 1818), northern two-lined salamander." In *Amphibian Declines: The Conservation Status of United State Species*, edited by M. J. Lannoo, 735-738. Berkeley: University of California Press.

Sever, D. M., and W. A. Hopkins. 2004. "Oviductal sperm storage in the ground skink *Scincella laterale* Holbrook (Reptilia: Scincidae)." *Journal of Experimental Zoology* 301A:599-601.

Sexton, O. J., J. Bizer, D. C. Gayou, P. Freiling, and M. Moutseous. 1986. "Field studies of breeding spotted salamanders, *Ambystoma maculatum*, in eastern Missouri, U.S.A." *Milwaukee Public Museum, Contributions in Biology and Geology* 67:1-19.

Sexton, O. J., C. Phillips, and J. E. Bramble. 1990. "The effects of temperature and precipitation on the breeding migration of the spotted salamander (*Ambystoma maculatum*)." *Copeia* 1990: 781-787.

Shaffer, L. L. 1991. *Pennsylvania Amphibians and Reptiles*. Harrisburg: Pennsylvania Fish Commission.

Shoop, C. R. 1968. "Migratory orientation of *Ambystoma maculatum*: movements near breeding ponds and displacements of migrating individuals." *Biological Bulletin* 134:230-238.

Shoop, C. R. 1974. "Yearly variation in larval survival of *Ambystoma maculatum*." *Ecology* 55:440-444.

Shufeldt, R. W. 1919. "Exhibition of a young specimen of the wood tortoise [*Clemmys insculpta* (Le Conte)]." *Journal of the Washington Academy of Sciences* 9:656.

Shufeldt, R. W. 1921. "Snake lore for forest lovers." *American Forestry* 27:445-454.

Sias, J. 2006. "Natural history and distribution of the upland chorus frog, *Pseudacris feriarum* Baird, in West Virginia." MS thesis. Marshall University, Huntington, WV.

Simmons, R. S., and J. B. Hanzley. 1952. "New county records for the wood turtle (*Clemmys inscripta*) in Maryland." *Maryland Naturalist* 22:22-23.

Simon, J. A., J. W. Snodgrass, R. E. Casey, and D. W. Sparling. 2009. "Spatial correlates of amphibian use of constructed wetlands in an urban landscape." *Landscape Ecology* 24:361-373.

Simpson, D. S. 2009. "The relative role of pH in structuring depression wetland amphibian communities: implications for conservation and management." MS thesis. Towson University, Towson, MD.

Sipple, W. S. 2008. "Note on box turtle (*Terrapene c. carolina*) aestivating along a marshy stream bank." *Bulletin of the Maryland Herpetological Society* 44:136-137.

Sites, J. W., Jr. 1978. "The foraging strategy of the dusky salamander, *Desmognathus fuscus* (Amphibia: Urodela: Plethodontidae): an empirical approach to predation theory." *Journal of Herpetology* 12:373-383.

Skinner, J. S. 1819. "Practical instructions for naturalists." *American Farmer* 1:188-190.

Slater, J. R. 1955. "Distribution of Washington amphibians." *Department of Biology, College of Puget Sound, Occasional Papers* 16: 120-154.

Slaughter, T. P. 1996. *The Natures of John and William Bartram*. Philadelphia: University of Pennsylvania Press.

Smallwood, W. M., and M. S. C. Smallwood. 1941. *Natural History and the American Mind*. New York: AMS Press.

Smith, C. K. 1983. "Notes on breeding period, incubation period, and egg masses of *Ambystoma jeffersonianum* (Green) (Amphibia: Caudata) from the southern limits of its range." *Brimleyana* 9: 135-140.

Smith, C. K., and J. W. Petranka. 1987. "Prey size distributions and size-specific foraging success of *Ambystoma* larvae." *Oecologia* 71:239-244.

Smith, D. G. 1997. "Ecological factors influencing the antipredator behaviors of the ground skink, *Scincella lateralis*." *Behavioral Ecology* 8:622-629.

Smith, E. M., and F. H. Pough. 1994. "Intergeneric aggression in salamanders." *Journal of Herpetology* 28:41-45.

Smith, E. R., J. C. Reed, and I. L. Delwiche. 2016. *The Atlantic Coast of Maryland, Sediment Budget Update: Tier 2, Assateague Island and Ocean City Inlet*. ERDC/CHL CHETN-XIV-48. Vicksburg, MI: US Army Engineer Research and Development Center.

Smith, G. R., A. Todd, J. E. Rettig, and F. Nelson. 2003. "Microhabitat selection by northern cricket frogs (*Acris crepitans*) along a west-central Missouri creek: field and experimental observations." *Journal of Herpetology* 37:383-385.

Smith, H. M. 1946. *Handbook of Lizards: Lizards of the United States and of Canada*. Ithaca, NY: Cornell University Press.

Smith, P. W. 1961. "The amphibians and reptiles of Illinois." *Illinois Natural History Survey Bulletin* 28:1-298.

Smith, S. A. 2004. "Status of the bog turtle and conservation efforts in Maryland." In *Bog Turtle Conservation in Maryland: Use of the Public and Private Sectors in Protection and Management of Small Isolated Wetlands*, edited by D. S. Lee, C. W. Swarth, and K. A. Buhlmann, 8-12. Lothian, MD: Jug Bay Wetlands Sanctuary.

Smith, S. [A.]. 2013. "Alas, poor Randall, I knew him well." Unpublished account in the *Maryland Amphibian & Reptile Atlas Newsletter*, January 2013, 2-4.

Smith, S. [A.]. 2014. "A tale of two sirens." Unpublished account in the *Maryland Amphibian & Reptile Atlas Newsletter*, July 2014, 2-9.

Smith, W. H., H. R. Cunningham, and A. Hall. 2011. "*Plethodon glutinosus*: arboreal behavior." *Herpetological Review* 42:582.

Smithberger, S. 1995. "Unusual *Nerodia sipedon* found at Jug Bay Wetlands Sanctuary." *Bulletin of the Maryland Herpetological Society* 31:100-101.

Smithberger, S. I., and C. W. Swarth. 1993. "Reptiles and amphibians of the Jug Bay Wetlands Sanctuary." *Maryland Naturalist* 37: 28-46.

Snider, A. T., and J. K. Bowler. 1992. "Longevity of reptiles and amphibians in North American collection." *Society for the Study of Amphibians and Reptiles, Herpetological Circular* 21:1-40.

Snodgrass, J. W., R. E. Casey, D. Joseph, and J. A. Simon. 2008. "Microcosm investigations of stormwater pond sediment toxicity to embryonic and larval amphibians: variation in sensitivity among species." *Environmental Pollution* 154:291-297.

Snodgrass, J. W., D. C. Forester, M. Lehti, and E. Lehman. 2007. "Dusky salamander nest-site selection over multiple spatial scales." *Herpetologica* 63:441-449.

Snodgrass, J. W., M. J. Komoroski, A. L. Bryan, Jr., and J. Burger. 2000. "Relationships among isolated wetland size, hydroperiod, and amphibian species richness: implications for wetland regulations." *Conservation Biology* 14:414-419.

Snyder, D. H. 1973. "Some adaptive value of brooding behavior in *Aneides aeneus*." *Herpetological Review* 1:63.

Somers, A. B., J. Mansfield-Jones, and J. Braswell. 2007. "In stream, streamside, and under stream bank movements of a bog turtle, *Glyptemys muhlenbergii*." *Chelonian Conservation and Biology* 6: 286-288.

Somers, A. B., and C. E. Mathews. 2006. *The Box Turtle Connection: A Passageway into the Natural World*. Greensboro: University of North Carolina.

Somma, L. A., A. Foster, and P. Fuller. 2018. "*Trachemys scripta elegans* (Weid-Neuwied, 1838)." US Geological Survey, Nonindigenous Aquatic Species Database, Gainesville, FL. Online at https://nas.er.usgs.gov/queries/FactSheet.aspx?speciesID=1261.

Southerland, M. T., R. E. Jung, D. P. Baxter, I. C. Chellman, G. Mercurio, and J. H. Völstad. 2004. "Stream salamanders as indicators of stream quality in Maryland, USA." *Applied Herpetology* 2:23-46.

Spotila, J. R. 2004. *Sea Turtles: A Complete Guide to Their Biology, Behavior, and Conservation*. Baltimore: Johns Hopkins University Press.

Spotila, J. R., and B. A. Bell. 2008. "Thermal ecology and feeding of the snapping turtle, *Chelydra serpentina*." In *Biology of the Snapping Turtle* (Chelydra serpentina), edited by A. C. Steyermark, M. S. Finkler, and R. J. Brooks, 71-79. Baltimore: Johns Hopkins University Press.

Standaert, W. F. 1967. "Growth, maturation, and population ecology of the carpenter frog (*Rana virgatipes*, Cope)." PhD dissertation. Rutgers University, New Brunswick, NJ.

Stearns, R. P. 1953. "James Petiver, promoter of natural science." *American Antiquarian Society Proceedings* 62:243-365.

Steen, D. A., J. M. Lineham, and L. L. Smith. 2010. "Multiscale habitat selection and refuge use of common kingsnakes, *Lampropeltis getula*, in southwestern Georgia." *Copeia* 2010:227-231.

Stein, B. A. 2002. *States of the Union: Ranking America's Biodiversity*. Arlington, VA: NatureServe.

Stein, B. A., L. S. Kutner, and J. S. Adams, editors. 2000. *Precious Heritage: The Status of Biodiversity in the United States*. New York: Oxford University Press.

Stein, R. A. 1977. "Selective predation, optimal foraging, and the predator-prey interaction between fish and crayfish." *Ecology* 58:1237-1253.

Stejneger, L., and T. Barbour. 1923. *A Check List of North American Amphibians and Reptiles*, 2nd ed. Cambridge, MA: Museum of Comparative Zoology.

Steuart, R. D. 1931. "A letter to the editor of the Journal of the Maryland Academy of Sciences, Col. R. B. Owens, D.S.O., D.Sc., F.R.S.C., relative to the Academy's needs and to its present activities; from Richard Steuart, Esq., Chairman, Membership Committee, The Maryland Academy of Sciences." *The Baltimore News/The Baltimore American*, January 24.

Stickel, L. F. 1950. "Populations and home range relationships of the box turtle, *Terrapene c. carolina*. (Linnaeus)." *Ecological Monographs* 20:351-378.

Stickel, L. F. 1978. "Changes in a box turtle population during three decades." *Copeia* 1978:221-225.

Stickel, L. F. 1989. "Home range behavior among box turtles (*Terrapene c. carolina*) of a bottomland forest in Maryland." *Journal of Herpetology* 23:40-44.

Stickel, L. F., and C. M. Bunck. 1989. "Growth and morphometrics of the box turtle, *Terrapene c. carolina*." *Journal of Herpetology* 23: 216-223.

Stickel, L. F., W. H. Stickel, and F. C. Schmid. 1980. "Ecology of a Maryland population of black rat snakes (*Elaphe o. obsoleta*)." *American Midland Naturalist* 103:1-14.

Stille, W. T. 1952. "The nocturnal amphibian fauna of the southern Lake Michigan beach." *Ecology* 33:149-162.

Stine, C. J. 1953. "Maryland salamanders of the genus *Ambystoma* (part 1) distribution." *Maryland Naturalist* 23:75-78.

Stine, C. J. 1984. "The life history and status of the eastern tiger salamander, *Ambystoma tigrinum tigrinum* (Green) in Maryland." *Bulletin of the Maryland Herpetological Society* 20:65-108.

Stine, C. J., and J. A. Fowler. 1956. "The ravine salamander in Maryland." *Maryland Naturalist* 26:1-4, 10-13.

Stine, C. J., J. A. Fowler, and R. S. Simmons. 1954. "Occurrence of the eastern tiger salamander, *Ambystoma tigrinum tigrinum* (Green) in Maryland, with notes on its life history." *Annals of the Carnegie Museum* 33:145-148.

Stine, C. J., and R. S. Simmons. 1952. "A new county record for Jefferson's salamander in Maryland." *Maryland Naturalist* 22:45-47.

Stine, C. J., R. S. Simmons, and J. A. Fowler. 1956. "New records for the eastern spadefoot toad in Maryland." *Herpetologica* 12: 295-296.

Stokes, G. D., and W. A. Dunson. 1982. "Permeability and channel structure of reptilian skin." *American Journal of Physiology—Renal Physiology* 242:681-689.

Stone, W. 1899. "Some Philadelphia ornithological collections and collectors, 1784-1850." *Auk* 16:166-177.

Storey, K. B., and J. M. Storey. 1986. "Freeze tolerance and intolerance as strategies of winter survival in terrestrially hibernating amphibians." *Comparative Biochemistry and Physiology* 83A: 613-617.

Storey, K. B., and J. M. Storey. 1987. "Persistence of freeze tolerance in terrestrially hibernating frogs after spring emergence." *Copeia* 1987:720-726.

Stratmann, T. S. M. 2015. "Finding the needle and the haystack: new insights into locating bog turtles (*Glyptemys muhlenbergii*) and their habitat in the southeastern United States." MS thesis. Clemson University, Clemson, SC.

Stuart, S. N., J. S. Chanson, N. A. Cox, B. E. Young, A. S. L. Rodrigues, D. L. Fischman, and R. W. Waller. 2004. "Status and trends of amphibian declines and extinctions worldwide." *Science* 306:1783-1786.

Sullivan, B. K. 1992. "Sexual selection and calling behavior in the American toad *Bufo americanus*." *Copeia* 1992:1-7.

Sun. 1838. "Rattlesnakes again." Anonymous. *The Sun* (1837-1991), October 8. ProQuest Historical Newspapers: *The Baltimore Sun*.

Sun. 1840. "Considerable of a snake." Anonymous. *The Sun* (1837-1991), August 19. ProQuest Historical Newspapers: *The Baltimore Sun*.

Sun. 1845. "Look out." Anonymous. *The Sun* (1837-1991), June 2. ProQuest Historical Newspapers: *The Baltimore Sun*.

Sun. 1856. "Rattle snakes killed." Anonymous. *The Sun* (1837-1991), August 8. ProQuest Historical Newspapers: *The Baltimore Sun*.

Sun. 1859. "Local matters: a nice souvenir." Anonymous. *The Sun* (1837-1991), August 20. ProQuest Historical Newspapers: *The Baltimore Sun*.

Sun. 1867. "Affairs in Frederick County." Anonymous. *The Sun* (1837-1991), September 6. ProQuest Historical Newspapers: *The Baltimore Sun*.

Sun. 1875. "That rattlesnake." Anonymous. *The Sun* (1837-1991), September 15. ProQuest Historical Newspapers: *The Baltimore Sun*.

Sun. 1877a. "Maryland items." Anonymous. *The Sun* (1837-1991), February 15. ProQuest Historical Newspapers: *The Baltimore Sun*.

Sun. 1877b. Anonymous. *The Sun* (1837-1991), June 9. ProQuest Historical Newspapers: *The Baltimore Sun*.

Sun. 1882. "Maryland affairs: Allegany." Anonymous. *The Sun* (1837-1991), August 14. ProQuest Historical Newspapers: *The Baltimore Sun*.

Sun. 1886. "The Maryland legislature senate, new bills introduced." Anonymous. *The Sun* (1837-1991), March 5. ProQuest Historical Newspapers: *The Baltimore Sun*.

Sun. 1890. "Maryland at a glance." Anonymous. *The Sun* (1837-1991), July 9. ProQuest Historical Newspapers: *The Baltimore Sun*.

Sun. 1892. "Wicomico County: extensive losses by lightning, copperheads and rattlesnakes." Anonymous. *The Sun* (1837-1991), August 4. ProQuest Historical Newspapers: *The Baltimore Sun*.

Sun. 1895. "Maryland." Anonymous. *The Sun* (1837-1991), June 17. ProQuest Historical Newspapers: *The Baltimore Sun*.

Sun. 1897a. "Dorchester County: a rattlesnake killed." Anonymous. *The Sun* (1837-1991), June 28. ProQuest Historical Newspapers: *The Baltimore Sun*.

Sun. 1897b. "Garrett County: a man who captures many rattles is bitten." Anonymous. *The Sun* (1837-1991), July 1. ProQuest Historical Newspapers: *The Baltimore Sun*.

Sun. 1897c. "Snakes in Dorchester." Editorial. *The Sun* (1837-1991), June 29. ProQuest Historical Newspapers: *The Baltimore Sun*.

Sun. 1905. "Killed rattlesnake in yard." Anonymous. *The Sun* (1837-1991), August 30. ProQuest Historical Newspapers: *The Baltimore Sun*.

Sun. 1906. "Mrs. Snelling as snake killer." Anonymous. *The Sun* (1837-1991), August 23. ProQuest Historical Newspapers: *The Baltimore Sun*.

Sun. 1909. "Boy bitten by rattlesnake." Anonymous. *The Sun* (1837-1991), July 14. ProQuest Historical Newspapers: *The Baltimore Sun*.

Sun. 1916. "Rattlesnake rides on train." Anonymous. *The Sun* (1837-1991), May 14. ProQuest Historical Newspapers:" *The Baltimore Sun*.

Sun. 1957. "Two deadly rattlesnakes killed in east Baltimore." Anonymous. *The Sun* (1837-1991), August 12. ProQuest Historical Newspapers: *The Baltimore Sun*.

Sun. 1991. "Gallimaufry." *The Baltimore Sun*, July 22, 6A.

Sutton, W. B., Y. Wang, and C. J. Schweitzer. 2010. "Habitat relationships of reptiles in pine beetle disturbed forests of Alabama, USA with guidelines for a modified drift-fence sampling method." *Current Zoology* 56:411-420.

Swarth, C. W. 1998a. *The Ecology and Population Status of Turtles at Jug Bay, Patuxent River*. Technical Report of the Jug Bay Wetlands Sanctuary. Lothian, MD: Jug Bay Wetlands Sanctuary.

Swarth, C. W. 1998b. "Monitoring turtles in wetlands." *The Volunteer Monitor* 10:20-21.

Swarth, C. W. 2004. "Natural history and reproductive biology of the red-bellied turtle (*Pseudemys rubriventris*)." In *Conservation and Ecology of Turtles of the Mid-Atlantic Region*, edited by C. W. Swarth, W. M. Roosenburg, and E. Kiviat, 73-84. Salt Lake City, UT: Bibliomania.

Swarth, C. W. 2005a. "Box turtle conservation: can we save them before it's too late?" *Audubon Naturalist News*, March, 4-6.

Swarth, C. W. 2005b. "Box turtle predation on songbirds." *Herpetologica* 36:315.

Swarth, C. W. 2005c. "Home range characteristics of box turtles." In *Summary of the Eastern Box Turtle Regional Conservation Workshop: Recommendations for Action*, edited by C. W. Swarth and S. Hagood, 9-10. Washington, DC: The Humane Society of the United States.

Swarth, C., and E. Friebele. 2008. *Seasonal Emergence Patterns in Red-bellied Turtle* (Pseudemys rubriventris) *hatchlings*. Technical Report of the Jug Bay Wetlands Sanctuary.Lothian, MD: Jug Bay Wetlands Sanctuary.

Swarth, C. W., and S. Hagood, editors. 2005. *Summary of the Eastern Box Turtle Regional Conservation Workshop: Recommendations for Action*. Washington, DC: The Humane Society of the United States.

Swarth, C. W., and E. Kiviat. 2009. "Animal communities of North American tidal freshwater wetlands." In *Tidal Freshwater Wetlands*, edited by A. Barendregt, D. Whigham, and A. Baldwin, 71-88. Leiden, The Netherlands: Backhuys Publishers.

Swarth, C. W., S. Matthews, L. Hollister, and E. Friebele. 2010. *Summary Report of the 2009 Jug Bay Bioblitz*. Lothian, MD: Jug Bay Wetlands Sanctuary.

Swarth, C. W., M. Quinlan, and J. Snodgrass. 2011. "Sex and age differences in home range and habitat use of eastern box turtles (*Terrapene carolina carolina*)." Abstract. Society for the Study of Amphibians and Reptiles, Minneapolis, MN.

Szlavecz, K., A. Terzis, R. Musaloiu, C. J. Liang, J. Cogan, A. Szalay, J. Gupchup, J. Klofas, L. Xia, C. Swarth, and S. Matthews. 2007. "Turtle nest monitoring with wireless sensor networks." Abstract. In *Proceedings of the American Geophysical Union, Fall Meeting, San Francisco, Cal*. Washington, DC: American Geophysical Union.

Taylor, G. J. 1984. "The Maryland Endangered Species Program: a history." In *Threatened and Endangered Plants and Animals of Maryland*, edited by A. W. Norden, D. C. Forester, and G. H. Fenwick, 43-49. Natural Heritage Program Special Publication 84-1. Annapolis: Maryland Department of Natural Resources.

Taylor, G. J., S. A. Dawson, S. A. Beall, and J. E. Schaeffer. 1984. "Distribution and habitat description of the Muhlenberg (bog) turtle (*Clemmys muhlenbergi*) in Maryland." *Transactions of the 1984 Northeast Fish and Wildlife Conference* 41:46-58.

Taylor, N., and J. Mays. 2006. "The salamanders *Eurycea longicauda* and *Plethodon glutinosus* in Gregorys Cave, TN: monitoring and observations on ecology and natural history." *Southeastern Naturalist* 5:435-442.

Teece, M., and C. Swarth. 1998. "Embryonic and fetal development of turtles: a stable isotope approach." *Marsh Notes: Newsletter of Jug Bay Wetlands Sanctuary* 13(4):5.

Teece, M., C. W. Swarth, N. Tuross, and M. L. Fogel. 2004. "Determination of the essential amino acid requirements of the red-bellied turtle, *Pseudemys rubriventris*, during hatchling development." In *Conservation and Ecology of Turtles of the Mid-Atlantic Region*, edited by C. W. Swarth, W. M. Roosenburg, and E. Kiviat, 85-91. Salt Lake City, UT: Bibliomania.

Therres, G. D., C. A. Davis, and C. W. Swarth. 2015. "Grid-based amphibian and reptile atlas using active searching: a pilot project." *Maryland Naturalist* 53:33-51.

Thompson, E. L. 1980. "Breeding site ecology of Ambystomatid salamanders in Maryland." MS thesis. Frostburg State University, Frostburg, MD.

Thompson, E. L. 1984a. "A report on the status and distribution of the Jefferson salamander, green salamander, mountain earth snake, and northern coal skink in Maryland." In *Threatened and Endangered Plants and Animals of Maryland*, edited by A. W. Norden, D. C. Forester, and G. H. Fenwick, 338-351. Natural Heritage Program Special Publication 84-1. Annapolis: Maryland Department of Natural Resources.

Thompson, E. L. 1984b. "Wehrle's salamander (*Plethodon wehrlei*) in Maryland." In *Threatened and Endangered Plants and Animals of Maryland*, edited by A. W. Norden, D. C. Forester, and G. H. Fenwick, 336-337. Natural Heritage Program Special Publication 84-1. Annapolis: Maryland Department of Natural Resources.

Thompson, E. 2000. "An amphibian survey of the C&O Canal National Historical Park in Allegeny and Washington counties, Maryland." Report to the National Park Service.

Thompson, E. L., and J. A. Chapman. 1978. "The first record of Wehrle's salamander from Maryland." *Proceedings of the Pennsylvania Academy of Science* 52:103.

Thompson, E. L., and J. E. Gates. 1979. "Geographic distribution: *Ambystoma jeffersonianum*." *Herpetological Review* 10:101.

Thompson, E. L., and J. E. Gates. 1982. "Breeding pool aggregations by the mole salamanders, *Ambystoma jeffersonianum* and *A. maculatum*, in a region of sympatry." *Oikos* 38:273-279.

Thompson, E. L., J. E. Gates, and G. J. Taylor. 1980. "Distribution and breeding habitat selection of the Jefferson salamander, *Ambystoma jeffersonianum*, in Maryland." *Journal of Herpetology* 14: 113-120.

Thompson, E. L., and G. J. Taylor. 1985. "Notes on the green salamander, *Aneides aeneus*, in Maryland." *Bulletin of the Maryland Herpetological Society* 21:107-114.

Thurow, G. R. 1976. "Aggression and competition in eastern *Plethodon* (Amphibia, Urodela, Plethodontidae)." *Journal of Herpetology* 10:277-291.

Thurston, L. 2005. *Nest Site Characteristics of the Eastern Box Turtle* (Terrapene carolina carolina): *Vegetation, Temperature, Insolation and Substrate*. Technical Report of the Jug Bay Wetlands Sanctuary. Lothian, MD: Jug Bay Wetlands Sanctuary.

Tilley, S. G., and M. J. Mahoney. 1996. "Patterns of genetic differentiation in salamanders of the *Desmognathus ochrophaeus* complex (Amphibia: Plethodontidae)." *Herpetological Monographs* 10: 1-42.

Tiner, R. W., and D. G. Burke. 1995. *Wetlands of Maryland*. Hadley, MA: US Fish and Wildlife Service, and Annapolis: Maryland Department of Natural Resources.

Tinkle, D. W., and R. E. Ballinger. 1972. "*Sceloporus undulatus*: a study of the intraspecific comparative demography of a lizard." *Ecology* 53:570-584.

Tobey, F. J. 1962. "Notes on Maryland herpetology." *Virginia Herpetological Society Bulletin*, no. 28.

Todd, B. D., J. D. Willson, C. T. Winne, and J. W. Gibbons. 2008. "Aspects of the ecology of the earth snakes (*Virginia valeriae* and *V. striatula*) in the Upper Coastal Plain." *Southeastern Naturalist* 7:349-358.

Todd, J., J. Amiel, and R. Wassersug. 2009. "Factors influencing the emergence of a northern population of eastern ribbon snakes (*Thamnophis sauritis*) from artificial hibernacula." *Canadian Journal of Zoology* 87:1221-1226.

Trauth, S. E. 1983. "Nesting habitat and reproductive characteristics of the lizard *Cnemidophorus sexlineatus* (Lacertilia: Teiidae)." *American Midland Naturalist* 109:289-299.

Trauth, S. E. 1991. "Distribution, scutellation, and reproduction in the queen snake, *Regina septemvittata* (Serpentes: Colubridae), from Arkansas." *Proceedings of the Arkansas Academy of Science* 45:103-106.

Trauth, S. E., and C. T. McAllister. 1995. "Vertebrate prey of selected Arkansas snakes." *Proceedings of the Arkansas Academy of Science* 49:188-192.

Treanor, R. R., and S. J. Nichola. 1972. *A Preliminary Report of the Commercial and Sporting Utilization of the Bullfrog,* Rana catesbeiana *Shaw, in California*. Inland Fish Administrative Report 72-4. Sacramento: California Department of Fish and Game.

True, F. W. 1887. "The turtle and terrapin fisheries." In *The Fisheries and Fishery Industries of the United States. Section V. History and Methods of the Fisheries. Volume II. A Geographical Review of the Fisheries Industries and Fishing Communities for the Year 1880*, edited by G. B. Goode, 495-503. Washington, DC: Government Printing Office.

Tyning, T. F. 1990. *A Guide to the Amphibians and Reptiles*. Boston: Little, Brown and Company.

Tyron, B. W., and G. Carl. 1980. "Reproduction in the mole kingsnake, *Lampropeltis calligaster rhombomaculata* (Serpentes, Colubridae)." *Transactions of the Kansas Academy of Science* 83: 66-73.

Uhler, F. M., C. Cottam, and T. E. Clarke. 1939. "Food of snakes of the George Washington National Forest, Virginia." *Transactions of the North American Wildlife Conference* 4:605-622.

Uhler, P. 1888. "Maryland Academy of Sciences: sketch of the history of the Maryland Academy of Sciences." *Johns Hopkins University, University Circulars* 7(64):49-51.

Ultsch, G. R. 2006. "The ecology of overwintering among turtles: where turtles overwinter and its consequences." *Biological Reviews* 81:339-367.

Unger, S. D., and R. N. Williams. 2012. *Eastern Hellbender: North America's Giant Salamander*. Purdue University Cooperative Extension Service, FNR-471-W. West Lafayette, IN: Purdue University.

USDA (US Department of Agriculture). 2013. *Forests of Maryland*. Newton Square, PA: US Forest Service, Northern Research Station.

USFWS (US Fish and Wildlife Service). 2001. *Bog Turtle* (Clemmys muhlenbergii), *Northern Population, Recovery Plan*. Hadley, MA: US Fish and Wildlife Service.

USGS (US Geological Survey). 1999. *Maryland and the District of Columbia: Surface Water Resources*. Water Supply Paper no. 2300. Baltimore: US Geological Survey.

Van Buskirk, J., and S. A. McCollum. 2000. "Functional mechanisms of an inducible defense in tadpoles: morphology and behaviour influence mortality risk from predation." *Journal of Evolutionary Biology* 13:336-347.

van Deusen, M., and R. H. Johnson. 1980. "Herpetological survey of the Elms site, St. Mary's County, Maryland." *Bulletin of the Maryland Herpetological Society* 16:1-8.

van Dijk, P. P., J. Harding, and G. A. Hammerson. 2011. "*Trachemys scripta* (errata version published in 2016)." The IUCN Red List of Threatened Species 2011: e.T22028A97429935. Online at http://dx.doi.org/10.2305/IUCN.UK.2011-1.RLTS.T22028A9347395.en.

Versar. 2011. *Results from Round 3 of the Maryland Biological Stream Survey (2007-2009)*. Final report to the Maryland Department of Natural Resources, Resource Assessment Service. Columbia, MD: Versar.

VertNet. 2016. "On-line database of vertebrate collections." Online at http://vertnet.org.

VHS (Virginia Herpetological Society). 1959. "Keep your eyes on the Interior Department's Outdoor Program." *Virginia Herpetological Society Bulletin* 13:2.

VHS (Virginia Herpetological Society). 1960. "The occurrence of poisonous snakes in Virginia, Maryland and D.C." *Virginia Herpetological Society Bulletin* 19:1-2.

VHS (Virginia Herpetological Society). 1962. "Notes on Maryland herpetology." *Virginia Herpetological Society Bulletin* 28:5-6.

Vitt, L. J., and W. E. Cooper, Jr. 1985a. "The evolution of sexual dimorphism in the skink *Eumeces laticeps*: an example of sexual selection." *Canadian Journal of Zoology* 63:995-1002.

Vitt, L. J., and W. E. Cooper, Jr. 1985b. "The relationship between reproduction and lipid cycling in the skink *Eumeces laticeps* with comments on brooding ecology." *Herpetologica* 41: 419-432.

Vitt, L. J., and W. E. Cooper, Jr. 1986. "Foraging and diet of a diurnal predator (*Eumeces laticeps*) feeding on hidden prey." *Journal of Herpetology* 20:408–415.

Volpe, E. P. 1952. "Physiological evidence for natural hybridization of *Bufo americanus* and *Bufo fowleri*." *Evolution* 6:393–406.

Waldman, B., and K. Adler. 1979. "Toad tadpoles associate preferentially with siblings." *Nature* 282:611–613.

Waldron, J. L., and W. J. Humphries. 2005. "Arboreal habitat use by the green salamander, *Aneides aeneus*, in South Carolina." *Journal of Herpetology* 39:486–492.

Waldron, J. L., and T. K. Pauley. 2007. "Green salamander (*Aneides aeneus*) growth and age at reproductive maturity." *Journal of Herpetology* 41:638–644.

Walker, C. F. 1946. "The amphibians of Ohio. Part I: frogs and toads." *Ohio State Museum of Science Bulletin* 1:1–109.

Walker, C. F., and W. Goodpaster. 1941. "The green salamander, *Aneides aeneus*, in Ohio." *Copeia* 1941:178.

Walton, M. 1988. "Relationships among metabolic, locomotory, and field measures of organismal performance in the Fowler's toad (*Bufo woodhousei fowleri*)." *Physiological Zoology* 61:107–118.

Walton, M., and B. D. Anderson. 1988. "The aerobic cost of saltatory locomotion in the Fowler's toad (*Bufo woodhousei fowleri*)." *Journal of Experimental Biology* 136:273–288.

Warden, D. B. 1816. *A Chorographical and Statistical Description of the District of Columbia, the Seat of the General Government of the United States, with an Engraved Plan of the District, and View of the Capital.* Paris: Smith, Rue Montmorency.

Warner, C. D. 2006. "Captain John Smith." *Project Gutenberg EBook*, no. 3130.

Wasserman, A. O. 1970. "Polyploidy in the common tree toad *Hyla versicolor* Le Conte." *Science* 167:385–386.

Watermolen, D. 1991. "*Storeria occipitomaculata occipitomaculata* (northern red-belly snake): behavior." *Herpetological Review* 22:61.

Watson, C. W. 2004. "The effects of controlled burning on ground skink populations in a mixed pine-hardwood habitat of East Texas." MS thesis. University of Texas at Arlington, Arlington.

Weir, L. A., J. A. Royle, K. D. Gazenski, and O. Villena. 2014. "Northeast regional and state trends in anuran occupancy from calling survey data (2001–2011) from the North American Amphibian Monitoring Program." *Herpetological Conservation and Biology* 9:223–245.

Wells, K. D. 2007. *The Ecology and Behavior of Amphibians.* Chicago: University of Chicago Press.

Wemple, P. 1971. "The eastern spiny soft-shelled turtle *Trionyx spinifer spinifer* [*sic*] Le Sueur in Maryland." *Bulletin of the Maryland Herpetological Society* 7:35–37.

Werler, J. E., and J. McCallion. 1951. "Notes on a collection of reptiles and amphibians from Princess Anne County, Virginia." *American Midland Naturalist* 45:245–252.

Whitaker, J. O., Jr. 1971. "A study of the western chorus frog, *Pseudacris triseriata*, in Vigo County, Indiana." *Journal of Herpetology* 5:127–150.

Whitaker, J. O., Jr., D. Rubin, and J. R. Munsee. 1977. "Observations on food habits of four species of spadefoot toads, genus *Scaphiopus*." *Herpetologica* 33:468–475.

White, J. F., Jr. 1987. "New frog species is discovered in Delaware." *Delaware Nature Education Society News* 23(4):3.

White, J. F., Jr., and A. W. White. 2007. *Amphibians and Reptiles of Delmarva*, rev. ed. Centreville, MD: Tidewater Publishers.

Wilbur, H. M. 1977a. "Density-dependent aspects of growth and metamorphosis in *Bufo americanus*." *Ecology* 58:196–200.

Wilbur, H. M. 1977b. "Propagule size, number, and dispersion pattern in *Ambystoma* and *Asclepias*." *American Naturalist* 111:43–68.

Wilder, I. W. 1913. "The life history of *Desmognathus fusca*." *Biological Bulletin* 24:251–342.

Wilhoft, D. C., E. Hotaling, and P. Franks. 1983. "Effects of temperature on sex determination in embryos of the snapping turtle, *Chelydra serpentina*." *Journal of Herpetology* 17:38–42.

Williams, J. 1995. "Another Atlantic leatherback turtle observed in the upper Chesapeake Bay." *Maryland Naturalist* 39:23–24.

Williams, K. L. 1978. "Systematics and natural history of the American milk snake, *Lampropeltis triangulum*." *Milwaukee Public Museum Publications in Biology and Geology*, no. 2.

Williams, K. L. 1988. *Systematics and Natural History of the American Milk Snake,* Lampropeltis triangulum, 2nd ed. Milwaukee, WI: Milwaukee Public Museum.

Williams, R. D., J. E. Gates, and C. H. Hocutt. 1981a. "An evaluation of known and potential sampling techniques for the hellbender, *Cryptobranchus alleganiensis*." *Journal of Herpetology* 15:23–27.

Williams, R. D., J. E. Gates, C. H. Hocutt, and G. J. Taylor. 1981b. "The hellbender: a nongame species in need of management." *Wildlife Society Bulletin* 9:94–100.

Willson, J. D., and M. E. Dorcas. 2004. "Aspects of the ecology of small fossorial snakes in the western Piedmont of North Carolina." *Southeastern Naturalist* 3:1–12.

Wolfe-Arnovits, J. 2000. "Columbia man warns about snakes in grass." *The Baltimore Sun*, July 20.

Wood, J. T. 1949. "Observations on *Natrix septemvittata* (Say) in southwestern Ohio." *American Midland Naturalist* 42:744–750.

Wood, J. T. 1953. "Observations on the complements of ova and nesting of the four-toed salamander in Virginia." *American Naturalist* 87:77–86.

Wood, J. T. 1955. "The nesting of the four-toed salamander, *Hemidactylium scutatum* (Schlegel), in Virginia." *American Midland Naturalist* 53:381–389.

Wood, J. T., and H. N. McCutcheon. 1954. "Ovarian egg complements and nest of the two-lined salamander, *Eurycea b. bislineata × cirrigera*, from southeastern Virginia." *American Midland Naturalist* 52:433–36.

Wood, R. C. 1997. "The impact of commercial crab traps on northern diamondback terrapins, *Malaclemys terrapin terrapin*." In *Proceedings: Conservation, Restoration, and Management of Tortoises and Turtles—An International Conference*, edited by J. Van Abbema, 21–27. New York: New York Turtle and Tortoise Society/Wildlife Conservation Society Turtle Recovery Program.

Woodward, T. 2005. "Hibernation behavior of eastern box turtles." Senior thesis. Roosevelt High School, Greenbelt, MD.

Woolcott, W. S. 1959. "Notes on the eggs and young of the scarlet snake, *Cemophora coccinea* Blumenbach." *Copeia* 1959:263.

Worthington, R. D. 1968. "Observations on the relative sizes of three species of salamander larvae in a Maryland pond." *Herpetologica* 24:242–246.

Worthington, R. D. 1969. "Additional observations on sympatric species of salamander larvae in a Maryland pond." *Herpetologica* 25:227-229.

Wright, A. H. 1914. *Life-History of the Anura of Ithaca, New York*. Washington, DC: Carnegie Institute of Washington.

Wright, A. H. 1932. *Life Histories of the Frogs of the Okefenokee Swamp, Georgia: North American Salientia (Anura)*, no. 2. New York: Macmillan Press.

Wright, A. H., and A. A. Wright. 1949. *Handbook of the Frogs and Toads of the United States and Canada*, 3rd ed. Ithaca, NY: Comstock Publishing Company.

Wright, A. H., and A. A. Wright. 1957. *Handbook of the Snakes of the United States and Canada*, vol. II. Ithaca, NY: Cornell University Press.

Wund, M. A., M. E. Torocco, R. T. Zappalorti, and H. K. Reinert. 2007. "Activity ranges and habitat use of *Lampropeltis getula getula* (eastern kingsnakes)." *Northeastern Naturalist* 14:343-360.

Young, B. E., S. N. Stuart, J. S. Chanson, N. A. Cox, and T. M. Boucher. 2004. *Disappearing Jewels: The Status of New World Amphibians*. Arlington, VA: NatureServe.

Zampella, R. A., and J. F. Bunnell. 2000. "The distribution of anurans in two river systems of a Coastal Plain watershed." *Journal of Herpetology* 34:210-221.

Zamudio, K. R., and A. M. Wieczorek. 2007. "Fine-scale spatial genetic structure and dispersal among spotted salamander (*Ambystoma maculatum*) breeding populations." *Molecular Ecology* 16: 257-274.

Zumer, B. 2013. "Camper bitten by copperhead still recovering; toxicity expert says bites more common than many people think." *The Baltimore Sun*, July 18.

Zweifel, R. G. 1968. "Effects of temperature, body size, and hybridization on mating calls of toads, *Bufo a. americanus* and *Bufo woodhousii fowleri*." *Copeia* 1968:269-285.

Zwinenberg, A. J. 1975. "The green turtle (*Chelonia mydas*), one of the reptiles most consumed by man, needs immediate protection." *Bulletin of the Maryland Herpetological Society* 11:45-61.

Index